Springer Tracts in Modern Physics

Volume 286

Springer Tracts in Modern Physics provides comprehensive and critical reviews of topics of current interest in physics. The following fields are emphasized:

– Particle and Nuclear Physics
– Condensed Matter Physics
– Light Matter Interaction
– Atomic and Molecular Physics

Suitable reviews of other fields can also be accepted. The Editors encourage prospective authors to correspond with them in advance of submitting a manuscript. For reviews of topics belonging to the above mentioned fields, they should address the responsible Editor as listed in "Contact the Editors".

V. T. Davis

Introduction to Photoelectron Angular Distributions

Theory and Applications

 Springer

V. T. Davis
University of Nevada, Reno
Reno, NV, USA

ISSN 0081-3869   ISSN 1615-0430   (electronic)
Springer Tracts in Modern Physics
ISBN 978-3-031-08026-5   ISBN 978-3-031-08027-2   (eBook)
https://doi.org/10.1007/978-3-031-08027-2

This Springer imprint is published by the registered company Springer Nature Switzerland AG
The registered company address is: Gewerbestrasse 11, 6330 Cham, Switzerland

For my parents

Acknowledgments

The author gratefully acknowledges the help of Dr. Kiattichart Chartkunchand, Professor Till Jahnke, and Professor Joshua Williams, all of whom contributed directly to the preparation of this manuscript by providing the results of their research. Professor Till Jahnke and Professor Joshua Williams further assisted by graciously inviting me to participate as a member of their research teams. Special thanks go to Professor Jeff Thompson, who first suggested this project, and to Professor Aaron Covington who provided support along the way. In addition, Professor Jeff Thompson and Professor Aaron Covington have allowed me to participate as a member of their research teams for over 20 years now. I would also like to thank Nemul Khan, Dinesh Vinayagam, Dr. Ute Heuser, and Dr. Sam Harrison of the editorial team for helping to bring this book to fruition. Finally, most of the credit for any success that I may have goes to my wife, Anne, who has been at my side for 40 years and counting.

Contents

Chapter 1
Introduction

The interaction of electromagnetic radiation with matter provides the primary physical process by which physicists and chemists study atoms and molecules. In some cases, the absorption of a photon (or photons) by atoms or molecules leads to bound-free transitions in which one or more electrons ("photoelectrons") are emitted in a photodissociation (pd) reaction. In photoelectron spectroscopy experiments, these photoelectrons are collected and analyzed. Analysis of collected photoelectrons provides information on fundamental atomic and molecular properties such as electronic structure and photoionization (or photodetachment) cross-sections. As it turns out, photoelectrons are not emitted randomly but instead in particular directions determined by the nature and dynamics of the material under study, and these directions are calculable by employing the appropriate physical theory. Photoelectron angular distribution (PAD) measurements are important because they can shed light on possible atomic and molecular electronic configurations and system dynamics that are not obtainable from total cross-sections. The measurement of differential cross-sections in particular can provide information not only on the magnitude of quantum transition amplitudes but also on their relative phases, as well as information on the nature of the interaction between light and matter. For example, recent measurements of molecular-frame photoelectron angular distributions (MFPADs) have been used to extract photoelectron emission delays in the attosecond range, delays which can provide ultrasensitive, time-dependent maps of molecular potentials. Also, photoelectron angular distribution measurements are particularly useful in the study of negative ions. The masking of the nuclear-electron interaction due to increased screening in negative ions elevates short-range electron-electron correlation forces to a dominant position within these ions, forces without which (for example) atomic anions could not exist. PAD measurements, being particularly sensitive to these types of many-body forces, are an essential tool for understanding these systems.

Before engaging in photoelectron angular distribution measurements, experimentalists should understand the basic analytical models involved because they reveal the conservation laws, dynamical processes, and geometrical effects that ultimately determine the shape of the measured angular distributions. It is often found that, after laborious computations, relatively simple expressions remain. This is no accident, as these simple expressions reveal deeper properties of the underlying physical processes involved in atomic or molecular photoionization (or photodetachment) under rotational and inversion symmetry [1]. The various physical influences on photoelectron angular distributions are revealed in analytical models primarily through the use of angular momentum coupling algebra and the mathematics of spherical tensor operators, as will be amply demonstrated. Key derivations are presented in (sometimes gruesome) detail in hopes of increasing the understanding of ultimate results. This manuscript breaks no new ground but instead attempts only to arrange the relevant information in such a way as to make these important theories more accessible (at an introductory level) to those engaged in experimental photoelectron spectroscopy studies in the laboratory. The information contained herein may also be useful to those interested in deepening their knowledge of the interaction of atoms and molecules with light, so this book may serve as a useful supplement in a standard course on atomic and molecular physics. In addition to the basic theory of PADs, common laboratory techniques used to measure PADs and areas of current research are briefly described.

Chapter 2
Angular Momentum in Quantum Mechanics

Since angular momentum plays a prominent role in the dynamics of photon-atom/ molecule interactions, a brief review of the formalism of quantum angular momentum is warranted. To be clear, angular momentum in quantum mechanics is a vast topic about which entire tomes can be (and have been) written. Here we concentrate on those aspects of quantum angular momentum that are most directly applicable to an examination of the theory of photoelectron angular distributions. We develop all our results from first principles starting with the definition of angular momentum operators in quantum mechanics.

2.1 Commutation Relations of Angular Momentum Operators

In standard quantum theory, angular momentum operators are vector operators whose Cartesian components are the Hermitian operators J_x, J_y, J_z, all of which obey the commutation relations:

$$\left[J_l, J_j\right] = i\hbar \sum_k \varepsilon_{ljk} J_k; \quad l, j, k = x, y, z \tag{2.1}$$

The operator for the square of the total angular momentum is

Primary references for this chapter: [2–9]

The original version of this chapter was revised. The correction to this chapter is available at https://doi.org/10.1007/978-3-031-08027-2_11

V. T. Davis, *Introduction to Photoelectron Angular Distributions*, Springer Tracts in Modern Physics 286, https://doi.org/10.1007/978-3-031-08027-2_2

$$J^2 = J_x^2 + J_y^2 + J_z^2 \tag{2.2}$$

This operator commutes with any of the component operators J_i. The proof is simple:

$$[J_i, J^2] = \sum_j \{[J_i, J_j]J_j + J_j[J_i, J_j]\} = i\hbar \sum_m \sum_j \varepsilon_{ijm}[J_m J_j + J_j J_m];$$

$$i, j, m = x, y, z \tag{2.3}$$

where (2.1) and the commutator relation:

$$[A, BC] = B[A, C] + [A, B]C \tag{2.4}$$

were used. The Levi-Civita tensor ε_{ijm} is defined in Problem 3.8.

In (2.3), i, j, m must be different from one another, else $\varepsilon_{ijm} = 0$. If they are all different, then the sums over m and j will generate an even number of terms which will subtract in pairs due to the antisymmetric nature of ε_{ijm}. Therefore,

$$[J_i, J^2] = 0 \tag{2.5}$$

which completes the proof.

It will prove convenient to introduce the so-called ladder operators:

$$J_\pm = J_x \pm iJ_y \tag{2.6}$$

Note that these two operators are not Hermitian, although they are Hermitian conjugates of each other.

We can now construct some additional commutation relations, as shown in the following example:

Example 2.1 Evaluate the commutators $[J_z, J_\pm]$ and $[J_+, J_-]$

$$[J_z, J_\pm] = [J_z, J_x \pm iJ_y] = [J_z, J_x] \pm i[J_z, J_y] = i\hbar J_y \pm (i)^2 \hbar(-1)J_x$$

$$\Rightarrow [J_z, J_\pm] = \pm\hbar J_\pm \tag{2.7}$$

$$[J_+, J_-] = [J_x + iJ_y, J_x - iJ_y]$$
$$= [J_x, J_x] + [iJ_y, J_x] + [J_x, -iJ_y] + [iJ_y, -iJ_y]$$
$$= i[J_y, J_x] - i[J_x, J_y] \Rightarrow [J_+, J_-] = 2\hbar J_z \tag{2.8}$$

∎

2.2   Construction of Eigenstates and the Spectrum of Eigenvalues

The ladder operators are useful in finding alternate expressions for J^2:

$$J_\mp J_\pm = (J_x \mp iJ_y)(J_x \pm iJ_y) = J_x^2 \mp iJ_y J_x \pm iJ_x J_y + J_y^2 = J_x^2 \mp i[J_y, J_x] + J_y^2$$
$$= J_x^2 + J_y^2 \mp \hbar J_z$$
$$\Rightarrow J_\mp J_\pm + J_z^2 \pm \hbar J_z = J_x^2 + J_y^2 + J_z^2 \tag{2.9}$$
$$\Rightarrow J^2 = J_\mp J_\pm + J_z^2 \pm \hbar J_z$$

The fact that the square of the angular momentum commutes with each of the Cartesian angular momentum components [c.f. (2.5)] means that we can choose a basis in which we simultaneously diagonalize the matrix representations of J^2 and any one component J_i. By convention, we choose the component J_z. Thus, we can construct the states $|j, m\rangle$ which are simultaneously eigenstates of J^2 and J_z.

Let

$$J^2|j, m\rangle = \hbar^2 \lambda |j, m\rangle \tag{2.10a}$$

$$J_z|j, m\rangle = \hbar m |j, m\rangle \tag{2.10b}$$

where j, m label the simultaneous eigenstates of J^2 and J_z, and the factors of $\hbar^2$ and $\hbar$ are chosen by appealing to dimensional analysis. Because the $|j, m\rangle$ are the eigenfunctions of Hermitian operators, they form a complete set. We also assume that we can make the $|j, m\rangle$ orthonormal. Note the inequality

$$\lambda \hbar^2 = \langle j, m|J^2|j, m\rangle = \langle j, m|J_x^2|j, m\rangle + \langle j, m|J_y^2|j, m\rangle + \langle j, m|J_z^2|j, m\rangle$$
$$= \|J_x|j, m\rangle\|^2 + \|J_y|j, m\rangle\|^2 + \hbar^2 m^2 \geq 0 \tag{2.11}$$

The first two terms on the RHS of (2.11) are positive or zero. Hence,

$$\lambda \geq m^2 \tag{2.12}$$

Consider the state $(J_\pm|j, m\rangle)$ acted on by the operator J_z,

$$J_z J_\pm|j, m\rangle = ([J_z, J_\pm] + J_\pm J_z)|j, m\rangle = (\pm \hbar J_\pm + J_\pm \hbar m)|j, m\rangle$$
$$= \hbar(m \pm 1)J_\pm|j, m\rangle \tag{2.13}$$

where (2.7) and (2.10b) were used. It is clear that the state $(J_\pm|j, m\rangle)$ is an eigenfunction of J_z with eigenvalues $\hbar(m \pm 1)$. Also, since (2.5) implies that J^2 commutes with the ladder operators, we have

$$J^2 J_\pm |j, m\rangle = J_\pm J^2 |j, m\rangle = \hbar^2 \lambda J_\pm |j, m\rangle \tag{2.14}$$

where (2.10a) was used. Equation (2.14) tells us that the state $(J_\pm |j, m\rangle)$ is an eigenfunction of J^2 with eigenvalues $\hbar^2 \lambda$.

The fact that the $(J_\pm |j, m\rangle)$ are simultaneous eigenfunctions of J_z and J^2, combined with the structure of (2.10b) and (2.13) allow us to infer that

$$J_\pm |j, m\rangle = C_\pm |j, m \pm 1\rangle \tag{2.15}$$

where $C_\pm$ are constants that depend on the quantum numbers j and m and where we have assumed that the eigenfunctions $|j, m \pm 1\rangle$ are normalized. Clearly, the ladder operators $J_\pm$ act on the states $|j, m\rangle$ to raise or lower the value of m by one while leaving the value of j unchanged. Thus, (2.12) and (2.15) tell us that, for a given value of j, we must have a finite sequence of values of m separated by integers and bounded by a minimum value m_{min} and a maximum value m_{max}. We must be able to reach $|j, m_{max}\rangle$ by applying J_+ to $|j, m_{min}\rangle$ a finite number of (integer) times or, conversely, we must be able to reach $|j, m_{min}\rangle$ by applying J_- to $|j, m_{max}\rangle$ a finite number of times. Physically, we expect $m_{max} = -m_{min}$ by the symmetry between z and $-z$, but we can also show this explicitly. To do this, first note that we can expect

$$J_+ |j, m_{max}\rangle = J_- |j, m_{min}\rangle = 0 \tag{2.16}$$

Then, using (2.9),

$$\left(J^2 = J_\mp J_\pm + J_z^2 \pm \hbar J_z \right) \left| j, m \begin{matrix} max \\ min \end{matrix} \right\rangle \Rightarrow \left\{ \begin{matrix} \lambda = m_{max}(m_{max} + 1) \\ \lambda = m_{min}(m_{max} + 1) \end{matrix} \right\}$$

$$\Rightarrow m_{max}^2 + m_{max} - m_{min}^2 + m_{min} = 0$$

$$\Rightarrow m_{max} = \frac{1}{2} \left(-1 \pm \sqrt{1 + 4m_{min}^2 - 4m_{min}} \right)$$

$$= \frac{1}{2}[-1 \pm (2m_{min} - 1)] = \left\{ \begin{matrix} m_{min} - 1 \\ -m_{min} \end{matrix} \right. \tag{2.17}$$

The upper result leads to a contradiction, leaving us to conclude

$$m_{max} = -m_{min} \tag{2.18}$$

Because successive values of m differ by integer units, the quantity $m_{max} - m_{min}$ is a positive definite integer which we set equal to $2j$, where j is a positive integer or half-integer. So, from (2.18), we now have

$$\left.\begin{array}{l} m_{\max} - m_{\min} = 2j \\ m_{\max} + m_{\min} = 0 \end{array}\right\} \Rightarrow \left\{\begin{array}{l} m_{\max} = j \\ m_{\min} = -j \end{array}\right. \tag{2.19}$$

and if we let $m_{\max} = j$, then (for a given value of j) the m-values run in a finite sequence as

$$-m_{\max}, -m_{\max} + 1, \cdots, m_{\max} - 1, m_{\max} \Leftrightarrow -j, -j+1, \cdots, j-1, j \tag{2.20}$$

Substitution of $m_{\max} = j$ into (2.17) also leads to the result

$$\lambda = j(j+1) \tag{2.21}$$

To summarize,

$$J^2|j, m\rangle = \hbar^2 j(j+1)|j, m\rangle; \quad j = \frac{1}{2}, 1, \frac{3}{2}, 2, \frac{5}{2}, \cdots \tag{2.22a}$$

$$J_z|j, m\rangle = \hbar m|j, m\rangle; \quad m = -j, -j+1, \cdots, j-1, j \tag{2.22b}$$

Finally, we see that there are $2j + 1$ values of m (m is the projection of the angular momentum j onto the z-axis) for a given value of j.

2.3   Matrix Elements of Angular Momentum Operators

To find the matrix elements of the angular momentum operators, first recall that the ladder operators are Hermitian conjugates and again use (2.9)

$$\langle j, m|(J^2 = J_- J_+ + J_z^2 + \hbar J_z)|j, m\rangle \Rightarrow \hbar^2 j(j+1) = |C_+|^2 + \hbar^2 m^2 + \hbar^2 m$$
$$\Rightarrow |C_+|^2 = \hbar^2[j(j+1) - m(m+1)] \tag{2.23}$$
$$\Rightarrow C_+ = e^{i\varphi_+} \hbar \sqrt{j(j+1) - m(m+1)} = \hbar \sqrt{j(j+1) - m(m+1)}$$

where we used (2.15) and (2.22a and 2.22b) and adopted a phase convention of $e^{i\varphi_+} = 1$. We also used the fact that the $|j, m\rangle$ are normalized.

In a similar fashion, we would also find

$$|C_-|^2 = \hbar^2[j(j+1) - m(m-1)] \tag{2.24}$$

Now consider

$$|C_\pm|^2 = \langle j, m|J_\mp J_\pm|j, m\rangle \Rightarrow \left\{\begin{array}{l} C_+(j, m) = e^{i\varphi_+}\hbar[j(j+1) - m(m+1)]^{1/2} \\ C_-(j, m) = e^{i\varphi_-}\hbar[j(j+1) - m(m-1)]^{1/2} \end{array}\right. \tag{2.25}$$

But at the same time,

$$\langle j, m|J_+J_-|j, m\rangle = [C_-(j, m)]\,[C_+(j, m-1)]$$
$$= \left[e^{i\varphi_-}\hbar[j(j+1) - m(m-1)]^{1/2}\right]\left[e^{i\varphi_+}\hbar[j(j+1) - (m-1)m]^{1/2}\right]$$
$$= e^{i(\varphi_+ + \varphi_-)}\hbar^2[j(j+1) - m(m-1)]$$

$$(2.26a)$$

Or, alternatively,

$$((\langle j, m|J_+)(J_-|j, m\rangle)) = [C_-(j, m)^*]\,[C_-(j, m)] = |C_-|^2 = \hbar^2[j(j+1) - m(m-1)]$$

$$(2.26b)$$

We have already chosen the phase convention $\varphi_+ = 0$. Equating (2.26a) and (2.26b) forces us to conclude that, with the chosen phase convention, we must have $\varphi_- = 0$ as well. Thus,

$$C_- = \hbar[j(j+1) - m(m-1)]^{1/2} \qquad (2.27)$$

And so,

$$J_\pm|j, m\rangle = \hbar[j(j+1) - m(m\pm 1)]^{1/2}|j, m\pm 1\rangle \qquad (2.28)$$

The matrix elements for our three operators J_x, J_y, J_z (for a given value of j in the basis in which J_z is diagonal) are summarized below:

$$\langle j, m'|J_z|j, m\rangle = \hbar m\delta_{m'm} \qquad (2.29a)$$

$$\langle j, m'|J_x|j, m\rangle = \langle j, m'|\tfrac{1}{2}(J_+ + J_-)|j, m\rangle$$
$$= \frac{\hbar}{2}\left[\sqrt{j(j+1) - m(m+1)}\,\delta_{m',m+1} + \sqrt{j(j+1) - m(m-1)}\,\delta_{m',m-1}\right]$$

$$(2.29b)$$

$$\langle j, m'|J_y|j, m\rangle = \langle j, m'|\tfrac{i}{2}(J_- - J_+)|j, m\rangle$$
$$= \frac{i\hbar}{2}\left[-\sqrt{j(j+1) - m(m+1)}\,\delta_{m',m+1} + \sqrt{j(j+1) - m(m-1)}\,\delta_{m',m-1}\right]$$

$$(2.29c)$$

The Kronecker delta $\delta_{m'm}$ is defined in Problem 3.8.

Example 2.2 Using (2.29), construct the matrix representations for the operators J_x, J_y, J_z for angular momentum $j = 1$.

For angular momentum $j = 1$, (2.20) tells us that the possible values for the z-projection of the angular momentum range from -1 to 1 in integer steps.

Therefore, $m', m = -1, 0, 1$. The matrix representations for the angular momentum operators for angular momentum $j = 1$ are thus represented by 3x3 matrices.

From (2.29b) and (2.29c), we see that the Kronecker deltas allow for nonzero elements only on the off-diagonals in the matrix representation of J_x and J_y. Filling in the numbers,

$$J_x = \frac{\hbar}{2} \begin{pmatrix} 0 & \sqrt{2} & 0 \\ \sqrt{2} & 0 & \sqrt{2} \\ 0 & \sqrt{2} & 0 \end{pmatrix} \qquad J_y = \frac{\hbar}{2} \begin{pmatrix} 0 & -i\sqrt{2} & 0 \\ i\sqrt{2} & 0 & -i\sqrt{2} \\ 0 & i\sqrt{2} & 0 \end{pmatrix}$$

As for the matrix representation of J_z, (2.29a) allows for nonzero elements only along the diagonal (as expected)

$$J_z = \hbar \begin{pmatrix} 1 & 0 & 0 \\ 0 & 0 & 0 \\ 0 & 0 & -1 \end{pmatrix}$$

∎

2.4 Orbital Angular Momentum and the Spherical Harmonics

The orbital angular momentum operator $\vec{L}$ in quantum mechanics is defined by the relation

$$\vec{L} = \vec{r} \times \vec{p} \tag{2.30}$$

where $\vec{r}$ is the single-particle position operator in three-dimensional coordinate space and $\vec{p}$ is the single-particle momentum operator. $\vec{L}, \vec{r}$, and $\vec{p}$ are vector operators whose (orthogonal) components are Hermitian operators. The components of the position and momentum operators obey the fundamental commutation relations:

$$[r_i, p_j] = i\hbar\delta_{ij}; \quad i, j = x, y, z \tag{2.31a}$$

$$[r_i, r_j] = [p_i, p_j] = 0; \quad i, j = x, y, z \tag{2.31b}$$

In the position representation, the momentum operator is given by

$$\vec{p} = -i\hbar\vec{\nabla} \tag{2.32}$$

Using (2.30) and (2.31a and 2.31b), it is easy to show that the components of the orbital angular momentum operator obey the following commutation relations:

$$[L_l, r_j] = i\hbar \sum_k \varepsilon_{ljk} r_k; \quad l, j, k = x, y, z \tag{2.33a}$$

$$[L_l, p_j] = i\hbar \sum_k \varepsilon_{ljk} p_k; \quad l, j, k = x, y, z \tag{2.33b}$$

$$[L_l, L_j] = i\hbar \sum_k \varepsilon_{ljk} L_k; \quad l, j, k = x, y, z \tag{2.33c}$$

Example 2.3 Using (2.30) and (2.31a and 2.31b), evaluate the commutators $[L_x, L_y]$, $[L_x, y]$, and $[L_z, p_x]$.

$$\begin{aligned}
[L_x, L_y] &= (yp_z - zp_y)(zp_x - xp_z) - (zp_x - xp_z)(yp_z - zp_y) \\
&= xp_y(zp_z - p_z z) - yp_x(zp_z - p_z z) = (xp_y - yp_x)[z, p_z] \\
&= i\hbar(xp_y - yp_x) = i\hbar L_z \\
[L_x, y] &= (yp_z - zp_y)y - y(yp_z - zp_y) = z[y, p_y] - [y, y]p_z = i\hbar z \\
[L_z, p_x] &= (xp_y - yp_x)p_x - p_x(xp_y - yp_x) = y[p_x, p_x] - [p_x, x]p_y = i\hbar p_y
\end{aligned}$$

where (2.4) was also used. ∎

Because the components of the orbital angular momentum operator obey (2.33c), we are justified in identifying $\vec{L}$ as an angular momentum operator [c.f. (2.1)]. As such, we can immediately write the following results:

$$L^2|l, m\rangle = \hbar^2 l(l+1)|l, m\rangle \tag{2.34a}$$

$$L_z|l, m\rangle = \hbar m|l, m\rangle \tag{2.34b}$$

$$L_\pm|l, m\rangle = \hbar[l(l+1) - m(m \pm 1)]^{1/2}|l, m \pm 1\rangle \tag{2.34c}$$

where the $|l, m\rangle$ are the simultaneous eigenkets of the operators L^2 and L_z with the indicated eigenvalues.

Many physical systems are conveniently described in spherical coordinates. In spherical coordinates, the unit vector $\hat{n}$ designates a direction in space that is specified by the spherical angles (θ, ϕ). If we are going to be operating in three-dimensional coordinate space using spherical coordinates, we will need to define the direction eigenkets $|\hat{n}\rangle$. We can then project the $|l, m\rangle$ into the coordinate representation as follows:

$$\langle \hat{n}|l, m\rangle \equiv Y_{lm}(\theta, \phi) \tag{2.35}$$

For now, all we will say about $Y_{lm}(\theta, \phi)$ is that it is the amplitude for the state characterized by l,m to be oriented in the direction $\hat{n}$ specified by the angles θ and ϕ.

To construct the form of the angular momentum operators in three-dimensional coordinate space, we start with the transformation equations between Cartesian coordinates and spherical coordinates:

$$x = r \cos \phi \sin \theta \tag{2.36a}$$

$$y = r \sin \phi \sin \theta \tag{2.36b}$$

$$z = r \cos \theta \tag{2.36c}$$

and the inverse transformations

$$r = \sqrt{x^2 + y^2 + z^2} \tag{2.37a}$$

$$\tan \theta = \sqrt{x^2 + y^2}/z \tag{2.37b}$$

$$\tan \phi = y/x \tag{2.37c}$$

The Cartesian components of the spherical unit vectors are

$$\hat{r} = \hat{x} \cos \phi \sin \theta + \hat{y} \sin \phi \sin \theta + \hat{z} \cos \theta \tag{2.38a}$$

$$\hat{\theta} = \hat{x} \cos \phi \cos \theta + \hat{y} \sin \phi \cos \theta - \hat{z} \sin \theta \tag{2.38b}$$

$$\hat{\phi} = -\hat{x} \sin \phi + \hat{y} \cos \phi \tag{2.38c}$$

Using the above transformation equations, the form of the gradient operator in spherical coordinates is found to be

$$\vec{\nabla} = \hat{r} \frac{\partial}{\partial r} + \hat{\theta} \frac{1}{r} \frac{\partial}{\partial \theta} + \hat{\phi} \frac{1}{r \sin \theta} \frac{\partial}{\partial \phi} \tag{2.39}$$

We now have enough information to find the orbital angular momentum operator and the eigenvalue equations in spherical coordinates:

$$\vec{L} = \frac{\hbar}{i} \vec{r} \times \vec{\nabla} = \frac{\hbar}{i} \left(\hat{r} \times \hat{\theta} \frac{\partial}{\partial \theta} + \hat{r} \times \hat{\phi} \frac{1}{\sin \theta} \frac{\partial}{\partial \phi} \right)$$

$$= \frac{\hbar}{i} \left(\hat{\phi} \frac{\partial}{\partial \theta} - \hat{\theta} \frac{1}{\sin \theta} \frac{\partial}{\partial \phi} \right) \tag{2.40}$$

where (2.32) was also used.

Example 2.4 Verify (2.39).

From (2.38a, 2.38b and 2.38c)

$$\frac{\partial \hat{r}}{\partial r} = 0$$

$$\frac{\partial \hat{r}}{\partial \theta} = \hat{x}\cos\phi\cos\theta + \hat{y}\sin\phi\cos\theta - \hat{z}\sin\theta = \hat{\theta}$$

$$\frac{\partial \hat{r}}{\partial \phi} = -\hat{x}\sin\phi\sin\theta + \hat{y}\cos\phi\sin\theta = \hat{\phi}\sin\theta$$

Also, since $\hat{r} = \hat{r}(\theta, \phi)$,

$$d\vec{r} = d(r\hat{r}) = \hat{r}dr + rd\hat{r} = \hat{r}dr + r\left(\frac{\partial \hat{r}}{\partial r}dr + \frac{\partial \hat{r}}{\partial \theta}d\theta + \frac{\partial \hat{r}}{\partial \phi}d\phi\right)$$

$$= dr\hat{r} + rd\theta\hat{\theta} + r\sin\theta d\phi\hat{\phi}$$

Suppose $u = u\left(\vec{r}\right) = u(r, \theta, \phi)$. Then,

$$du = \frac{\partial u}{\partial r}dr + \frac{\partial u}{\partial \theta}d\theta + \frac{\partial u}{\partial \phi}d\phi$$

By definition, we also have

$$du = \vec{\nabla}u \cdot d\vec{r}$$

Therefore,

$$\frac{\partial u}{\partial r}dr + \frac{\partial u}{\partial \theta}d\theta + \frac{\partial u}{\partial \phi}d\phi = \left(\vec{\nabla}u\right)_r dr + \left(\vec{\nabla}u\right)_\theta rd\theta + \left(\vec{\nabla}u\right)_\phi r\sin\theta d\phi$$

giving us the gradient in spherical polar coordinates

$$\vec{\nabla}u = \hat{r}\left(\vec{\nabla}u\right)_r + \hat{\theta}\left(\vec{\nabla}u\right)_\theta + \hat{\phi}\left(\vec{\nabla}u\right)_\phi = \hat{r}\frac{\partial u}{\partial r} + \hat{\theta}\frac{1}{r}\frac{\partial u}{\partial \theta} + \hat{\phi}\frac{1}{r\sin\theta}\frac{\partial u}{\partial \phi}$$

Thus, verifying (2.39). ∎

Combining (2.38a, 2.38b and 2.38c) and (2.40), and taking (2.30) into account gives us the forms of the component operators:

$$L_x = \left(yp_z - zp_y\right) = i\hbar\left(\sin\phi\frac{\partial}{\partial \theta} + \frac{\cos\phi}{\tan\theta}\frac{\partial}{\partial \phi}\right) \qquad (2.41a)$$

$$L_y = \left(zp_x - xp_z\right) = i\hbar\left(\frac{\sin\phi}{\tan\theta}\frac{\partial}{\partial\phi} - \cos\phi\frac{\partial}{\partial\theta}\right) \tag{2.41b}$$

$$L_z = \left(xp_y - yp_x\right) = -i\hbar\frac{\partial}{\partial\phi} \tag{2.41c}$$

We can also construct the form of the orbital angular momentum ladder operators, and the operator for the square of the orbital angular momentum:

$$L_\pm = -i\hbar e^{\pm i\phi}\left(\pm i\frac{\partial}{\partial\theta} - \cot\theta\frac{\partial}{\partial\phi}\right) = \hbar e^{\pm i\phi}\left(\pm\frac{\partial}{\partial\theta} + i\cot\theta\frac{\partial}{\partial\phi}\right) \tag{2.42a}$$

$$L^2 = L_x^2 + L_y^2 + L_z^2 = -\hbar^2\left[\frac{1}{\sin\theta}\frac{\partial}{\partial\theta}\left(\sin\theta\frac{\partial}{\partial\theta}\right) + \frac{1}{\sin^2\theta}\frac{\partial^2}{\partial\phi^2}\right] \tag{2.42b}$$

In the position representation, the orbital angular momentum operator acts on the states $|l, m\rangle$ as follows:

$$\langle\hat{n}|\vec{L}|l, m\rangle = -i\hbar\vec{r} \times \vec{\nabla}Y_{lm}(\theta, \phi) \tag{2.43}$$

To see this more clearly, let the following operator

$$1 - i\left(\frac{\delta\phi}{\hbar}\right)L_z = 1 - i\left(\frac{\delta\phi}{\hbar}\right)\left(xp_y - yp_x\right) \tag{2.44}$$

act on a position eigenket $|\vec{r}\rangle = |x, y, z\rangle$. Keeping in mind that the momentum operator is the generator of translations in quantum mechanics, we get

$$\left[1 - i\left(\frac{\delta\phi}{\hbar}\right)L_z\right]|x, y, z\rangle = \left[1 - i\left(\frac{\delta\phi}{\hbar}\right)\left(xp_y - yp_x\right)\right]|x, y, z\rangle$$

$$= |x - y\delta\phi, y + x\delta\phi, z\rangle \tag{2.45}$$

By the way, this result is exactly what we would expect if the operator of (2.44) rotated the position eigenket by an infinitesimal angle $\delta\phi$ about the z-axis. Furthermore, we see that L_z is the generator of this rotation (more on this later). Now consider the ket $|l, m\rangle$ projected into the coordinate representation in spherical coordinates:

$$\langle x, y, z|l, m\rangle \rightarrow \langle r, \theta, \phi|l, m\rangle \tag{2.46}$$

After an infinitesimal rotation about the z-axis, we have

$$\langle x, y, z| \left[1 - i\left(\frac{\delta\phi}{\hbar}\right) L_z \right] |l, m\rangle = \langle x + y\delta\phi, y - x\delta\phi, z|l, m\rangle$$

$$\rightarrow \langle r, \theta, \phi - \delta\phi|l, m\rangle \qquad (2.47)$$

Expanding to first-order in $\delta\phi$,

$$\langle r, \theta, \phi - \delta\phi|l, m\rangle \approx \langle r, \theta, \phi|l, m\rangle - \delta\phi \frac{\partial}{\partial\phi}\langle r, \theta, \phi|l, m\rangle \qquad (2.48)$$

and substituting into (2.47) gives,

$$\langle r, \theta, \phi| \left[1 - i\left(\frac{\delta\phi}{\hbar}\right) L_z \right] |l, m\rangle = \langle r, \theta, \phi|l, m\rangle - \delta\phi \frac{\partial}{\partial\phi}\langle r, \theta, \phi|l, m\rangle \qquad (2.49)$$

Comparing both sides of (2.49) allows us to conclude

$$\langle r, \theta, \phi|L_z|l, m\rangle = -i\hbar\frac{\partial}{\partial\phi}\langle r, \theta, \phi|l, m\rangle \qquad (2.50)$$

We could perform the same analysis on the other components of the orbital angular momentum operator, thereby validating (2.43). We now strip the radial dependance out of (2.50) by first letting $\langle \vec{r}| = \langle r, \theta, \phi| \rightarrow \langle \hat{n}| = \langle \theta, \phi|$ and then multiplying both sides of (2.34b) from the left by the bra $\langle \hat{n}|$,

$$\langle \hat{n}|L_z|l, m\rangle = \hbar m\langle \hat{n}|l, m\rangle \Rightarrow -i\hbar\frac{\partial}{\partial\phi}\langle \hat{n}|l, m\rangle = \hbar m\langle \hat{n}|l, m\rangle$$

$$\Rightarrow -i\frac{\partial}{\partial\phi}Y_{lm}(\theta, \phi) = mY_{lm}(\theta, \phi) \qquad (2.51)$$

This equation implies that the ϕ-dependance of $Y_{lm}(\theta, \phi)$ goes as $e^{im\phi}$. The requirement that $Y_{lm}(\theta, \phi)$ be single-valued means that m must be an integer. It then follows from (2.20) that l must also be an integer.

From (2.34a) and (2.42b) we could construct, in a similar way,

$$\left[\frac{1}{\sin\theta}\frac{\partial}{\partial\theta}\left(\sin\theta\frac{\partial}{\partial\theta}\right) + \frac{1}{\sin^2\theta}\frac{\partial^2}{\partial\phi^2} + l(l+1) \right] Y_{lm}(\theta, \phi) = 0 \qquad (2.52)$$

And from (2.34c) and (2.42a),

$$-ie^{\pm i\phi}\left(\pm i\frac{\partial}{\partial\theta} - \cot\theta\frac{\partial}{\partial\phi}\right) Y_{lm}(\theta, \phi) = [l(l+1) - m(m \pm 1)]^{1/2}Y_{l,m\pm 1}(\theta, \phi)$$

$$(2.53)$$

The solutions to (2.51), (2.52) and (2.53) are the well-known spherical harmonics:

$$Y_{lm}(\theta,\phi) = \begin{cases} \dfrac{(-1)^l}{2^l l!}\sqrt{\dfrac{(2l+1)(l+m)!}{4\pi(l-m)!}}\, e^{im\phi}\, \dfrac{1}{\sin^m\theta}\dfrac{d^{l-m}}{d(\cos\theta)^{l-m}}\sin^{2l}\theta; \, m \geq 0 \\ (-1)^m Y^*_{l,-m}(\theta,\phi); \, m < 0 \end{cases}$$

(2.54)

where $l = 0, 1, 2, \ldots$ and $-l \leq m \leq l$.

Equations (2.54) and (2.55) reflect standard phase conventions and are consistent with the phase conventions of (2.29b) and (2.29c).

Below are listed some properties of the spherical harmonics, which may be deduced from (2.54):

$$Y^*_{lm}(\theta,\phi) = (-1)^m Y_{l,-m}(\theta,\phi) \quad \text{(complex conjugation)} \tag{2.55}$$

$$Y_{lm}(\pi-\theta, \phi+\pi) = (-1)^l Y_{lm}(\theta,\phi) \quad \text{(space inversion)} \tag{2.56}$$

$$Y_{lm}(\theta,-\phi) = (-1)^l Y_{lm}(\theta,\phi) = Y_{lm}(\theta, 2\pi-\phi) = Y^*_{lm}(\theta,\phi) \tag{2.57}$$

$$Y_{lm}(\pi,\phi) = (-1)^l\sqrt{\frac{2l+1}{4\pi}}\delta_{m0} \quad \text{for } l+m = \text{even} \tag{2.58}$$

$$Y_{lm}(\pi/2,0) = (-1)^{\frac{l-m}{2}}\sqrt{\frac{2l+1}{4\pi}}\frac{[(l-m)!(l+m)!]^{1/2}}{(l-m)!!(l+m)!!} \quad \text{for } l+m = \text{even} \tag{2.59}$$

$$Y_{l0}(\theta,\phi) = \sqrt{\frac{2l+1}{4\pi}}P_l(\cos\theta) \tag{2.60}$$

where the $P_l(\cos\theta)$ are the Legendre polynomials (see App. A). Continuing,

$$Y_{lm}(0,\phi) = \sqrt{\frac{2l+1}{4\pi}}\delta_{m0} \tag{2.61}$$

From the orthogonality relation,

$$\langle l', m' | l, m \rangle = \delta_{ll'}\delta_{mm'} \tag{2.62}$$

and the completeness relation for the position eigenkets,

$$\int d\Omega |\hat{n}\rangle\langle\hat{n}| = 1 \tag{2.63}$$

we get the orthonormality condition for the spherical harmonics:

$$\int\limits_{0}^{2\pi} d\phi \int\limits_{-1}^{1} d(\cos\theta) Y^{*}_{l'm'}(\theta,\phi) Y_{lm}(\theta,\phi) = \delta_{ll'}\delta_{mm'} \qquad (2.64)$$

Since the spherical harmonics are the eigenfunctions of a Hermitian operator, they form a complete set. We can see this directly from the properties of the position eigenkets $|\theta,\phi\rangle$:

$$\langle\theta,\phi|\theta',\phi'\rangle = \delta(\Omega-\Omega') \Rightarrow \sum_{l}\sum_{m}\langle\theta,\phi|l,m\rangle\langle l,m|\theta',\phi'\rangle = \delta(\Omega-\Omega')$$

$$\Rightarrow \sum_{l}\sum_{m} Y^{*}_{lm}(\theta,\phi) Y_{lm}(\theta',\phi') = \delta(\Omega-\Omega') = \frac{\delta(\theta-\theta')\delta(\phi-\phi')}{\sin\theta} \qquad (2.65)$$

where we also used the completeness relation

$$\sum_{l}\sum_{m}|l,m\rangle\langle l,m| = 1 \qquad (2.66)$$

The following is a list of the first few spherical harmonics in their spherical and Cartesian forms:

$$Y_{00}(\theta,\phi) = \frac{1}{\sqrt{4\pi}} \qquad (2.67a)$$

$$Y_{10}(\theta,\phi) = \sqrt{\frac{3}{4\pi}}\cos\theta = \sqrt{\frac{3}{4\pi}}\frac{z}{r} \qquad (2.67b)$$

$$Y_{1,\pm1}(\theta,\phi) = \mp\sqrt{\frac{3}{8\pi}}e^{\pm i\phi}\sin\theta = \mp\sqrt{\frac{3}{8\pi}}\frac{(x\pm iy)}{r} \qquad (2.67c)$$

$$Y_{20}(\theta,\phi) = \sqrt{\frac{5}{16\pi}}(3\cos^2\theta - 1) = \sqrt{\frac{5}{16\pi}}\frac{2z^2 - x^2 - y^2}{r^2} \qquad (2.67d)$$

$$Y_{2,\pm1}(\theta,\phi) = \mp\sqrt{\frac{15}{8\pi}}e^{\pm i\phi}\cos\theta\sin\theta = \mp\sqrt{\frac{15}{8\pi}}\frac{(x\pm iy)z}{r^2} \qquad (2.67e)$$

$$Y_{2,\pm2}(\theta,\phi) = \sqrt{\frac{15}{32\pi}}e^{\pm 2i\phi}\sin^2\theta = \sqrt{\frac{15}{32\pi}}\frac{(x\pm iy)^2}{r^2} \qquad (2.67f)$$

$$Y_{30}(\theta,\phi) = \sqrt{\frac{7}{4\pi}}\left(\frac{5}{2}\cos^3\theta - \frac{3}{2}\cos\theta\right) = \sqrt{\frac{7}{4\pi}}z\frac{\left(\frac{5}{2}z^2 - \frac{3}{2}r^2\right)}{r^3} \qquad (2.67g)$$

$$Y_{3,\pm 1}(\theta, \phi) = \mp\sqrt{\frac{7}{48\pi}} e^{\pm i\phi}\left(\frac{15}{2}\cos^2\theta - \frac{3}{2}\right)\sin\theta$$

$$= \mp\sqrt{\frac{7}{48\pi}}\frac{\left(\frac{15}{2}z^2 - \frac{3}{2}r^2\right)(x \pm iy)}{r^3} \tag{2.67h}$$

$$Y_{3,\pm 2}(\theta, \phi) = \sqrt{\frac{7}{480\pi}} e^{\pm 2i\phi}\left(15\cos\theta\sin^2\theta\right)$$

$$= \sqrt{\frac{7}{480\pi}}\frac{15z(x^2 - y^2 \pm 2ixy)}{r^3} \tag{2.67i}$$

$$Y_{3,\pm 3}(\theta, \phi) = \mp\sqrt{\frac{7}{2880\pi}} e^{\pm 3i\phi} 15\sin^3\theta$$

$$= \mp\sqrt{\frac{7}{2880\pi}}\frac{15[x^3 - 3xy^2 \pm i(3x^2y - y^3)]}{r^3} \tag{2.67j}$$

A recurrence relation for the spherical harmonics is given in (2.68). This relation will be required later. For a proof of this relation, see App. A (see also Problem 4.9).[1]

$$\cos\theta Y_{lm} = \left[\frac{(l - m + 1)(l + m + 1)}{(2l + 1)(2l + 3)}\right]^{1/2} Y_{l+1,m}$$

$$+ \left[\frac{(l - m)(l + m)}{(2l - 1)(2l + 1)}\right]^{1/2} Y_{l-1,m} \tag{2.68}$$

Example 2.5 Equation (2.51) implies that we can write the eigenfunctions of the orbital angular momentum operators as $Y_{lm}(\theta, \phi) = N_l e^{im\phi} P_l^m(\theta)$. Use (2.42a and 2.42b), (2.53), and the appropriate normalization integrals to find the general form of $Y_{ll}(\theta, \phi)$.

From (2.42a and 2.42b) and (2.53), we operate on $Y_{ll}(\theta, \phi)$ with the raising operator, knowing that the result will vanish:

$$he^{i\phi}\left[i\cot\theta\frac{\partial}{\partial\phi} + \frac{\partial}{\partial\theta}\right]Y_{ll}(\theta, \phi) = 0 \Rightarrow \left[i\cot\theta\frac{\partial}{\partial\phi} + \frac{\partial}{\partial\theta}\right]e^{il\phi}P_l^l(\theta) = 0$$

$$\Rightarrow \left(-l\cot\theta + \frac{\partial}{\partial\theta}\right)P_l^l(\theta) = 0 \Rightarrow \frac{\partial P_l^l(\theta)/\partial\theta}{P_l^l(\theta)} = l\cot\theta \Rightarrow \frac{\partial}{\partial\theta}\ln P_l^l(\theta) = l\frac{\partial}{\partial\theta}\ln(\sin\theta)$$

$$\Rightarrow P_l^l(\theta) = (\sin\theta)^l$$

We use (2.64) to find the normalization constant N_l

---

[1] For an exhaustive list of formulas involving the spherical harmonics, see also [10].

$$\int_0^{2\pi} d\phi \int_{-1}^1 d(\cos\theta)Y_{ll}^*(\theta,\phi)Y_{ll}(\theta,\phi) = 1 \Rightarrow 2\pi|N_l|^2 \int_{-1}^1 d(\cos\theta)(\sin\theta)^{2l}$$

$$= 2\pi|N_l|^2 \int_{-1}^1 dx(1-x^2)^{2l} = 1$$

$$\Rightarrow |N_l| = \frac{1}{2^l l!}\sqrt{\frac{(2l+1)!}{4\pi}} \Rightarrow N_l = \frac{(-1)^l}{2^l l!}\sqrt{\frac{(2l+1)!}{4\pi}}$$

where we have used the indicated phase convention and the result

$$\int_{-1}^1 dx(1-x^2)^{2l} = \frac{2^{2l+1}(l!)^2}{(2l+1)!}$$

This gives us the general form for $Y_{ll}(\theta,\phi)$,

$$Y_{ll}(\theta,\phi) = \frac{(-1)^l}{2^l l!}\sqrt{\frac{(2l+1)!}{4\pi}}e^{il\phi}(\sin\theta)^l$$

■

2.5   The Addition Theorem for Spherical Harmonics

Suppose we have two vectors $\vec{x}$ and $\vec{x}'$ whose orientations with respect to a fixed set of axes are described by the spherical angles as shown in Fig. 2.1.

Fig. 2.1 The angle γ is the angle between the vectors $\vec{x}$ and $\vec{x}'$

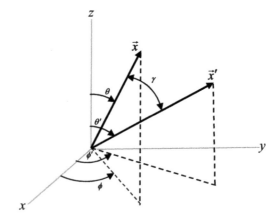

Recall the completeness relation for the spherical harmonics from (2.65) and note that the term $\delta(\Omega - \Omega')$ in that equation depends only on the angle γ, which is the same as the angle between the two vectors shown in Fig. 2.1. If we imagine the vector $\vec{x}$ as the polar axis of a new set of coordinate axes, then it is natural to want to expand $\delta(\Omega - \Omega')$ in the Legendre polynomials $P_l(\cos\gamma)$:

$$\delta(\Omega - \Omega') = \sum_l b_l P_l(\cos\gamma) \qquad (2.69)$$

where

$$b_l = \frac{2l+1}{2} \int_{-1}^{1} \delta(\Omega - \Omega')P_l(\cos\gamma)d(\cos\gamma)$$

$$= \frac{2l+1}{2(2\pi)} \int_0^{2\pi} d\psi \int_{-1}^{1} \delta(\Omega - \Omega')P_l(\cos\gamma)d(\cos\gamma) \qquad (2.70)$$

and where we have introduced an integration over azimuthal angle ψ about "axis $\vec{x}$" (γ would be the polar angle about this axis), and we used the orthogonality integral for the Legendre polynomials:

$$\int_{-1}^{1} dx P_l(x)P_{l'}(x) = \frac{2}{2l+1}\delta_{ll'} \qquad (2.71)$$

We further recognize that

$$d\Omega = d(\cos\gamma)d\psi \qquad (2.72)$$

We have (in our double integral) a solid-angle integration over a unit sphere. The integration was constructed for some arbitrary γ, so we are free to choose any γ. So, for purposes of the integration, we choose $\gamma = 0 \Rightarrow \cos\gamma = 1$:

$$\Rightarrow b_l = \frac{2l+1}{2(2\pi)}P_l(1)\int_{\substack{unit \\ sphere}} \delta(\Omega - \Omega')d\Omega = \frac{2l+1}{4\pi} \qquad (2.73)$$

where we used the boundary condition

$$P_l(1) = 1 \qquad (2.74)$$

Since we are integrating over the entire unit sphere, it does not matter which set of angles we use. Hence,

$$\int_{\substack{unit \\ sphere}} \delta(\Omega - \Omega')d\Omega = \int_{\substack{unit \\ sphere}} \delta(\Omega - \Omega')d\Omega' = \int_{\substack{unit \\ sphere}} \delta(\Omega' - \Omega)d\Omega' = 1 \quad (2.75)$$

Combining (2.65), (2.69), and (2.73)

$$\Rightarrow \sum_{lm} Y^*_{lm}(\Omega')Y_{lm}(\Omega) = \sum_l \left(\frac{2l+1}{4\pi}\right)P_l(\cos\gamma) \qquad (2.76)$$

We are not quite finished until we satisfy ourselves that we can equate terms in l on both sides of (2.76).

First, it is always possible to define a new set of axes via some rotation. All the $Y_{lm}(\Omega')$ measured in the new coordinate system can then be expressed as linear combinations of the $Y_{lm}(\Omega)$ from the old coordinate system, all having the same l:

$$Y_{lm}(\Omega') = \sum_{m'=-l}^{l} D^l_{m'm} Y_{lm'}(\Omega) \qquad (2.77)$$

where $D^l_{mm'}$ describes the rotation (see the section on rotations below). Remember that Ω and Ω' refer to the same direction in space but are expressed in different coordinate systems. Furthermore, from (2.60), note that $P_l(\cos\gamma)$ can be similarly expressed:

$$P_l(\cos\gamma) = \sqrt{\frac{4\pi}{2l+1}}Y_{l0}(\Omega') = \sum_{m'=-l}^{l} D^l_{0m'} Y_{lm'}(\Omega) \qquad (2.78)$$

Thus, we are satisfied that we can equate terms in l on both sides of (2.76):

$$P_l(\cos\gamma) = \frac{4\pi}{2l+1} \sum_{m=-l}^{l} Y^*_{lm}(\Omega')Y_{lm}(\Omega) \qquad (2.79)$$

This is the addition theorem for spherical harmonics. This theorem is valid for any γ as indicated by the symmetry in the expression between the prime and the unprimed coordinates. Either function on the RHS could be the complex conjugate, since the sum is over positive and negative values of m, and spherical harmonics are related to their complex conjugates by (2.55).

Example 2.6 As an example of the usefulness of addition theorem for spherical harmonics, let us apply it under the condition $\gamma = 0$:

$$\Rightarrow \frac{2l+1}{4\pi} = \sum_{m=-l}^{l} |Y_{lm}(\theta,\phi)|^2 \tag{2.80}$$

This result is called the sum rule for spherical harmonics. ∎

Example 2.7 As another example, we apply the addition theorem for the case $l = 1$. There are three terms in the summation:

$$P_1(\cos\gamma) = \cos\gamma = \frac{1}{2}\left(\sin\theta'e^{-i\phi'}\right)^*\left(\sin\theta e^{-i\phi}\right) + \cos\theta'\cos\theta$$
$$+ \frac{1}{2}\left(-\sin\theta'e^{i\phi'}\right)^*\left(-\sin\theta e^{i\phi}\right)$$
$$\Rightarrow \cos\gamma = \cos\theta'\cos\theta + \frac{1}{2}\sin\theta\sin\theta'\left[e^{i(\phi'-\phi)} + e^{i(\phi-\phi')}\right]$$
$$\Rightarrow \cos\gamma = \cos\theta'\cos\theta + \sin\theta\sin\theta'\cos(\phi-\phi') \tag{2.81}$$

thereby verifying a trigonometric result that can be obtained using elementary methods (but only after considerable effort). ∎

2.6 Rotations in Quantum Mechanics

In the study of classical mechanics, one often encounters rotations. In classical mechanics, rotations are intimately connected with angular momentum, and it is no different in quantum mechanics. Just as the components of a vector change when the vector is rotated, so do state kets of rotated systems differ from their unrotated progenitors. We will confine ourselves to the formalism of the active transformation in which the state itself is rotated as opposed to passive transformation in which the basis kets (i.e., the axes) are rotated. We define a rotation operator $\mathcal{R}$ in the appropriate ket space as follows:

$$|\alpha'\rangle = \mathcal{R}|\alpha\rangle \Leftrightarrow \langle\alpha'| = \langle\alpha|\mathcal{R}^\dagger \tag{2.82}$$

where $|\alpha'\rangle/\langle\alpha'|$ and $|\alpha\rangle/\langle\alpha|$ represent the rotated and unrotated kets/bras, respectively.

The first observation we make is that probabilities should be unaffected by rotations,

$$\langle\alpha'|\alpha'\rangle = \langle\alpha|\alpha\rangle \Rightarrow \langle\alpha|\mathcal{R}^\dagger\mathcal{R}|\alpha\rangle = \langle\alpha|\alpha\rangle \Rightarrow \mathcal{R}^\dagger\mathcal{R} = 1 \tag{2.83}$$

The conclusion is that the rotation operator is unitary.

Our next observation is that the expectation of an observable A should not change under rotation:

$$\langle \alpha'|A'|\alpha'\rangle = \langle \alpha|A|\alpha\rangle \Rightarrow \langle \alpha|\mathcal{R}^\dagger A' \mathcal{R}|\alpha\rangle = \langle \alpha|A|\alpha\rangle \Rightarrow A = \mathcal{R}^\dagger A' \mathcal{R}$$

$$\Rightarrow A' = \mathcal{R}A\mathcal{R}^\dagger \tag{2.84}$$

Equation (2.84) tells us how operators transform under rotations. We have already seen [c.f. (2.44)] that the operator $\mathcal{R}_{\hat{z}}(\delta\phi) = 1 - i(\frac{\delta\phi}{\hbar})L_z$ rotated an orbital angular momentum state by an infinitesimal angle $\delta\phi$ about the z-axis. Furthermore, we saw that if the operator $\vec{p}$ is a generator of translations, then the orbital angular momentum operator $\vec{L} = \vec{r} \times \vec{p}$ is the generator of rotations in three-dimensional coordinate space. There are, however, other types of angular momentum—such as spin angular momentum—that have nothing to do with $\vec{r}$ and $\vec{p}$. In quantum mechanics, we therefore define the rotation operator $\mathcal{R}_{\hat{n}}(\delta\alpha)$ so that it is the operator responsible for a rotation about an axis oriented in the $\hat{n}$−direction by an infinitesimal angle $\delta\alpha$ and is given by

$$\mathcal{R}_{\hat{n}}(\delta\alpha) \equiv 1 - i\left(\frac{\vec{J}\cdot\hat{n}}{\hbar}\right)\delta\alpha \tag{2.85}$$

Since $\vec{J}$ is a Hermitian operator, $\mathcal{R}_{\hat{n}}(\delta\alpha)$ will be unitary, as required. We also expect that the rotation operator should approach the identity operator in the limit $\delta\alpha \to 0$. The operator of (2.85) meets this requirement.

Example 2.8 Equation (2.84) shows us how operators transform under rotations. Find the condition under which an operator is invariant under a rotation.

First recall that the operator responsible for a rotation about an axis oriented in the $\hat{n}$−direction by an infinitesimal angle $\delta\alpha$ is given by

$$\mathcal{R}_{\hat{n}}(\delta\alpha) = 1 - i\left(\frac{\vec{J}\cdot\hat{n}}{\hbar}\right)\delta\alpha = 1 - \frac{i}{\hbar}J_n\delta\alpha$$

Apply to (2.84):

$$A' = \mathcal{R}(\delta\alpha)A\mathcal{R}^\dagger(\delta\alpha) = \left(1 - \frac{i}{\hbar}J_n\delta\alpha\right)A\left(1 + \frac{i}{\hbar}J_n\delta\alpha\right) = A - \frac{i}{\hbar}\delta\alpha(J_nA - AJ_n) + \cdots$$

$$\approx A - \frac{i}{\hbar}\delta\alpha[J_n, A]$$

So, the operator A is invariant under a rotation about an axis oriented in the $\hat{n}$−direction if $[J_n, A] = 0$.

<u>Comment</u>: We know from quantum theory that the operator A is a constant of the motion under the Hamiltonian H if $[A, H] = 0$. If $[J_n, H] = 0$, then J_n is a constant of the motion, and we see that angular momentum conservation corresponds to invariance under rotation. ∎

A finite rotation (about a specified axis u) through an angle α can be built out of a succession of n equal infinitesimal rotations of angle $\delta\alpha$ such that $\delta\alpha = \alpha/n$. The operator that describes this finite rotation is

$$\mathcal{R}_u(\alpha) = [\mathcal{R}_u(\delta\alpha)]^n = \lim_{n\to\infty}\left(1 - \frac{i}{\hbar}\frac{\alpha}{n}J_u\right)^n = \exp\left(-\frac{i\alpha}{\hbar}J_u\right)$$

$$\Rightarrow \mathcal{R}_u(\alpha) = e^{-i\alpha\vec{J}\cdot\widehat{u}/\hbar} \tag{2.86}$$

The form of the rotation operator of (2.86) and the fact that angular momentum operators obey the commutation rules of (2.1) together guarantee that rotations about different axes will not commute. This is as it should be, because physical successive finite rotations about axes in different directions generally do not commute.

In three-dimensional coordinate space, one could uniquely characterize a finite rotation of angle α about the $\widehat{n}-$ axis by specifying the three components of the vector $\alpha\widehat{n}$. A more convenient characterization is achieved by making use of the Euler-angle description of rotations. The Euler angles can be used to specify the orientation of a rotated body with respect to a set of (unrotated) axes that remain fixed. The two sets of coordinates have a common origin. In Fig. 2.2, the "old" (unrotated) axes are symbolized by the (red) unit vectors $\left\{\widehat{X}, \widehat{Y}, \widehat{Z}\right\}$, and the "new" (rotated) axes are symbolized by the (green) unit vectors $\{\widehat{x}, \widehat{y}, \widehat{z}\}$. One transitions from the old coordinates to the new coordinates via a series of three sequential rotations as follows:

Fig. 2.2 The Euler angles and the relative placements of the $\left\{\widehat{X}, \widehat{Y}, \widehat{Z}\right\}$-axes and the $\{\widehat{x}, \widehat{y}, \widehat{z}\}$-axes. The angle γ in this application is not necessarily the same as the angle γ in Fig. 2.1 and (2.79) and (2.81)

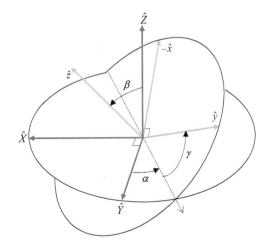

- Rotate counterclockwise about axis $\widehat{Z}$ through an angle α. The $\widehat{Y}$ axis is thus brought into alignment with the "line of nodes," denoted by a dotted line in the figure. The angle α is called the precession angle.
- Rotate about the line of nodes counterclockwise through an angle β. This action brings the $\widehat{Z}$-axis to its final orientation. The $\widehat{Z}$-axis now becomes the $\widehat{z}$-axis. The angle β is called the nutation angle.
- Rotate about the $\widehat{z}$-axis counterclockwise through an angle γ. Now the axes are all in their final resting places. The angle γ is called the body angle.

The rotation operator that describes these three successive rotations is

$$\mathscr{R}(\alpha,\beta,\gamma) = \exp\left(-\frac{i\gamma}{\hbar}J_z\right)\exp\left(-\frac{i\beta}{\hbar}J_N\right)\exp\left(-\frac{i\alpha}{\hbar}J_z\right) \quad (2.87)$$

The operator J_N (the angular momentum along the line of nodes) is actually the operator J_Y transformed by the rotation operator $\exp(-i\alpha J_z/\hbar)$. Therefore, according to (2.84),

$$J_N = \exp\left(-\frac{i\alpha}{\hbar}J_z\right)J_Y\exp\left(\frac{i\alpha}{\hbar}J_z\right) \quad (2.88)$$

The same transformation applies to all powers of J:

$$
\begin{aligned}
\exp\left(-\frac{i\beta}{\hbar}J_N\right) &= \sum_n (-\tfrac{i\beta}{\hbar})^n \left(\tfrac{1}{n!}\right)(J_N)^n \\
&= \sum_n (-\tfrac{i\beta}{\hbar})^n \left(\tfrac{1}{n!}\right)\left[\exp\left(-\frac{i\alpha}{\hbar}J_z\right)J_Y^n\exp\left(\frac{i\alpha}{\hbar}J_z\right)\right] \\
&= \exp\left(-\frac{i\alpha}{\hbar}J_z\right)\exp\left(-\frac{i\beta}{\hbar}J_Y\right)\exp\left(\frac{i\alpha}{\hbar}J_z\right)
\end{aligned}
\quad (2.89)
$$

Similarly,

$$\exp\left(-\frac{i\gamma}{\hbar}J_z\right) = \exp\left(-\frac{i\beta}{\hbar}J_N\right)\exp\left(-\frac{i\gamma}{\hbar}J_z\right)\exp\left(\frac{i\beta}{\hbar}J_N\right) \quad (2.90)$$

So, substituting (2.90) into (2.87) gives

$$\mathscr{R}(\alpha,\beta,\gamma) = \exp\left(-\frac{i\beta}{\hbar}J_N\right)\exp\left(-\frac{i\gamma}{\hbar}J_z\right)\exp\left(\frac{i\beta}{\hbar}J_N\right)\exp\left(-\frac{i\beta}{\hbar}J_N\right)\exp\left(-\frac{i\alpha}{\hbar}J_z\right)$$

$$= \exp\left(-\frac{i\beta}{\hbar}J_N\right)\exp\left(-\frac{i\gamma}{\hbar}J_z\right)\exp\left(-\frac{i\alpha}{\hbar}J_z\right) \quad (2.91)$$

Example 2.9 Put $\mathcal{R}(\alpha,\beta,\gamma)$ into a form that consists entirely of rotations in the same frame.

We use (2.89) to substitute for the first term on the RHS of (2.91)

$$\mathcal{R}(\alpha,\beta,\gamma) = \exp\left(-\frac{i\alpha}{\hbar}J_Z\right)\exp\left(-\frac{i\beta}{\hbar}J_Y\right)\exp\left(\frac{i\alpha}{\hbar}J_Z\right)\exp\left(-\frac{i\gamma}{\hbar}J_Z\right)\exp\left(-\frac{i\alpha}{\hbar}J_Z\right)$$

$$= \exp\left(-\frac{i\alpha}{\hbar}J_Z\right)\exp\left(-\frac{i\beta}{\hbar}J_Y\right)\exp\left(-\frac{i\gamma}{\hbar}J_Z\right)$$

$$(2.92)$$

where we used the fact that J_Z commutes with itself to cancel some terms. Equation (2.92) has the advantage in that it consists entirely of rotations in the same frame. It also shows that the Euler rotations may be carried out in the fixed frame if the order of the rotations is reversed. ∎

2.7 Matrix Elements of the Rotation Operators

The matrix elements of the rotation operator that describes a rotation through an angle α about an axis whose direction is specified by the unit vector $\hat{u}$ (in the basis that diagonalizes J^2 and J_z) are defined by

$$D^j_{m'm}(\alpha) \equiv \langle j,m'|\mathcal{R}_u(\alpha)|j,m\rangle = \langle j,m'|e^{-i\alpha\vec{J}\cdot\hat{u}/\hbar}|j,m\rangle \qquad (2.93)$$

We do not need to consider matrix elements between states of different j because the rotation operator commutes with J^2, and so those elements will vanish trivially. This result reflects the fact that the angular momentum of any system should depend neither on the direction in which we are looking nor the orientation of any set of axes. In this circumstance, the completeness relation for the angular momentum eigenstates reduces to

$$\sum_j \sum_{m'} |j,m'\rangle\langle j,m'| = 1 \rightarrow \sum_{m'} |j,m'\rangle\langle j,m'| = 1 \qquad (2.94)$$

Making use of (2.94), we perform a rotation on the state $|j,m\rangle$:

$$\mathcal{R}_u(\alpha)|j,m\rangle = \left(\sum_{m'}|j,m'\rangle\langle j,m'|\right)\mathcal{R}_u(\alpha)|j,m\rangle = \sum_{m'}D^j_{m'm}(\alpha)|j,m'\rangle \qquad (2.95)$$

So, although the rotated states $\mathcal{R}_u(\alpha)|j,m\rangle$ are eigenfunctions of J^2, they do not (generally) form a basis that diagonalizes J_z. In fact, (2.95) tells us that the rotated state is a linear superposition of eigenstates $|j, m'\rangle$, each with a different value of m'. The coefficients of the expansion in (2.95) are the $D^j_{m'm}$ (also known as the Wigner D-functions), which are the elements of a $(2j + 1) \times (2j + 1)$ unitary matrix.

In the case of a rotation characterized by the Euler angles, we have

$$D^j_{m'm}(\alpha, \phi, \gamma) = \langle j, m' | e^{-\frac{i\alpha}{\hbar}J_z} e^{-\frac{i\beta}{\hbar}J_y} e^{-\frac{i\gamma}{\hbar}J_z} | j, m \rangle = e^{-i\alpha m'} d^j_{m'm}(\beta) e^{-i\gamma m} \qquad (2.96)$$

where we have defined

$$d^j_{m'm}(\beta) \equiv \langle j, m' | e^{-\frac{i\beta}{\hbar}J_y} | j, m \rangle \qquad (2.97)$$

Note that

$$\left| D^j_{m'm}(\alpha, \beta, \gamma) \right|^2 = \left| d^j_{m'm}(\beta) \right|^2 \qquad (2.98)$$

A closed-form expression for the $d^j_{m'm}(\beta)$ is given by the Wigner formula [6]:

$$d^j_{m'm}(\beta) = \sum_n (-1)^n \frac{\sqrt{(j+m)!(j-m)!(j+m')!(j-m')!}}{n!(j-m'-n)!(j+m-n)!(m+m'-n)!}$$
$$\times \left(\cos \frac{\beta}{2} \right)^{2j+m-m'-2n} \left(-\sin \frac{\beta}{2} \right)^{m'-m+2n} \qquad (2.99)$$

where the summation is over values of n for which the arguments of the factorials in the denominator are not negative. Eq. (2.99) can be used to find a number of useful relations for the $d^j_{m'm}(\beta)$. For example, (2.99) is invariant under the replacements $m \rightleftarrows -m'$. As a result, we have

$$d^j_{m'm}(\beta) = d^j_{-m,-m'}(\beta) \qquad (2.100)$$

As another example, we replace β by $-\beta$ in (2.99):

$$d^j_{m'm}(-\beta) = d^j_{mm'}(\beta) \qquad (2.101)$$

Combining,

$$d^j_{m'm}(\beta) = d^{j*}_{m'm}(\beta) = d^j_{mm'}(-\beta) = (-1)^{m'-m} d^j_{-m',-m}(\beta)$$
$$= (-1)^{m'-m} d^j_{mm'}(\beta) \qquad (2.102)$$

For the case $\beta = 0$, only the $n = 0$ term in the summation of (2.99) contributes, and then only if $m' = m$,

$$d^j_{m'm}(0) = \delta_{m'm} \tag{2.103}$$

Similarly,

$$d^j_{m'm}(\pm\pi) = (-1)^{j\pm m'}\delta_{m',-m} \tag{2.104}$$

We are now in the position of being able to relate the spherical harmonics to the rotation matrices. Let us apply the rotation operator of (2.92) to the position eigenket $|\hat{z}\rangle$ to achieve the general position eigenket $|\hat{n}\rangle$. Since we are in three-dimensional coordinate space, we are obviously talking about orbital angular momentum with integer values of l. The symbolism is adjusted accordingly. With (2.92) in mind, we imagine first rotating about the Y-axis through an angle θ, and then about the Z-axis through an angle ϕ:

$$|\hat{n}\rangle = \mathcal{R}(\alpha = \phi, \beta = \theta, \gamma = 0)|\hat{z}\rangle \tag{2.105}$$

Applying the completeness relation for the orbital angular momentum eigenfunctions,

$$\sum_l \sum_m |l,m\rangle\langle l,m| = 1 \tag{2.106}$$

to (2.105) gives

$$|\hat{n}\rangle = \mathcal{R}(\phi,\theta,\gamma=0)\sum_l\sum_m |l,m\rangle\langle l,m|\hat{z}\rangle = \sum_l\sum_m \mathcal{R}(\phi,\theta,0)|l,m\rangle\langle l,m|\hat{z}\rangle$$

$$\Rightarrow \langle l',m'|\hat{n}\rangle = \sum_m D^l_{m'm}(\phi,\theta,0)\langle l,m|\hat{z}\rangle \tag{2.107}$$

where $D^l_{m'm}(\phi,\theta,\gamma)$ are the matrix elements defined in (2.96) for orbital angular momentum and for the angles indicated. From (2.35), (2.54), and (2.61), we have

$$\langle l,m|\hat{z}\rangle = Y_{lm}(\theta=0, \phi = \text{undetermined}) = \sqrt{\frac{2l+1}{4\pi}}\delta_{m0} \tag{2.108}$$

Combining (2.107) and (2.108),

$$\Rightarrow Y^*_{lm'}(\theta,\phi) = \sqrt{\frac{2l+1}{4\pi}}D^l_{m'0}(\phi,\theta,0) \Rightarrow D^l_{m0}(\alpha=\phi,\beta=\theta,0)$$

$$= \sqrt{\frac{4\pi}{2l+1}}Y^*_{lm}(\theta,\phi)\bigg|_{\theta=\beta,\phi=\alpha} \tag{2.109a}$$

Taking the complex conjugate and using (2.55) and example 2.10, we find

$$D^l_{0m}(\alpha = \phi, \beta = \theta, 0) = \sqrt{\frac{4\pi}{2l+1}} Y_{lm}(\theta, \phi) \Big|_{\theta=\beta, \phi=\alpha} \qquad (2.109b)$$

For the case $m = 0$,

$$D^l_{00}(\beta = \theta) = d^l_{00}(\beta) = P_l(\cos\theta) \qquad (2.110)$$

where we also used (2.96).[2]

Example 2.10 Find an expression for $D^{j*}_{m'm}(\alpha, \beta, \gamma)$. Also show that
$\sum_m D^{j*}_{km}(\alpha, \beta, \gamma) D^j_{k'm}(\alpha, \beta, \gamma) = \delta_{kk'}$.

$$\begin{aligned}
D^{j*}_{m'm}(\alpha, \beta, \gamma) &= \langle j, m' | \mathcal{R}(\alpha, \beta, \gamma) | j, m \rangle^* \\
&= \langle j, m | \mathcal{R}^\dagger(\alpha, \beta, \gamma) | j, m' \rangle = \langle j, m | \mathcal{R}^{-1}(\alpha, \beta, \gamma) | j, m' \rangle \\
&= D^j_{mm'}(-\alpha, -\beta, -\gamma) = (-1)^{m'-m} D^j_{-m',-m}(\alpha, \beta, \gamma)
\end{aligned}$$

where we also used the fact that the rotation operator is unitary.

Continuing,

$$\begin{aligned}
\sum_m D^{j*}_{nm}(\alpha, \beta, \gamma) D^j_{mn'}(\alpha, \beta, \gamma) &= \sum_m \langle j, n | \mathcal{R}^{-1}(\alpha, \beta, \gamma) | j, m \rangle \langle j, m | \mathcal{R}(\alpha, \beta, \gamma) | j, n' \rangle \\
&= \langle j, n | \mathcal{R}^{-1}(\alpha, \beta, \gamma) \mathcal{R}(\alpha, \beta, \gamma) | j, n' \rangle = \langle j, n | j, n' \rangle = \delta_{nn'}
\end{aligned}$$

∎

2.8   The Coupling of Two Angular Momenta

When examining a physical system, often we find that we must add two or more
independent angular momenta. For example, to find the total orbital angular momentum of an atom, we must add the orbital angular momenta of all the electrons in the
atom. To find the total angular momentum of an individual atomic electron, we add the
electron's spin angular momentum to its orbital angular momentum. In the latter case,
the total angular momentum $\vec{J}$ is the vector sum of the two individual angular
momentum vectors:

$$\vec{J} = \vec{L} + \vec{S} \qquad (2.111)$$

---

[2]For an exhaustive list of properties of the Wigner D-functions, see also [10].

where $\vec{L}$ is the orbital angular momentum operator we first saw in (2.30) and $\vec{S}$ is the single-particle spin operator. This example leads us to first consider a more general case of adding two angular momenta from different subspaces, $\vec{J}_1$ and $\vec{J}_2$. The total angular momentum for such a system is

$$\vec{J} = \vec{J}_1 + \vec{J}_2 \qquad (2.112)$$

Equation (2.112) implies

$$J_k = J_{1k} + J_{2k}; \quad k = x, y, z \qquad (2.113)$$

It is important to remember that $\vec{J}_1$ and $\vec{J}_2$ obey the angular-momentum commutation relations within their respective subspaces:

$$\left[J_{1l}, J_{1j}\right] = i\hbar \sum_k \epsilon_{ljk} J_{1k}; \quad l, j, k = x, y, z \qquad (2.114a)$$

$$\left[J_{2l}, J_{2j}\right] = i\hbar \sum_k \epsilon_{ljk} J_{2k}; \quad l, j, k = x, y, z \qquad (2.114b)$$

but that angular momenta from different subspaces commute

$$\left[J_{1l}, J_{2j}\right] = 0; \quad l, j = x, y, z \qquad (2.115)$$

A direct consequence of (2.114a and 2.114b) and (2.115) is that the Cartesian components of the total angular momentum $\vec{J}$ obey the angular-momentum commutation relations:

$$\left[J_l, J_j\right] = i\hbar \sum_k \epsilon_{ljk} J_k; \quad l, j, k = x, y, z \qquad (2.116)$$

Thus, $\vec{J}$ is, by definition, an angular momentum operator in the sense described in Sect. 2.1. We can therefore (for example) define the ladder operators J_+ and J_- for the composite system and the operator J^2 for the square of the total angular momentum of the composite system. $\vec{J}$ has the standard set of eigenvectors, eigenvalues, and matrix elements and is the generator of rotations for the entire system. We can also derive the following commutation relations:

$$\left[J^2, J_1^2\right] = \left[J^2, J_2^2\right] = \left[J_z, J_1^2\right] = \left[J_z, J_2^2\right] = 0 \qquad (2.117)$$

We therefore have two complete sets of mutually commuting observables that we can use to describe the composite system. We can either use the set of observables $J_1^2, J_{1z}, J_2^2, J_{2z}$ with their simultaneous eigenkets $\{ |j_1, m_1\rangle \otimes |j_2, m_2\rangle \Leftrightarrow |j_1, j_2; m_1, m_2\rangle \}$ as a basis set, or we can use the set of observables J^2, J_z, J_1^2, J_2^2 with their simultaneous set of eigenkets $\{ |j_1, j_2; j, m\rangle \Leftrightarrow |j, m\rangle \}$ as a basis set. The first representation is called the uncoupled representation or the uncoupled basis. The second is called the coupled representation or the coupled basis. Note that, for the coupled representation, the symbol for the eigenket $|j, m\rangle$ is shorthand for $|j_1, j_2; j, m\rangle$, since the quantities j_1, j_2 are understood (or given). In the same vein, we could also shorten the symbol for the eigenket in the uncoupled basis: $|j_1, j_2; m_1, m_2\rangle \rightarrow |m_1, m_2\rangle$, and this is sometimes done in the literature.

2.9   The Clebsch-Gordan Coefficients

The two representations described above are equivalent descriptions that span a composite space of dimensionality N, where

$$N = \sum_{\substack{j=j_1-j_2 \\ j_1 \geq j_2}}^{j=j_1+j_2} (2j+1) = (2j_1+1)(2j_2+1) \tag{2.118}$$

We connect the two descriptions using the Clebsch-Gordan (C-G) coefficients. The C-G coefficients $\langle j_1, j_2; m_1, m_2 | j, m \rangle$ are the elements of a unitary transformation which connect the uncoupled angular momentum basis $\{ |j_1, j_2; m_1, m_2\rangle \}$ with the coupled basis $\{ |j, m\rangle \}$. They arise in the expansion of the coupled basis on the uncoupled basis as follows:

$$\begin{aligned} |j, m\rangle &= \left[\sum_{m_1=-j_1}^{j_1} \sum_{m_2=-j_2}^{j_2} |j_1, j_2; m_1, m_2\rangle \langle j_1, j_2; m_1, m_2| \right] |j, m\rangle \\ &= \sum_{m_1} \sum_{m_2} \langle j_1, j_2; m_1, m_2 | j, m \rangle |j_1, j_2; m_1, m_2\rangle \end{aligned} \tag{2.119}$$

The C-G coefficients are sometimes also known as the Wigner coefficients or as the vector-coupling coefficients.

Because $J_z = J_{1z} + J_{2z}$, we can say,

$$\langle j_1, j_2; m_1, m_2 | (J_z - J_{1z} - J_{2z}) | j, m\rangle = 0 \Rightarrow (m - m_1 - m_2) \langle j_1, j_2; m_1, m_2 | j, m\rangle = 0$$

$$\Rightarrow \langle j_1, j_2; m_1, m_2 | j, m\rangle = 0 \text{ unless } m = m_1 + m_2 \tag{2.120}$$

Therefore, the double sum in (2.119) is actually a single sum, the sum over m_2 being superfluous:

$$|j, m\rangle = \sum_{m_1} \langle j_1, j_2; m_1, m - m_1 | j, m\rangle |j_1, j_2; m_1, m - m_1\rangle \tag{2.121}$$

In addition, the C-G coefficients must also satisfy the triangle selection rule(s). That is, the C-G coefficients are all identically zero unless the following conditions are simultaneously satisfied:

$$m = m_1 + m_2 \tag{2.122a}$$

$$|j_1 - j_2| \leq j \leq j_1 + j_2 \tag{2.122b}$$

It is clear that the (so-called) magnetic quantum numbers m add algebraically, while the angular momenta add as vectors. The general tenants of angular momentum theory derived above require that, for the ket $|j, m\rangle$ to exist, we must also have

$$m = j, j - 1, j - 2, \cdots, -j \tag{2.123a}$$

$$m_1 = j_1, j_1 - 1, j_1 - 2, \cdots, -j_1 \tag{2.123b}$$

$$m_2 = j_2, j_2 - 1, j_2 - 2, \cdots, -j_2 \tag{2.123c}$$

If the conditions in (2.122) and (2.123a, 2.123b and 2.123c) are not met, then the C-G coefficients are not defined.

The C-G coefficients are further defined by the initial condition:

$$\langle j_1, j_2; m_1, m_2 | j_1 + j_2, j_1 + j_2 \rangle = 1 \tag{2.124}$$

and by the phase convention that the C-G coefficients are all real. Finally, the C-G coefficients are sometimes represented by the abbreviated notations $\langle m_1, m_2 | j, m \rangle$, $C_{j_1, m_1; j_2, m_2}^{j, m}$, or $C_{m_1, m_2}^{j, m}$. We will use some of these alternate notations later when it proves convenient.

If the $|j, m\rangle$ are normalized, then

$$\langle j', m' | j, m \rangle = \delta_{j'j} \delta_{m'm}$$
$$\Rightarrow \langle j', m' | j, m \rangle = \sum_{m_1', m_2'} \langle j', m' | j_1, j_2; m_1', m_2' \rangle$$
$$\times \langle j_1, m_1' | \langle j_2, m_2' | \sum_{m_1, m_2} \langle j_1, j_2; m_1, m_2 | j, m \rangle | j_1, m_1 \rangle | j_1, m_1 \rangle$$
$$= \delta_{j'j} \delta_{m'm}$$
$$= \sum_{m_1, m_1', m_2, m_2'} \langle j', m' | j_1, j_2; m_1', m_2' \rangle \langle j_1, j_2; m_1, m_2 | j, m \rangle \overbrace{\langle j_1, m_1' | j_1, m_1 \rangle}^{\delta_{m_1' m_1}} \overbrace{\langle j_2, m_2' | j_2, m_2 \rangle}^{\delta_{m_2' m_2}}$$
$$= \delta_{j'j} \delta_{m'm}$$
$$\Rightarrow \sum_{m_1, m_2} \langle j', m' | j_1, j_2; m_1, m_2 \rangle \langle j_1, j_2; m_1, m_2 | j, m \rangle = \delta_{j'j} \delta_{m'm} \tag{2.125}$$

where we have used the fact that the C-G coefficients are real. Note again that, strictly speaking, we need only one index in the summation, since m_1 and m_2 are constrained by (2.122a).

Example 2.11 Find the normalization condition for the C-G coefficients.

If we set $j = j'$ and $m' = m = m_1 + m_2$ in (2.125), we get,

$$\sum_{m_1, m_2 = m - m_1} |\langle j_1, j_2; m_1, m_2 | j, m \rangle|^2 = 1 \qquad (2.126)$$

which is the normalization condition for C-G coefficients. ∎

Example 2.12 Invert (2.119) and use the result to find a second orthogonality condition for the C-G coefficients.

Using (2.122a and 2.122b), (2.123a, 2.123b and 2.123c), and (2.124),

$$|j, m\rangle = \sum_{j' = |j_1 - j_2|}^{j_1 + j_2} \sum_{m' = -j}^{j} \delta_{j'j} \delta_{m'm} |j', m'\rangle = \sum_{j' = |j_1 - j_2|}^{j_1 + j_2} \sum_{m' = -j}^{j} \langle j', m' | j, m \rangle |j', m'\rangle$$

$$= \sum_{j', m'} \left(\sum_{m_1, m_2} \langle j', m' | j_1, j_2; m_1, m_2 \rangle \langle j_1, j_2; m_1, m_2 | j, m \rangle \right) |j', m'\rangle$$

$$\Rightarrow |j, m\rangle = \sum_{j', m'} \sum_{m_1, m_2} \langle j_1, j_2; m_1, m_2 | j', m' \rangle \langle j_1, j_2; m_1, m_2 | j, m \rangle |j', m'\rangle \qquad (2.127)$$

But, by definition, we have

$$|j, m\rangle = \sum_{m_1, m_2} \langle j_1, j_2; m_1, m_2 | j, m \rangle |j_1, m_1\rangle |j_2, m_2\rangle \qquad (2.128)$$

Equating (2.127) and (2.128),

$$\sum_{m_1, m_2} \langle j_1, j_2; m_1, m_2 | j, m \rangle \left(\sum_{j', m'} \langle j_1, j_2; m_1, m_2 | j', m' \rangle |j', m'\rangle - |j_1, m_1\rangle |j_2, m_2\rangle \right) = 0$$

$$\Rightarrow |j_1, j_2; m_1, m_2\rangle = \sum_{j, m} \langle j_1, j_2; m_1, m_2 | j, m \rangle |j, m\rangle \qquad (2.129)$$

Equation (2.129) is the inverse of (2.119). We can now deduce a second orthogonality condition from (2.129) as follows:

$$\underbrace{\langle j_1, m_1' | j_1, m_1 \rangle}_{\delta_{m_1 m_1'}} \underbrace{\langle j_2, m_2' | j_2, m_2 \rangle}_{\delta_{m_2 m_2'}} = \sum_{j, m} \sum_{j', m'} \langle j_1, j_2; m_1, m_2 | j, m \rangle$$

$$\times \langle j', m' | j_1, j_2; m_1', m_2' \rangle \underbrace{\langle j', m' | j, m \rangle}_{\delta_{j'j} \delta_{m'm}}$$

$$\Rightarrow \sum_{j, m} \langle j_1, j_2; m_1, m_2 | j, m \rangle \langle j_1, j_2; m_1', m_2' | j, m \rangle = \delta_{m_1 m_1'} \delta_{m_2 m_2'} \qquad (2.130)$$

Of course, we already knew this because the C-G coefficients form a real unitary matrix, and a real unitary matrix is an orthogonal matrix. ∎

Since the $|j_1, j_2; m_1, m_2\rangle$ form a standard angular momentum basis, we have

$$J_{1\pm}|j_1, j_2; m_1, m_2\rangle = \hbar\sqrt{j_1(j_1+1) - m_1(m_1 \pm 1)}|j_1, j_2; m_1 \pm 1, m_2\rangle \quad (2.131a)$$

$$J_{2\pm}|j_1, j_2; m_1, m_2\rangle = \hbar\sqrt{j_2(j_2+1) - m_2(m_2 \pm 1)}|j_1, j_2; m_1, m_2 \pm 1\rangle \quad (2.131b)$$

And, by construction, the kets $|j, m\rangle$ satisfy

$$J_{\pm}|j, m\rangle = \hbar\sqrt{j(j+1) - m(m \pm 1)}|j, m \pm 1\rangle \quad (2.132)$$

for

$$J_{\pm} = J_{1\pm} + J_{2\pm} \quad (2.133)$$

So, for $m > -j$ we have, applying (2.131a, 2.131b, 2.132 and 2.133) to (2.119):

$$\sqrt{j(j+1) - m(m \pm 1)}|j, m \pm 1\rangle =$$
$$\sum_{m'_1, m'_2} \langle j_1, j_2; m'_1, m'_2 | j, m \rangle \left[\begin{array}{l} \sqrt{j_1(j_1+1) - m'_1(m'_1 \pm 1)}|j_1, j_2; m'_1 \pm 1, m'_2\rangle \\ + \sqrt{j_2(j_2+1) - m'_2(m'_2 \pm 1)}|j_1, j_2; m'_1, m'_2 \pm 1\rangle \end{array} \right]$$
$$(2.134)$$

Multiplying by the bra $\langle j_1, j_2; m_1, m_2 |$ gives

$$\sqrt{j(j+1) - m(m \pm 1)}\langle j_1, j_2; m_1, m_2 | j, m \pm 1\rangle$$
$$= \sqrt{j_1(j_1+1) - m_1(m_1 \mp 1)}\langle j_1, j_2; m_1 \mp 1, m_2 | j, m\rangle \quad (2.135)$$
$$+ \sqrt{j_2(j_2+1) - m_2(m_2 \mp 1)}\langle j_1, j_2; m_1, m_2 \mp 1 | j, m\rangle$$

A more convenient form of (2.135) is

$$\sqrt{(j \mp m)(j \pm m + 1)}\langle j_1, j_2; m_1, m_2 | j, m \pm 1\rangle$$
$$= \sqrt{(j_1 \mp m_1 + 1)(j_1 \pm m_1)}\langle j_1, j_2; m_1 \mp 1, m_2 | j, m\rangle \quad (2.136)$$
$$+ \sqrt{(j_2 \mp m_2 + 1)(j_2 \pm m_2)}\langle j_1, j_2; m_1, m_2 \mp 1 | j, m\rangle$$

Similarly, we could calculate $\langle j_1, j_2; m_1, m_2 | J_{\pm} | j, m \mp 1\rangle$ to get the recursion relation

$$\sqrt{(j \pm m)(j \mp m + 1)}\langle j_1, j_2; m_1, m_2 | j, m\rangle$$
$$= \sqrt{(j_1 \mp m_1 + 1)(j_1 \pm m_1)}\langle j_1, j_2; m_1 \mp 1, m_2 | j, m \mp 1\rangle \quad (2.137)$$
$$+ \sqrt{(j_2 \mp m_2 + 1)(j_2 \pm m_2)}\langle j_1, j_2; m_1, m_2 \mp 1 | j, m \mp 1\rangle$$

Equations (2.124), (2.126), and (2.136), along with the given phase convention are all that are needed to uniquely determine all the C-G coefficients.

That notwithstanding, the following closed-form expression for the C-G coefficients was developed by Racah [2]:

$$\langle j_1, j_2; m_1, m_2 | j_3, m_3 \rangle = \delta_{m_1 + m_2, m} \left[\frac{(2j+1)(j_1 + j_2 - j)!(j_1 - j_2 + j)!(j + j_2 - j_1)!}{(j_1 + j_2 + j + 1)!} \right]^{\frac{1}{2}}$$

$$\times \sum_z (-1)^z \frac{\sqrt{(j_1 + m_1)!(j_1 - m_1)!(j_2 + m_2)!(j_2 - m_2)!(j - m)!(j + m)!}}{z!(j_1 + j_2 - j - z)!(j_1 - m_1 - z)!(j_2 + m_2 - z)!(j - j_2 + m_1 + z)!(j - j_1 - m_2 + z)!}$$

$$(2.138)$$

To put all the angular momentum indices on a completely equal footing [and thereby completely highlight the symmetric nature of (2.138)], let $j \rightarrow j_3$; $m \rightarrow m_3$, and let $s \equiv j_1 + j_2 + j_3$, which gives [3]

$$\langle j_1, j_2; m_1, m_2 | j_3, m_3 \rangle = \delta_{m_1 + m_2, m_3} \left[\frac{(2j_3 + 1)(s - 2j_3)!(s - 2j_2)!(s - 2j_1)!}{(s + 1)!} \right]^{\frac{1}{2}}$$

$$\times \sum_z (-1)^z \frac{\sqrt{(j_1 + m_1)!(j_1 - m_1)!(j_2 + m_2)!(j_2 - m_2)!(j_3 + m_3)!(j_3 - m_3)!}}{z!(j_1 + j_2 - j_3 - z)!(j_1 - m_1 - z)!(j_2 + m_2 - z)!(j_3 - j_2 + m_1 + z)!(j_3 - j_1 - m_2 + z)!}$$

$$(2.139)$$

Many symmetry relations for the C-G coefficients can be readily obtained from (2.138) or (2.139) [10].

Example 2.13 Use (2.139) to find an expression for $\langle j_2, j_3; -m_2, m_3 | j_1, m_1 \rangle$ in terms of $\langle j_1, j_2; m_1, m_2 | j_3, m_3 \rangle$.

Making the substitution $n = (j_2 + m_2) - z$ and summing over n in (2.139),

$$\langle j_1, j_2; m_1, m_2 | j_3, m_3 \rangle = \delta_{m_1 + m_2, m_3} (-1)^{j_2 + m_2} \left[\frac{(2j_3 + 1)}{(2j_1 + 1)} \right]^{\frac{1}{2}} \left[\frac{(2j_1 + 1)(s - 2j_3)!(s - 2j_2)!(s - 2j_1)!}{(s + 1)!} \right]^{\frac{1}{2}}$$

$$\times \sum_n (-1)^n \frac{\sqrt{(j_1 + m_1)!(j_1 - m_1)!(j_2 + m_2)!(j_2 - m_2)!(j_3 + m_3)!(j_3 - m_3)!}}{\left[(j_2 + m_2 - n)!(j_1 - j_3 - m_2 + n)! \left(j_1 - j_2 \underbrace{- m_1 - m_2}_{-m_3} + n \right)! n! \right.}$$

$$\left. \times \left(j_3 \underbrace{+ m_1 + m_2}_{+m_3} - n \right)!(j_3 - j_1 + j_2 - n)! \right]$$

$$= \delta_{m_1+m_2,m_3}(-1)^{j_2+m_2}\left[\frac{(2j_3+1)}{(2j_1+1)}\right]^{\frac{1}{2}}\left[\frac{(2j_1+1)(s-2j_3)!(s-2j_2)!(s-2j_1)!}{(s+1)!}\right]^{\frac{1}{2}}\times$$

$$\sum_n(-1)^n\frac{\sqrt{(j_1+m_1)!(j_1-m_1)!(j_2+m_2)!(j_2-m_2)!(j_3+m_3)!(j_3-m_3)!}}{\left[\begin{array}{c}n!(j_3-j_1+j_2-n)!(j_2+m_2-n)!(j_3+m_3-n)!\\ \times(j_1-j_3-m_2+n)!(j_1-j_2-m_3+n)!\end{array}\right]}$$

$$\Rightarrow \langle j_1,j_2;m_1,m_2|j_3,m_3\rangle = (-1)^{j_2+m_2}\left[\frac{(2j_3+1)}{(2j_1+1)}\right]^{\frac{1}{2}}\langle j_2,j_3; -m_2,m_3|j_1,m_1\rangle$$

$$(2.140)$$

∎

Similarly, by making the substitution $n=(j_1+j_2-j_3)-z$ and summing over n, one can easily verify that

$$\langle j_1,j_2;m_1,m_2|j_3,m_3\rangle = (-1)^{j_1+j_2-j_3}\langle j_1,j_2; -m_1, -m_2|j_3, -m_3\rangle$$
$$= (-1)^{j_1+j_2-j_3}\langle j_2,j_1;m_2,m_1|j_3,m_3\rangle \qquad (2.141)$$

And by making the substitution $n=j_1-m_1-z$ and summing over n, one could also show

$$\langle j_1,j_2;m_1,m_2|j_3,m_3\rangle = (-1)^{j_1-m_1}\sqrt{\frac{(2j_3+1)}{(2j_2+1)}}\langle j_1,j_3;m_1, -m_3|j_2, -m_2\rangle \quad (2.142)$$

and so forth.

We must be cautious when interpreting (2.138) and (2.139). In these equations, the summation index z ranges over all integral values for which the factorial arguments in the denominator of the last term on the RHS are positive; the factorial of a negative number being undefined in this context. This is demonstrated below explicitly for (2.139) as follows [3]:

$$\langle j_1,j_2;m_1,m_2|j_3,m_3\rangle =$$

$$\delta_{m_1+m_2,m_3}\left[\begin{array}{c}(2j_3+1)\dfrac{(s-2j_3)!(s-2j_2)!(s-2j_1)!}{(s+1)!}\\ \times(j_1+m_1)!(j_1-m_1)!(j_2+m_2)!(j_2-m_2)!(j_3+m_3)!(j_3-m_3)!\end{array}\right]^{\frac{1}{2}}$$

$$\times\sum_{z_{low}}^{z_{up}}\frac{(-1)^z}{z!\underbrace{(j_1+j_2-j_3-z)!}_{\geq 0}\underbrace{(j_1-m_1-z)!}_{\geq 0}\underbrace{(j_2+m_2-z)!}_{\geq 0}\underbrace{(j_3-j_2+m_1+z)!}_{\geq 0}\underbrace{(j_3-j_1-m_2+z)!}_{\geq 0}}$$

gives an upper limit for z gives an upper limit for z gives an upper limit for z gives a lower limit for z gives a lower limit for z

$$\Rightarrow z_{up}=j_1+j_2-j_3 \quad \Rightarrow z_{up}=j_1-m_1 \quad \Rightarrow z_{up}=j_2+m_2 \quad \Rightarrow z_{low}=j_2-j_3-m_1 \quad \Rightarrow z_{low}=j_1-j_3+m_2$$

$$\underbrace{\phantom{\Rightarrow z_{up}=j_1+j_2-j_3 \quad \Rightarrow z_{up}=j_1-m_1 \quad \Rightarrow z_{up}=j_2+m_2}}_{z_{up}=\text{minimum of these}} \qquad \underbrace{\phantom{\Rightarrow z_{low}=j_2-j_3-m_1 \quad \Rightarrow z_{low}=j_1-j_3+m_2}}_{z_{low}=\text{maximum of these}}$$

$$(2.143)$$

For those who are curious, the Racah formula for the C-G coefficients is derived in App. B.

For some parts of this document, we use not the C-G coefficients, but the Wigner 3-j symbols, which are defined as

$$\begin{pmatrix} j_1 & j_2 & j_3 \\ m_1 & m_2 & m_3 \end{pmatrix} = (-1)^{j_1-j_2-m_3}(2j_3+1)^{-\frac{1}{2}}\langle j_1,j_2;m_1,m_2|j_3,-m_3\rangle \quad (2.144a)$$

$$\langle j_1,j_2;m_1,m_2|j_3,m_3\rangle = (-1)^{j_1-j_2+m_3}(2j_3+1)^{\frac{1}{2}}\begin{pmatrix} j_1 & j_2 & j_3 \\ m_1 & m_2 & -m_3 \end{pmatrix} \quad (2.144b)$$

We use the 3-j symbols because they tend to have better symmetry properties upon exchange of angular momentum quantum numbers. For example, the 3-j symbols have the advantage that an even permutation of the columns is cyclic, that is, leaves the numerical value of the 3-j symbol unchanged (the presence of the phase factors in (2.144a and 2.144b) ensures this). The proof is simple (showing one cyclic permutation as an example):

$$\langle j_1,j_2;m_1,m_2|j_3,m_3\rangle$$

$$= (-1)^{j_2+m_2}\left[\frac{(2j_3+1)}{(2j_1+1)!}\right]^{\frac{1}{2}}\langle j_2,j_3;-m_2,m_3|j_1,m_1\rangle$$

$$= (-1)^{j_2+m_2}\left[\frac{(2j_3+1)}{(2j_1+1)!}\right]^{\frac{1}{2}}(-1)^{j_1+j_2-j_3}\langle j_2,j_3;m_2,-m_3|j_1,-m_1\rangle$$

$$\Rightarrow (-1)^{j_1-j_2+m_3}(2j_3+1)^{\frac{1}{2}}\begin{pmatrix} j_1 & j_2 & j_3 \\ m_1 & m_2 & -m_3 \end{pmatrix}$$

$$= (-1)^{j_2+m_2}\left[\frac{(2j_3+1)}{(2j_1+1)!}\right]^{\frac{1}{2}}(-1)^{j_1+j_2-j_3}(-1)^{j_2-j_3-m_1}(2j_1+1)^{\frac{1}{2}}\begin{pmatrix} j_2 & j_3 & j_1 \\ m_2 & -m_3 & m_1 \end{pmatrix}$$

$$\Rightarrow \begin{pmatrix} j_1 & j_2 & j_3 \\ m_1 & m_2 & -m_3 \end{pmatrix} = \underbrace{(-1)^{4j_2-2j_3-m_3+m_2-m_1}}_{=(-1)^{4j_2-2j_3-2m_1}=1 \ (\text{for integer or half-integer})}\begin{pmatrix} j_2 & j_3 & j_1 \\ m_2 & -m_3 & m_1 \end{pmatrix}$$

$$= \begin{pmatrix} j_2 & j_3 & j_1 \\ m_2 & -m_3 & m_1 \end{pmatrix}$$

$$(2.145a)$$

where we used (2.122a and 2.122b), (2.140), (2.141), and (2.144a and 2.144b).

From (2.141) and (2.144a), we see that an odd permutation is equivalent to a multiplication by $(-1)^{j_1+j_2+j_3}$:

$$\begin{pmatrix} j_2 & j_1 & j_3 \\ m_2 & m_1 & m_3 \end{pmatrix} = (-1)^{j_1+j_2+j_3} \begin{pmatrix} j_1 & j_2 & j_3 \\ m_1 & m_2 & m_3 \end{pmatrix} \qquad (2.145b)$$

Again, from (2.141) and (2.144a), we have also

$$\begin{pmatrix} j_1 & j_2 & j_3 \\ -m_1 & -m_2 & -m_3 \end{pmatrix} = (-1)^{j_1+j_2+j_3} \begin{pmatrix} j_1 & j_2 & j_3 \\ m_1 & m_2 & m_3 \end{pmatrix} \qquad (2.145c)$$

Equation (2.145c) tells us that any 3-j symbol of the form $\begin{pmatrix} j_1 & j_2 & j_3 \\ 0 & 0 & 0 \end{pmatrix}$ must have $j_1 + j_2 + j_3$=even integer; otherwise, the 3-j symbol is zero.

The restriction that the summation index z in (2.143) ranges only over all integral values for which the factorial arguments in the denominator of the last term on the RHS are positive has consequences for the 3-j symbols.

For example, from this restriction, it follows that, for the 3-j symbol $\begin{pmatrix} j_1 & j_2 & j_3 \\ m_1 & m_2 & m_3 \end{pmatrix}$, we must have (for the 3-j symbol to be real):

$$j_1 + j_2 + j_3 = \text{integer}, m_1 + m_2 + m_3 = \text{integer, and } j_1 - j_2 - m_3$$
$$= \text{integer} \qquad (2.145d)$$

We must also have.

$$j_1 + j_2 \geq j_3, j_2 + j_3 \geq j_1, \text{and } j_3 + j_1 \geq j_2 \qquad (2.145e)$$

Since there is a sign change in the relation between the C-G coefficients and the 3-j symbols [c. f. (2.144)], the relation between the projection quantum numbers for the 3-j symbol $\begin{pmatrix} j_1 & j_2 & j_3 \\ m_1 & m_2 & m_3 \end{pmatrix}$ is

$$m_1 + m_2 + m_3 = 0 \qquad (2.145f)$$

Equations (2.145d, 2.145e and 2.145f) represent the triangle rules for the 3-j symbols. We will see that the triangle rules are often essential for evaluating sums involving 3-j symbols.

The utility of the Racah formula for evaluating 3-j symbols can be demonstrated by a few simple examples.

Example 2.14 Calculate $\begin{pmatrix} j & j & 0 \\ m & -m & 0 \end{pmatrix}$

$$
\begin{pmatrix} j & j & 0 \\ m & -m & 0 \end{pmatrix} = \left[\frac{(2j)!(j-m)!(j+m)!(j-m)!(j+m)!}{(2j+1)!} \right]^{\frac{1}{2}}
$$

$$
\times \sum_z \frac{(-1)^z}{z!(2j-z)!(j-m-z)!(j-m-z)!(-j+m+z)!(-j+m+z)!}
$$

$$
= (2j+1)^{-\frac{1}{2}}[(j-m)!(j+m)!] \sum_z \frac{(-1)^z}{z!(2j-z)![(j-m-z)!]^2[(-j+m+z)!]^2}
$$

$$
= (2j+1)^{-\frac{1}{2}}[(j-m)!(j+m)!] \left[\frac{(-1)^{j-m}}{(j-m)!(j+m)!} \right] = (2j+1)^{-\frac{1}{2}}(-1)^{j-m}
$$

$$(2.146)$$

where using the guidance from (2.143), we see that the summation over z collapsed to the range $j - m \leq z \leq j - m$ (a single value for z). ∎

Example 2.15

Calculate $\begin{pmatrix} j & j-\dfrac{1}{2} & \dfrac{1}{2} \\ m & -m-\dfrac{1}{2} & \dfrac{1}{2} \end{pmatrix}$

$$
\begin{pmatrix} j & j-\dfrac{1}{2} & \dfrac{1}{2} \\ m & -m-\dfrac{1}{2} & \dfrac{1}{2} \end{pmatrix} = \left[\frac{(2j-1)!(j-m)!(j+m)!(j+m)!(j-m-1)!}{(2j+1)!} \right]^{\frac{1}{2}}
$$

$$
\times \sum_z \frac{(-1)^z}{z!(2j-1-z)!(j-m-z)!(j-m-1-z)!(-j+1+m+z)!(-j+m+1+z)!}
$$

$$
= (j+m)!(j-m-1)! \left[\frac{(j-m)}{2j(2j+1)} \right]^{\frac{1}{2}} \left[\frac{(-1)^{j-m-1}}{(j-m-1)!(j+m)!} \right]
$$

$$
\Rightarrow \begin{pmatrix} j & j-\dfrac{1}{2} & \dfrac{1}{2} \\ m & -m-\dfrac{1}{2} & \dfrac{1}{2} \end{pmatrix} = \left[\frac{(j-m)}{2j(2j+1)} \right]^{\frac{1}{2}}(-1)^{j-m-1}
$$

$$(2.147)$$

where using the guidance from (2.143), we see that the summation over z collapsed to the range $j - m - 1 \leq z \leq j - m - 1$ (a single value for z). ∎

The Racah formula (and the symmetry relations derived from the Racah formula) can be used to compute the values of many of the C-G coefficients and 3-j symbols. Many tables of these algebraic expressions exist (see, e.g., [3, 5, 10]).

We finish this section by noting that to successfully navigate the many different available sources that discuss angular momentum in atomic and molecular applications, one must be able to move effortlessly between using C-G coefficients and 3-j symbols.

2.10 The Clebsch-Gordan Series

Recall from (2.129):

$$|j_1,j_2;m_1,m_2\rangle \equiv |j_1,m_1\rangle|j_2,m_2\rangle = \sum_{j,m}\langle j_1,j_2;m_1,m_2|j,m\rangle|j,m\rangle \qquad (2.129)$$

Apply a rotation $\mathscr{R}$ to both sides:

$$\sum_{m_1''}\sum_{m_2''}D^{j_1}_{m_1''m_1}D^{j_2}_{m_2''m_2}|j_1,m_1''\rangle|j_2,m_2''\rangle = \sum_{j,m}\sum_{m'}\langle j_1,j_2;m_1,m_2|j,m\rangle D^{j}_{m'm}|j,m'\rangle$$

$$\qquad (2.148)$$

where we used (2.95). Now multiply both sides from the left by $\langle j_1,m_1'|\langle j_2,m_2'|$,

$$\Rightarrow D^{j_1}_{m_1'm_1}D^{j_2}_{m_2'm_2} = \sum_j \langle j_1,j_2;m_1,m_2|j,m\rangle\langle j_1,j_2;m_1',m_2'|j,m'\rangle D^{j}_{m'm} \qquad (2.149)$$

Notice that the summations over m and m' have disappeared from the RHS of (2.149). This is because (2.122a) tells us that, for a given m_1 and m_2, m is determined, and for a given m_1' and m_2', m' is determined. Equation (2.149) is called the Clebsch-Gordan series. We can use the C-G series to solve useful integrals. Consider, for example,

$$\int d\Omega D^{j_1*}_{m_1'm_1}D^{j_2}_{m_2'm_2} = \int d\Omega(-1)^{m_1'-m_1}D^{j_1}_{-m_1',-m_1}D^{j_2}_{m_2'm_2} \qquad (2.150)$$

where we have used (2.102) which also applies to the $D^{j}_{m'm}$. Let us apply (2.149) to an integral over two rotation matrices $D^{j}_{m'm}(\alpha,\beta,\gamma)$ for which $d\Omega = d\alpha\sin\beta d\beta d\gamma$

$$\int d\Omega D^{j_1*}_{m_1'm_1}D^{j_2}_{m_2'm_2}$$

$$= \sum_{j_3}(-1)^{m_1'-m_1}\langle j_1,j_2;-m_1,m_2|j_3,-m_1+m_2\rangle\langle j_1,j_2;-m_1',m_2'|j_3,-m_1'+m_2'\rangle$$

$$\times \int_0^{2\pi}d\alpha e^{-i\alpha(-m_1'+m_2')}\int_0^{2\pi}d\gamma e^{-i\gamma(-m_1+m_2)}\int_0^{\pi}d\beta\sin\beta d^{j_3}_{-m_1'+m_2',-m_1+m_2}$$

where we also used (2.96). Examining the integrals,

$$
\int_0^{2\pi} d\alpha e^{i\alpha(-m_1'+m_2')} \int_0^{2\pi} d\gamma e^{-i\gamma(-m_1+m_2)} \int_0^{\pi} d\beta \sin\beta d^{j_3}_{-m_1'+m_2',-m_1+m_2}
$$

$$
= (2\pi)\delta_{m_1'm_2'}(2\pi)\delta_{m_1m_2} \int_0^{\pi} d\beta \sin\beta d^{j_3}_{-m_1'+m_2',-m_1+m_2}
$$

$$
= (2\pi)^2 \int_0^{\pi} d\beta \sin\beta d^{j_3}_{00}
$$

$$
= (2\pi)^2 \int_0^{\pi} d\beta \sin\beta P_{j_3}(\cos\beta)
$$

$$
= (2\pi)^2 \int_0^{\pi} d\beta \sin\beta P_{j_3}(\cos\beta)P_0(\cos\beta) = (2\pi)^2 \left(\frac{2}{2j_3+1}\right)\delta_{j_30} = 8\pi^2\delta_{j_30}
$$

where we used (2.110) and (2.71) and noted that $P_0(\cos\beta) = 1$. Putting everything together,

$$
\int d\Omega D^{j_1*}_{m_1'm_1} D^{j_2}_{m_2'm_2} = 8\pi^2(-1)^{m_1'-m_1}\langle j_1,j_2; -m_1,m_1|0,0\rangle\langle j_1,j_2; -m_1',m_1'|0,0\rangle
$$

$$
\tag{2.151}
$$

Using (2.144b) and (2.146) and realizing that (2.122b) implies that we must have $j_1 = j_2$:

$$
\langle j_1,j_2; -m_1,m_1|0,0\rangle = (-1)^{j_1-j_2}\begin{pmatrix} j_1 & j_2 & 0 \\ -m_1 & m_1 & 0 \end{pmatrix}
$$

$$
= (2j_1+1)^{-\frac{1}{2}}(-1)^{j_1+m_1}\delta_{j_1j_2} \tag{2.152a}
$$

$$
\langle j_1,j_2; -m_1',m_1'|0,0\rangle = (-1)^{j_1-j_2}\begin{pmatrix} j_1 & j_2 & 0 \\ -m_1' & m_1' & 0 \end{pmatrix} = (-1)^{j_1+m_1'}(2j_1+1)^{-\frac{1}{2}}\delta_{j_1j_2}
$$

$$
\tag{2.152b}
$$

giving us finally

$$
\int d\Omega D^{j_1*}_{m_1'm_1}(\alpha,\beta,\gamma)D^{j_2}_{m_2'm_2}(\alpha,\beta,\gamma) = \frac{8\pi^2}{2j_1+1}(-1)^{2(j_1+m_1')}\delta_{j_1j_2}\delta_{m_1'm_2'}\delta_{m_1m_2}
$$

$$
= \frac{8\pi^2}{2j_1+1}\delta_{j_1j_2}\delta_{m_1'm_2'}\delta_{m_1m_2} \tag{2.153}
$$

where we see that $(-1)^{2(j_1+m_1')} = 1$ for j_1, m_1' integer or half-integer. Equation (2.153) serves as a normalization condition for the $D^{j_2}_{m_2'm_2}(\alpha, \beta, \gamma)$.

Example 2.16

Use the C-G series to solve the integral over the spherical harmonic triple product:

$$\int\limits_0^{2\pi} d\phi \int\limits_{-1}^1 d(\cos\theta) Y_{lm}^*(\theta, \phi) Y_{l_1 m_1}(\theta, \phi) Y_{l_2 m_2}(\theta, \phi)$$

In (2.149), let $j_1 \to l_1, j_2 \to l_2, m_1 \to 0, m_2 \to 0, j \to l_3$,

$$\Rightarrow D^{l_1}_{m_1'0} D^{l_2}_{m_2'0} = \sum_{l_3} \langle l_1, l_2; 0, 0 | l_3, 0\rangle \langle l_1, l_2; m_1', m_2' | l_3, m'\rangle D^{l_3}_{m'0} \qquad (2.154)$$

where we used the fact that, since $m_1 = m_2 = 0$, (2.122a) tells us that we must also have $m = 0$. Now let $m_{1,2}' \to m_{1,2}, m' \to m_3$, and invoke (2.109a and 2.109b)

$$\Rightarrow Y_{l_1 m_1}^*(\theta, \phi) Y_{l_2 m_2}^*(\theta, \phi) = \sqrt{\frac{(2l_1 + 1)(2l_2 + 1)}{4\pi}} \sum_{l_3} (2l_3 + 1)^{-1/2} \langle l_1, l_2; 0, 0 | l_3, 0\rangle$$

$$\times \langle l_1, l_2; m_1, m_2 | l_3, m_3\rangle Y_{l_3 m_3}^*(\theta, \phi)$$

$$\Rightarrow (-1)^{m_1 + m_2 - m_3} Y_{l_1 m_1}(\theta, \phi) Y_{l_2 m_2}(\theta, \phi) =$$
$$\sqrt{\frac{(2l_1 + 1)(2l_2 + 1)}{4\pi}} \sum_{l_3} (2l_3 + 1)^{-1/2} \langle l_1, l_2; 0, 0 | l_3, 0\rangle \langle l_1, l_2; m_1, m_2 | l_3, m_3\rangle Y_{l_3 m_3}(\theta, \phi)$$

$$\Rightarrow Y_{l_1 m_1}(\theta, \phi) Y_{l_2 m_2}(\theta, \phi) = \sqrt{\frac{(2l_1 + 1)(2l_2 + 1)}{4\pi}} \sum_{l_3} (2l_3 + 1)^{-1/2} \langle l_1, l_2; 0, 0 | l_3, 0\rangle$$

$$\times \langle l_1, l_2; m_1, m_2 | l_3, m_3\rangle Y_{l_3 m_3}(\theta, \phi)$$

$$(2.155)$$

where we observed that $(-1)^{m_1 + m_2 - m_3} = 1$ since $m_1 + m_2 = m_3$.

Now multiply both sides by $Y_{lm}^*(\theta, \phi)$ and integrate

$$\int\limits_0^{2\pi} d\phi \int\limits_{-1}^1 d(\cos\theta) Y_{lm}^*(\theta, \phi) Y_{l_1 m_1}(\theta, \phi) Y_{l_2 m_2}(\theta, \phi)$$

$$= \sqrt{\frac{(2l_1 + 1)(2l_2 + 1)}{4\pi(2l + 1)}} \langle l_1, l_2; 0, 0 | l, 0\rangle \langle l_1, l_2; m_1, m_2 | l, m\rangle \qquad (2.156)$$

Notice that the spherical harmonic orthonormality condition (2.64) collapsed the summation over l_3 on the RHS of (2.155). ∎

Let us manipulate (2.156) to put the result in terms of 3-j symbols:

$$
\int_0^{2\pi} d\phi \int_{-1}^{1} d(\cos\theta)(-1)^m Y_{l,-m}(\theta,\phi) Y_{l_1 m_1}(\theta,\phi) Y_{l_2 m_2}(\theta,\phi)
$$

$$
= (-1)^{2(l_1-l_2)+m} \sqrt{\frac{(2l_1+1)(2l_2+1)(2l+1)}{4\pi}} \begin{pmatrix} l_1 & l_2 & l \\ 0 & 0 & 0 \end{pmatrix} \begin{pmatrix} l_1 & l_2 & l \\ m_1 & m_2 & -m \end{pmatrix}
$$

(2.157)

Relabel $m \to -m$ and note that $(-1)^{2(l_1-l_2)} = 1$ for l_1, l_2 integer or half-integer:

$$
\int_0^{2\pi} d\phi \int_{-1}^{1} d(\cos\theta) Y_{l,m}(\theta,\phi) Y_{l_1 m_1}(\theta,\phi) Y_{l_2 m_2}(\theta,\phi)
$$

$$
= \sqrt{\frac{(2l_1+1)(2l_2+1)(2l+1)}{4\pi}} \begin{pmatrix} l_1 & l_2 & l \\ 0 & 0 & 0 \end{pmatrix} \begin{pmatrix} l_1 & l_2 & l \\ m_1 & m_2 & m \end{pmatrix}
$$

(2.158)

We now use the C-G series and (2.153) to evaluate an integral over a triple product of rotation matrices (all with the same angular arguments):

$$
\int d\Omega D^{j_3*}_{m_3' m_3}(R) D^{j_2}_{m_2' m_2}(R) D^{j_1}_{m_1' m_1}(R) =
$$

$$
\Bigg[\sum_j \langle j_1, j_2; m_1, m_2 | j, m_1+m_2 \rangle \langle j_1, j_2; m_1', m_2' | j, m_1'+m_2' \rangle (R) \Bigg]
$$

$$
\Bigg[\times \int d\Omega D^{j_3*}_{m_3' m_3}(R) D^{j}_{m_1'+m_2', m_1+m_2}(R) \Bigg]
$$

$$
= \sum_j \langle j_1, j_2; m_1, m_2 | j, m_1+m_2 \rangle \langle j_1, j_2; m_1', m_2' | j, m_1'+m_2' \rangle \frac{8\pi^2}{2j_3+1} \delta_{j_3 j} \delta_{m_1'+m_2', m_3'} \delta_{m_1+m_2, m_3}
$$

$$
\Rightarrow \int d\Omega D^{j_3*}_{m_3' m_3}(R) D^{j_2}_{m_2' m_2}(R) D^{j_1}_{m_1' m_1}(R) = \frac{8\pi^2}{2j_3+1} \langle j_1, j_2; m_1, m_2 | j_3, m_3 \rangle \langle j_1, j_2; m_1', m_2' | j_3, m_3' \rangle
$$

(2.159a)

Or in terms of 3-j symbols,

$$
\int d\Omega D^{j_1}_{m_1' m_1}(R) D^{j_2}_{m_2' m_2}(R) D^{j_3}_{m_3' m_3}(R) = 8\pi^2 \begin{pmatrix} j_1 & j_2 & j_3 \\ m_1' & m_2' & m_3' \end{pmatrix} \begin{pmatrix} j_1 & j_2 & j_3 \\ m_1 & m_2 & m_3 \end{pmatrix}
$$

(2.159b)

Note that (2.156) is a special case of (2.159a).

2.11 The Coupling of Three Angular Momenta

In the previous sections, we found that the states $\{|j, m\rangle\}$ completely characterized the resultant system when two angular momenta were added together. However, when coupling together three angular momenta, the states $\{|j, m\rangle\}$ are no longer unique since there is more than one way to add three angular momenta together to get the same total angular momentum. That is, there are several different sets of mutually commuting angular momentum operators that can be combined to describe the same physical system.

For example, one may first add j_1 and j_2 and add the result to j_3. Or, one may first add j_2 to j_3 and add the result to j_1. In the first case, (2.121) gives

$$
\begin{aligned}
|j_{12}, m_{12}\rangle &= \sum_{m_1} C^{j_{12},m_{12}}_{j_1,m_1;j_2,m_{12}-m_1} |j_1, m_1\rangle |j_2, m_{12} - m_1\rangle \Rightarrow |(j_{12},j_3)j, m\rangle \\
&= \sum_{m_{12}} C^{j,m}_{j_{12},m_{12};j_3,m-m_{12}} |j_3, m - m_{12}\rangle \sum_{m_1} C^{j_{12},m_{12}}_{j_1,m_1;j_2,m_{12}-m_1} |j_1, m_1\rangle \\
&\quad \times |j_2, m_{12} - m_1\rangle
\end{aligned}
\tag{2.160}
$$

Notice that the resultant states for this coupling are symbolized by $|(j_{12}, j_3)j, m\rangle$ (here we are using alternate symbols for the C-G coefficients for ease of tracking).

Similarly, for the second case,

$$
\begin{aligned}
|j_{23}, m_{23}\rangle &= \sum_{m_2} C^{j_{23},m_{23}}_{j_2,m_2;j_3,m_{23}-m_2} |j_2, m_2\rangle |j_3, m_{23} - m_2\rangle \Rightarrow |(j_1,j_{23})j, m\rangle \\
&= \sum_{m_{23}} C^{j,m}_{j_1,m-m_{23};j_{23},m_{23}} |j_1, m - m_{23}\rangle \sum_{m_2} C^{j_{23},m_{23}}_{j_2,m_2;j_3,m_{23}-m_2} |j_2, m_2\rangle \\
&\quad \times |j_3, m_{23} - m_2\rangle
\end{aligned}
\tag{2.161}
$$

where the resultant states for this coupling are symbolized by $|(j_1, j_{23})j, m\rangle$.

These two representations of $|j, m\rangle$ are physically equivalent and so must be connected by a unitary transformation:

$$
|(j_{12},j_3)j, m\rangle = \sum_{j_{23}} \langle (j_1,j_{23})j, m|(j_{12},j_3)j, m\rangle \, |(j_1,j_{23})j, m\rangle
\tag{2.162}
$$

Substituting (2.160) and (2.161) into (2.162) gives

$$
\begin{aligned}
&\sum_{m_1,m_{12}} C^{j_{12},m_{12}}_{j_1,m_1;j_2,m_{12}-m_1} C^{j,m}_{j_{12},m_{12};j_3,m-m_{12}} |j_1, m_1\rangle |j_2, m_{12} - m_1\rangle |j_3, m - m_{12}\rangle \\
&= \sum_{m_2,m_{23}} \sum_{j_{23}} \langle (j_1,j_{23})j, m|(j_{12},j_3)j, m\rangle C^{j,m}_{j_1,m-m_{23};j_{23},m_{23}} C^{j_{23},m_{23}}_{j_2,m_2;j_3,m_{23}-m_2} \\
&\quad \times |j_1, m - m_{23}\rangle |j_2, m_2\rangle |j_3, m_{23} - m_2\rangle
\end{aligned}
\tag{2.163}
$$

The objects introduced in (2.162), $\langle (j_1, j_{23})j, m | (j_{12}, j_3)j, m \rangle$, are known as the recoupling coefficients.

Multiplying both sides by $\langle j_1, \mu_1 | \langle j_2, \mu_2 | \langle j_3, \mu_3 |$,

$$\sum_{m_1, m_{12}} C^{j_{12}, m_{12}}_{j_1, m_1; j_2, m_{12} - m_1} C^{j, m}_{j_{12}, m_{12}; j_3, m - m_{12}} \delta_{m_1, \mu_1} \delta_{m_{12} - m_1, \mu_2} \delta_{m - m_{12}, \mu_3}$$

$$= \sum_{m_2, m_{23}} \sum_{j_{23}} \underbrace{\langle (j_1, j_{23})j, m | (j_{12}, j_3)j, m \rangle}_{\equiv R_{j_{23}, j_{12}}} C^{j, m}_{j_1, m - m_{23}; j_{23}, m_{23}}$$

$$\times C^{j_{23}, m_{23}}_{j_2, m_2; j_3, m_{23} - m_2} \delta_{m - m_{23}, \mu_1} \delta_{m_2, \mu_2} \delta_{m_{23} - m_2, \mu_3} \tag{2.164}$$

$$\Rightarrow C^{j_{12}, m_{12}}_{j_1, \mu_1; j_2, \mu_2} C^{j, m}_{j_{12}, m_{12} = \mu_1 + \mu_2; j_3, \mu_3} \delta_{\mu_1 + \mu_2 + \mu_3, m} = \delta_{\mu_1 + \mu_2 + \mu_3, m}$$

$$\times \sum_{j_{23}} R_{j_{23}, j_{12}} C^{j_{23}, m_{23}}_{j_2, \mu_2; j_3, \mu_3 = m - \mu_1 - \mu_2} C^{j, m}_{j_1, \mu_1; j_{23}, m_{23} = \mu_2 + \mu_3}$$

$$\Rightarrow C^{j_{12}, m_{12}}_{j_1, \mu_1; j_2, \mu_2} C^{j, \mu_1 + \mu_2 + \mu_3}_{j_{12}, \mu_1 + \mu_2; j_3, \mu_3} = \sum_{j_{23}} R_{j_{23}, j_{12}} C^{j_{23}, m_{23}}_{j_2, \mu_2; j_3, \mu_3} C^{j, \mu_1 + \mu_2 + \mu_3}_{j_1, \mu_1; j_{23}, \mu_2 + \mu_3}$$

where we have labeled the recoupling coefficients as the unitary transformation $R_{j_{23}, j_{12}}$ (just as a reminder that the transformation is a unitary one).

Now multiply by $C^{j, m}_{j_2, \mu_2; j_3, \mu_3}$ and sum over μ_2 with $\mu_2 + \mu_3$ fixed

$$\sum_{\mu_2} C^{j_{12}, m_{12}}_{j_1, \mu_1; j_2, \mu_2} C^{j, \mu_1 + \mu_2 + \mu_3}_{j_{12}, \mu_1 + \mu_2; j_3, \mu_3} C^{j, m}_{j_2, \mu_2; j_3, \mu_3} = \sum_{j_{23}} C^{j_{23}, m_{23}}_{j_2, \mu_2; j_3, \mu_3} \underbrace{\sum_{\mu_2} C^{j, m}_{j_2, \mu_2; j_3, \mu_3} \sum_{j_{23}} R_{j_{23}, j_{12}} C^{j, \mu_1 + \mu_2 + \mu_3}_{j_1, \mu_1; j_{23}, \mu_2 + \mu_3}}_{\delta_{j_{23}, j} \delta_{m, m_{23}}}$$

$$\tag{2.165}$$

where (as indicated) we have used (2.125) with $\mu_2 + \mu_3$ fixed (which determines μ_3).

The $\delta_{j_{23}, j}$ collapses the sum over j_{23} to a single value j; and after relabeling $j \to j_{23}$ on both sides of (2.165), and letting $m \to m_{23}$,

$$\sum_{\mu_2} C^{j_{12}, m_{12}}_{j_1, \mu_1; j_2, \mu_2} C^{j, \mu_1 + \mu_2 + \mu_3}_{j_{12}, \mu_1 + \mu_2; j_3, \mu_3} C^{j_{23}, m_{23}}_{j_2, \mu_2; j_3, \mu_3} = R_{j_{23}, j_{12}} C^{j, \mu_1 + \mu_2 + \mu_3}_{j_1, \mu_1; j_{23}, \mu_2 + \mu_3} \tag{2.166}$$

Multiply both sides of (2.166) by $C^{J, M}_{j_1, \mu_1; j_{23}, \mu_2 + \mu_3}$ and sum over μ_1 (keeping $\mu_1 + \mu_2 + \mu_3$ fixed):

$$\sum_{\mu_1, \mu_2} C^{j_{12}, m_{12}}_{j_1, \mu_1; j_2, \mu_2} C^{j, \mu_1 + \mu_2 + \mu_3}_{j_{12}, \mu_1 + \mu_2; j_3, \mu_3} C^{j_{23}, m_{23}}_{j_2, \mu_2; j_3, \mu_3} C^{J, M}_{j_1, \mu_1; j_{23}, \mu_2 + \mu_3}$$

$$= R_{j_{23}, j_{12}} \underbrace{\sum_{\mu_1} C^{j, \mu_1 + \mu_2 + \mu_3}_{j_1, \mu_1; j_{23}, \mu_2 + \mu_3} C^{J, M}_{j_1, \mu_1; j_{23}, \mu_2 + \mu_3}}_{\delta_{j, J} \delta_{M, \mu_1 + \mu_2 + \mu_3}} \tag{2.167}$$

where (as indicated) we have used (2.125) with $\mu_1 + \mu_2 + \mu_3$ fixed (which determines $\mu_2 + \mu_3$).

So now we have

$$\sum_{\mu_1,\mu_2} C^{j_{12},m_{12}}_{j_1,\mu_1;j_2,\mu_2} C^{j,\mu_1+\mu_2+\mu_3}_{j_{12},\mu_1+\mu_2;j_3,\mu_3} C^{j_{23},m_{23}}_{j_2,\mu_2;j_3,\mu_3} C^{j,m}_{j_1,\mu_1;j_{23},\mu_2+\mu_3} = R_{j_{23},j_{12}} \tag{2.168}$$

Since both coupling schemes result in the same projection m of the total angular momentum, the recoupling coefficients must be independent of m, which allows us to write

$$\langle (j_1,j_{23})j, m | (j_{12},j_3)j, m \rangle \rightarrow \langle (j_1,j_{23})j | (j_{12},j_3)j \rangle \tag{2.169}$$

Rewriting in terms of the original variables gives

$$\langle (j_1,j_{23})j | (j_{12},j_3)j \rangle = \sum_{m_1,m_{12}} C^{j_{12},m_{12}}_{j_1,m_1;j_2,m_{12}-m_1} C^{j,m}_{j_{12},m_{12};j_3,m-m_{12}} C^{j_{23},m-m_1}_{j_2,m_{12}-m_1;j_3,m-m_{12}} C^{j,m}_{j_1,m_1;j_{23},m-m_1} \tag{2.170}$$

In (2.170), the sum is still over all six m-values, but only two summation indices are shown, since, by the preceding analysis, they serve to fix all the other m-values (i.e., only two of the six summation indices are independent). Equation (2.170) can be used to find all the recoupling coefficients. However, again for reasons of symmetry, we prefer to use the Wigner 6-j symbols, which are defined by the formula

$$\begin{Bmatrix} j_1 & j_2 & j_{12} \\ j_3 & j & j_{23} \end{Bmatrix} \equiv (-1)^{j_1+j_2+j_3+j}[(2j_{12}+1)(2j_{23}+1)]^{-\frac{1}{2}}\langle (j_1,j_{23})j | (j_{12},j_3)j \rangle \tag{2.171}$$

The 6-j symbols can also be computed via the Racah formula

$$\begin{Bmatrix} a & b & e \\ d & c & f \end{Bmatrix} \equiv \Delta(abe)\Delta(acf)\Delta(bdf)\Delta(cde)$$

$$\times \sum_{z'} \frac{(-1)^{z'}(z'+1)!}{\left[\begin{array}{c} (z'-a-b-e)!(z'-c-d-e)!(z'-a-c-f)!(z'-b-d-f)! \\ x(a+b+c+d-z')!(a+d+e+f-z')!(b+c+e+f-z')! \end{array} \right]} \tag{2.172}$$

where (for example)

$$\Delta(abe) \equiv \left[\frac{(a+b-e)!(a-b+e)!(-a+b+e)!}{(a+b+e+1)!} \right] \qquad (2.173)$$

In addition, we must remember that the $\Delta(abe)$ vanish unless the triangle condition of (2.122b) is met for a, b, and e. The same is true for all the like terms in (2.172). For those who want more visibility on this topic, the Racah formula for the 6-j symbols is developed in App. C.

2.12 Spherical Tensor Operators and the Wigner-Eckart Theorem

Earlier in this chapter, we described how operators transformed under rotations [c.f. (2.84) and example 2.8]. In this section, we explore this subject in more detail. Operators transform in various ways when subject to a rotation and can be classified according to their resultant behavior. Such a classification scheme allows us to create a convenient operator basis for the development of photoelectron angular distribution theory (among other things). As it turns out, Cartesian tensor operators are inconvenient for the purpose of describing spherically or axially symmetric systems because they can be decomposed ("reduced") into several different operators, each of which may transform differently under rotation (see example 2.19). Spherical tensor operators, on the other hand, are irreducible, making them excellent vehicles for the study of photoelectron angular distributions as well as a host of other topics in atomic and molecular physics. Operating with the language of spherical tensor operators also allows us to derive the most powerful theorem in our study of quantum angular momentum, the Wigner-Eckart theorem.

Example 2.17 A scalar operator A is one that is invariant under a rotation so that $A' = A$. Therefore, according to the result from example 2.8, a scalar operator obeys the following commutation relation:

$$[J_n, A] = 0 \qquad\qquad \blacksquare$$

A tensor operator is a natural generalization of a vector operator. A spherical tensor operator of rank k (k = integer) is actually a set of $2k + 1$ functions, $T_q^{(k)}; q = -k, -k+1, \cdots, k-1, k$, that transform under the rotation $\mathscr{R}$ as follows:

$$\mathscr{R} T_q^{(k)} \mathscr{R}^{-1} \equiv \sum_{q'} D_{q'q}^k T_{q'}^{(k)} \qquad (2.174)$$

An entirely equivalent definition of a spherical tensor operator is that they obey the following commutation relations with the (various manifestations of the) angular momentum operator $\vec{J}$:

$$\left[J_z, T_q^{(k)}\right] = \hbar q T_q^{(k)} \tag{2.175a}$$

$$\left[J_\pm, T_q^{(k)}\right] = \hbar \sqrt{k(k+1) - q(q \pm 1)}\, T_{q\pm 1}^{(k)} \tag{2.175b}$$

A spherical tensor operator of rank $k = 1$ corresponds to a vector operator. The operators $T_q^{(1)}$ are related to the Cartesian components of the vector operator $\vec{A} = A_x \hat{e}_x + A_y \hat{e}_y + A_z \hat{e}_z$ as follows:

$$T_{\pm 1}^{(1)} = \mp \frac{1}{\sqrt{2}}\left(A_x \pm iA_y\right); \quad T_0^{(1)} = A_z \tag{2.176}$$

According to (2.175a and 2.175b), spherical tensor operators of rank 1 (vector operators) also have the following commutator properties with the angular momentum operators:

$$\left[J_z, T_q^{(1)}\right] = \hbar q T_q^{(1)} \quad (q = -1, 0, \ 1) \tag{2.177a}$$

$$\left[J_\pm, T_q^{(1)}\right] = \hbar \sqrt{1(1+1) - q(q \pm 1)}\, T_{q\pm 1}^{(1)} \tag{2.177b}$$

Example 2.18 Show that (2.174) and (2.175a and 2.175b) are equivalent definitions of a spherical tensor operator.

From (2.29a) we have

$$\langle k, q' | J_z | k, q \rangle = \hbar q \delta_{q'q} \Rightarrow \sum_{q'=-k}^{k} T_{q'}^{(k)} \langle k, q' | J_z | k, q \rangle = \hbar q T_q^{(k)}$$

From (2.29b) and (2.29c), we can construct

$$\langle k, q' | J_\pm | k, q \rangle = \hbar \sqrt{k(k+1) - q(q \pm 1)}\, \delta_{q',q+1} \Rightarrow$$

$$\sum_{q'=-k}^{k} T_{q'}^{(k)} \langle k, q' | J_\pm | k, q \rangle = \hbar \sqrt{k(k+1) - q(q \pm 1)}\, T_{q\pm 1}^{(k)}$$

Combine these results with (2.177):

$$\sum_{q'=-k}^{k} T_{q'}^{(k)} \langle k, q'|J_z|k, q\rangle = \left[J_z, T_q^{(k)}\right]$$

$$\text{and} \sum_{q'=-k}^{k} T_{q'}^{(k)} \langle k, q'|J_\pm|k, q\rangle = \left[J_\pm, T_q^{(k)}\right]$$

If $\vec{J}$ is decomposed in a spherical basis [c.f. (2.185) below], then we can merge the last two results into one form:

$$\left[\vec{J}, T_q^{(k)}\right] = \sum_{q'=-k}^{k} T_{q'}^{(k)} \langle k, q'|\vec{J}|k, q\rangle \Rightarrow \left[\vec{J} \cdot \hat{n}, T_q^{(k)}\right]$$

$$= \sum_{q'=-k}^{k} T_{q'}^{(k)} \langle k, q'|\vec{J} \cdot \hat{n}|k, q\rangle$$

From example 2.8, we can say

$$\mathscr{R}(\delta\alpha)T_q^{(k)}\mathscr{R}^\dagger(\delta\alpha) = T_q^{(k)} - \frac{i}{\hbar}\delta\alpha\left[J_n, T_q^{(k)}\right]$$

Combining,

$$\mathscr{R}(\delta\alpha)T_q^{(k)}\mathscr{R}^\dagger(\delta\alpha) = T_q^{(k)} - \frac{i}{\hbar}\delta\alpha \sum_{q'=-k}^{k} T_{q'}^{(k)} \langle k, q'|\vec{J} \cdot \hat{n}|k, q\rangle$$

$$= \sum_{q'=-k}^{k} T_{q'}^{(k)} \langle k, q'|\left(1 - \frac{i}{\hbar}\delta\alpha\vec{J} \cdot \hat{n}\right)|k, q\rangle$$

$$= \sum_{q'=-k}^{k} T_{q'}^{(k)} \langle k, q'|e^{-i\delta\alpha\vec{J} \cdot \hat{n}/\hbar}|k, q\rangle$$

This result also holds in the limit (for finite rotations):

$$\mathscr{R}(\alpha)T_q^{(k)}\mathscr{R}^{-1}(\alpha) = \sum_{q'=-k}^{k} T_{q'}^{(k)} \langle k, q'|\exp\left(-\frac{i\alpha}{\hbar}J_n\right)|k, q\rangle$$

$$= \sum_{q'} D_{q'q}^k(\alpha)T_{q'}^{(k)}$$

∎

We now have all we need to derive the Wigner-Eckart (W-E) theorem. From (2.95) and (2.174), under a rotation, we have

$$\mathcal{R}\left[T_q^{(k)}|\alpha'j'm'\rangle\right] = \mathcal{R}T_q^{(k)}\mathcal{R}^{-1}\mathcal{R}|\alpha'j'm'\rangle = \sum_{q'}D_{q'q}^k(R)T_{q'}^{(k)}\sum_{\mu'}D_{\mu'm'}^{j}(R)|\alpha'j'\mu'\rangle$$

$$= \sum_{q',\mu'}D_{q'q}^k(R)D_{\mu'm'}^{j}(R)\left[T_{q'}^{(k)}|\alpha'j'\mu'\rangle\right]$$

$$(2.178)$$

where the label "α" stands for all other quantum numbers needed to uniquely identify the state.

Multiply the quantity $\left[T_q^{(k)}|\alpha'j'm'\rangle\right]$ from the left by the bra $\langle\alpha jm|$ and use (2.178)

$$\langle\alpha jm|\left[T_q^{(k)}|\alpha'j'm'\rangle\right] = \langle\alpha jm|\mathcal{R}^{-1}\mathcal{R}\left[T_q^{(k)}|\alpha'j'm'\rangle\right]$$

$$= \sum_{\mu}D_{\mu m}^{j*}(R)\langle\alpha j\mu|\sum_{q',\mu'}D_{q'q}^k(R)D_{\mu'm'}^{j}(R)\left[T_{q'}^{(k)}|\alpha'j'\mu'\rangle\right] \Rightarrow \langle\alpha jm|T_q^{(k)}|\alpha'j'm'\rangle$$

$$= \sum_{\mu}\langle\alpha j\mu|\sum_{\mu q',\mu'}D_{\mu m}^{j*}(R)D_{q'q}^k(R)D_{\mu'm'}^{j}(R)\langle\alpha j\mu|T_{q'}^{(k)}|\alpha'j'\mu'\rangle \qquad (2.179)$$

Integrate both sides of (2.179) over the Euler angles $d\Omega = d\phi\sin\theta\, d\theta d\chi$,

$$\langle\alpha jm|T_q^{(k)}|\alpha'j'm'\rangle\int d\Omega = \sum_{\mu,q',\mu'}\langle\alpha j\mu|T_{q'}^{(k)}|\alpha'j'\mu'\rangle\int d\Omega D_{\mu m}^{j}(R)^*D_{\mu'm'}^{j}(R)D_{q'q}^k(R)$$

$$\Rightarrow 8\pi^2\langle\alpha jm|T_q^{(k)}|\alpha'j'm'\rangle = \sum_{\mu,q',\mu'}\langle\alpha j\mu|T_{q'}^{(k)}|\alpha'j'\mu'\rangle\frac{8\pi^2}{(2j+1)}C_{k,q;j',m'}^{j,m}C_{k',q';j',\mu'}^{j,\mu}$$

$$\Rightarrow \langle\alpha jm|T_q^{(k)}|\alpha'j'm'\rangle = \frac{C_{k,q;j',m'}^{j,m}}{(2j+1)}\sum_{\mu,q',\mu'}\langle\alpha j\mu|T_{q'}^{(k)}|\alpha'j'\mu'\rangle C_{k',q';j',\mu'}^{j,\mu} \qquad (2.180)$$

where we used (2.159a). Now define the reduced matrix elements:

$$\langle\alpha j\|T^{(k)}\|\alpha'j'\rangle \equiv (-1)^{k-j'+j}(2j+1)^{-\frac{1}{2}}\sum_{\mu,q',\mu'}\langle\alpha j\mu|T_{q'}^{(k)}|\alpha'j'\mu'\rangle C_{k,q';j',\mu'}^{j,\mu} \qquad (2.181)$$

which gives

$$\langle \alpha j m | T_q^{(k)} | \alpha' j' m' \rangle = (-1)^{-k+j'-j} \frac{C_{k,q;j',m'}^{j,m}}{(2j+1)^{\frac{1}{2}}} \langle \alpha j \| T^{(k)} \| \alpha' j' \rangle$$

$$= (-1)^{-k+j'-j}(-1)^{k-j'+m} \begin{pmatrix} k & j' & j \\ q & m' & -m \end{pmatrix} \langle \alpha j \| T^{(k)} \| \alpha' j' \rangle$$

$$= (-1)^{-k+j'-j}(-1)^{k-j'+m} \begin{pmatrix} j & k & j' \\ -m & q & m' \end{pmatrix} \langle \alpha j \| T^{(k)} \| \alpha' j' \rangle$$

$$= (-1)^{-j+m} \begin{pmatrix} j & k & j' \\ -m & q & m' \end{pmatrix} \langle \alpha j \| T^{(k)} \| \alpha' j' \rangle$$

$$\Rightarrow \langle \alpha j m | T_q^{(k)} | \alpha' j' m' \rangle = (-1)^{j-m} \begin{pmatrix} j & k & j' \\ -m & q & m' \end{pmatrix} \langle \alpha j \| T^{(k)} \| \alpha' j' \rangle \tag{2.182}$$

where we used (2.144b) and (2.145b and 2.145c). We also used the fact that for any angular momentum quantum number j and its projection m, the quantity $j \pm m =$ integer, and for any integer k, $(-1)^k = (-1)^{-k}$. The theorem (2.182) is called the Wigner-Eckart (W-E) theorem, and it has profound physical implications. In a nutshell, the W-E theorem states that the matrix elements of a spherical tensor operator between the states $|j, m\rangle$ and $|j', m'\rangle$ can be written as the product of a C-G coefficient (or equivalently, a 3-j symbol), which contains all the information on the orientation (geometry) of the system (i.e., depends only on m', m, and q), and a "reduced" matrix element, $\langle j' \| \hat{T}^{(k)} \| j \rangle$ which contains all the information on the dynamics of the system (i.e., depends only on j, j', and k). Equation (2.182) is the usual form of the W-E theorem and reflects standard conventions [3, 5, 10].

We can also write the W-E theorem as follows:

$$\langle \alpha' j' m' | T_q^{(k)} | \alpha j m \rangle \propto (-1)^{-k+j-j'} C_{k,q;j,m}^{j',m'} \langle \alpha' j' \| T^{(k)} \| \alpha j \rangle$$

$$= (-1)^{-k+j'-j}(-1)^{k+j-j'} C_{j,m;k,q}^{j',m'} \langle \alpha' j' \| T^{(k)} \| \alpha j \rangle = C_{j,m;k,q}^{j',m'} \langle \alpha' j' \| T^{(k)} \| \alpha j \rangle$$

$$\tag{2.183}$$

The question now arises as to how we can create spherical tensor operators. One way to do so is by forming the spherical tensor operator $T_q^{(k)}$ from the product of two other spherical tensor operators $T_{q_1}^{(k_1)}$ and $T_{q_2}^{(k_2)}$ as follows:

$$T_q^{(k)} = \sum_{q_1} \sum_{q_2} \langle k_1, k_2; q_1, q_2 | k, q \rangle T_{q_1}^{(k_1)} T_{q_2}^{(k_2)} \tag{2.184}$$

To prove that $T_q^{(k)}$ is, in fact, a spherical tensor operator, we need to show that the indicated product transforms according to (2.174):

$$\mathscr{R}T_q^{(k)}\mathscr{R}^{-1} = \sum_{q_1}\sum_{q_2}\langle k_1,k_2;q_1,q_2|k,q\rangle\mathscr{R}T_{q_1}^{(k_1)}\mathscr{R}^{-1}\mathscr{R}T_{q_2}^{(k_2)}\mathscr{R}^{-1}$$

$$= \sum_{q_1}\sum_{q_2}\sum_{q_1'}\sum_{q_2'}\langle k_1,k_2;q_1,q_2|k,q\rangle D_{q_1'q_1}^{k_1}D_{q_2'q_2}^{k_2}T_{q_1'}^{(k_1)}T_{q_2'}^{(k_2)}$$

$$= \sum_{q_1'}\sum_{q_2'}\sum_{k''}\underbrace{\sum_{q_1}\sum_{q_2}\langle k_1,k_2;q_1,q_2|k'',q''\rangle\langle k_1,k_2;q_1,q_2|k,q\rangle}_{\delta_{kk''}\delta_{qq''}}\langle k_1,k_2;q_1',q_2'|k'',q'\rangle D_{q'q''}^{k''}T_{q_1'}^{(k_1)}T_{q_2'}^{(k_2)}$$

$$= \sum_{q_1'}\sum_{q_2'}\langle k_1,k_2;q_1',q_2'|kq'\rangle T_{k_1}^{(q_1')}T_{k_2}^{(q_2')}D_{q'q}^{k} \rightarrow \sum_{q'}\sum_{q_1'}\sum_{q_2'}\langle k_1,k_2;q_1',q_2'|k,q'\rangle T_{q_1'}^{(k_1)}T_{q_2'}^{(k_2)}D_{q_1'+q_2',q}^{k}$$

$$= \sum_{q'}\left(\sum_{q_1'}\sum_{q_2'}\langle k_1,k_2;q_1',q_2'|k,q'\rangle T_{q_1'}^{(k_1)}T_{q_2'}^{(k_2)}\right)D_{q'q}^{k} = \sum_{q'}D_{q'q}^{k}T_{q'}^{(k)}$$

where we used the Clebsch-Gordan series (2.149) and the orthogonality relation of the C-G coefficients (2.125). We also made the move, $\sum_{q_1'}\sum_{q_2'} \rightarrow \sum_{q'}\sum_{q_1'}\sum_{q_2'}$, which does not change the value of the equation since $q' = q_1' + q_2'$. We see that the RHS of (2.184) transforms correctly, thereby validating the fact that we can construct spherical tensor operators of higher or lower rank by multiplying two tensor operators together.

It is important to realize that the components L_i of the orbital angular momentum vector operators can themselves be written in the form of spherical tensor operators of rank one. If written in a spherical basis, the orbital angular momentum operators would appear as

$$L_1^{(1)} = -\sqrt{\frac{1}{2}}L_+; L_0^{(1)} = L_z; L_{-1}^{(1)} = \sqrt{\frac{1}{2}}L_- \tag{2.185}$$

We would then find that the orbital angular momentum operators obey the following commutation relations:

$$\left[L_z, L_q^{(k)}\right] = qL_q^{(k)} \tag{2.186a}$$

$$\left[L_\pm, L_q^{(k)}\right] = [k(k+1) - q(q\pm1)]^{\frac{1}{2}}L_{q\pm1}^{(k)} \tag{2.186b}$$

making them (by definition) spherical tensor operators of rank one.

Example 2.19 As an exercise in operating with angular momentum operators in their spherical tensor form, we will now use (2.184) to decompose the operator L_z^2 in terms of the rank-1 vector operators $\vec{U}$ and $\vec{V}$ as follows [3, 4, 6]. First consider

$$
T_0^{(2)} = \sum_{q_1}\sum_{q_2}\langle 1,1;q_1,q_2|2,0\rangle U_{(q_1)}^1 V_{(q_2)}^1
$$
$$
= \langle 1,1;0,0|2,0\rangle U_0 V_0 + \langle 1,1;-1,1|2,0\rangle U_{-1}V_{+1} + \langle 1,1;1,-1|2,0\rangle U_{+1}V_{-1}
$$
$$
\tag{2.187}
$$

There are only three non-zero terms in the expansion. All other terms in the double summation are zero because the C-G coefficients in those terms violate the triangle condition $0 = q = q_1 + q_2$.

Using standard tables of C-G coefficients [easier than trying to apply (2.143)],

$$
T_0^{(2)} = \sqrt{\frac{2}{3}}U_0 V_0 + \sqrt{\frac{1}{6}}U_{-1}V_{+1} + \sqrt{\frac{1}{6}}U_{+1}V_{-1}
$$
$$
= \sqrt{\frac{2}{3}}U_z V_z - \left(\frac{1}{2}\right)\sqrt{\frac{1}{6}}\left(U_x - iU_y\right)\left(V_x + iV_y\right) - \left(\frac{1}{2}\right)\sqrt{\frac{1}{6}}\left(U_x + iU_y\right)\left(V_x - iV_y\right)
$$
$$
= \sqrt{\frac{1}{6}}\left(2U_z V_z - U_x V_x - U_y V_y\right)
$$
$$
\tag{2.188}
$$

where we also used (2.176).

Now let $\vec{U} = \vec{V} = \vec{L}$,

$$
T_0^{(2)} \rightarrow L_0^{(2)} = \sqrt{\frac{1}{6}}\left(2L_z L_z - L_x L_x - L_y L_y\right) = \sqrt{\frac{1}{6}}\left(3L_z^2 - L^2\right)
$$
$$
\Rightarrow L_z^2 = \frac{\sqrt{6}L_0^{(2)} + L^2}{3}
\tag{2.189}
$$

$L_0^{(2)}$ is a spherical tensor angular momentum operator of rank 2, and L^2 is a scalar invariant. We say that we have "reduced" the Cartesian tensor L_z^2 to the sum of a second-rank spherical tensor and a rank-zero (scalar) spherical tensor. ∎

Example 2.20 Evaluate the sum: $\sum_m m^2 |Y_{lm}|^2$.

First, consider the following [4, 6]:

$$
\sum_m m^2 |d_{m'm}^l(\theta)|^2 = \sum_m m^2 \langle l,m'|e^{-i\theta L_y}|l,m\rangle\langle l,m|e^{i\theta L_y}|l,m'\rangle
$$
$$
= \sum_m \langle l,m'|e^{-i\theta L_y}L_z^2|l,m\rangle\langle l,m|e^{i\theta L_y}|l,m'\rangle
$$
$$
= \langle l,m'|e^{-i\theta L_y}L_z^2 e^{i\theta L_y}|l,m'\rangle
$$
$$
= \langle l,m'|\mathcal{R}(\alpha=0,\beta=\theta,\gamma=0)L_z^2\mathcal{R}^{-1}(\alpha=0,\beta=\theta,\gamma=0)|l,m'\rangle
$$
$$
\tag{2.190}
$$

Substituting (2.189) into (2.190) gives

$$\sum_m m^2 |d^l_{m'm}(\theta)|^2 = \langle l, m' | \mathcal{R} \left(\frac{\sqrt{6}L^{(2)}_0 + L^2}{3} \right) \mathcal{R}^{-1} |l, m' \rangle \qquad (2.191)$$

We apply the definition in (2.174) to the RHS interior term of (2.191):

$$\langle l, m' | \mathcal{R} \left(\frac{\sqrt{6}L^{(2)}_0 + L^2}{3} \right) \mathcal{R}^{-1} |l, m' \rangle$$

$$= \frac{\sqrt{6}}{3} \sum_{q'=-2}^{2} D^2_{q'0} \langle l, m' | L^{(2)}_{q'} l, m' | \rangle + \frac{1}{3} \langle l, m' | \mathcal{R} L^2 \mathcal{R}^{-1} |l, m' \rangle \qquad (2.192)$$

$$= \frac{\sqrt{6}}{3} D^2_{00} \langle l, m' | L^{(2)}_0 |l, m' \rangle + \frac{1}{3} \langle l, m' | L^2 |l, m' \rangle$$

$$= \frac{\sqrt{6}}{3} P_2(\cos\theta) \left[\frac{1}{\sqrt{6}} \langle l, m' | (3L^2_z - L^2) |l, m' \rangle \right] + \frac{1}{3} \langle l, m' | L^2 |l, m' \rangle$$

In (2.192) only the term $q' = 0$ contributed in the summation, since $\langle l, m' | L^{(2)}_{q' \neq 0} |l, m' \rangle = 0$, (which can be easily proved using the W-E theorem), and since L^2 was unchanged by the rotation because it is a scalar invariant. We also used (2.110).

Substituting the results from (2.192) into (2.191),

$$\sum_m m^2 |d^l_{m'm}(\theta)|^2 = \frac{\sqrt{6}}{3} P_2(\cos\theta) \langle l, m' | \left[\frac{1}{\sqrt{6}} (3L^2_z - L^2) \right] |l, m' \rangle$$

$$+ \frac{1}{3} \langle l, m' | L^2 |l, m' \rangle$$

$$= \frac{\left[3m'^2 - l(l+1) \right]}{3} P_2(\cos\theta) + \frac{1}{3} l(l+1) \qquad (2.193)$$

$$= m'^2 P_2(\cos\theta) + \frac{1}{2} l(l+1) \sin^2\theta$$

For $m' = 0$, we have

$$\sum_m m^2 |d^l_{0m}(\theta)|^2 = \frac{1}{2} l(l+1) \sin^2\theta \qquad (2.194)$$

Recalling (2.98) and (2.109) gives us finally [6]

$$\sum_m m^2 |Y_{lm}|^2 = \frac{l(l+1)(2l+1)}{8\pi} \sin^2\theta \qquad (2.195)$$

∎

The W-E theorem is essential for quantitatively evaluating operators that represent the interaction of atoms and molecules with photons. To use this theorem, however, we must be able to evaluate the reduced matrix elements that result from the use of the W-E theorem. Fortunately, there are ways to find reduced matrix elements that commonly appear. We start with the matrix elements for the normalized spherical harmonic tensor operators of (4.11).

According to the W-E theorem,

$$\langle l,m|C_q^{(k)}|l',m'\rangle = (-1)^{l-m}\begin{pmatrix} l & k & l' \\ -m & q & m' \end{pmatrix}\langle l\|C^{(k)}\|l'\rangle \tag{2.196}$$

But from (2.157),

$$\langle l,m|C_q^{(k)}|l',m'\rangle = \sqrt{\frac{4\pi}{2k+1}}\int_0^{2\pi}d\phi\int_{-1}^{1}d(\cos\theta)Y_{lm}^*(\theta,\phi)Y_{kq}(\theta,\phi)Y_{l'm'}(\theta,\phi)$$

$$= (-1)^{2(k-l')+m}\sqrt{(2l+1)(2l'+1)}\begin{pmatrix} l & k & l' \\ 0 & 0 & 0 \end{pmatrix}\begin{pmatrix} l & k & l' \\ m & q & -m' \end{pmatrix} \tag{2.197}$$

$$= (-1)^m\sqrt{(2l+1)(2l'+1)}\begin{pmatrix} l & k & l' \\ 0 & 0 & 0 \end{pmatrix}\begin{pmatrix} l & k & l' \\ m & q & -m' \end{pmatrix}$$

where we have used the fact that $(-1)^{2(k-l')} = 1$. Comparing (2.196) and (2.197),

$$\langle l\|C^{(k)}\|l'\rangle = (-1)^l\sqrt{(2l+1)(2l'+1)}\begin{pmatrix} l & k & l' \\ 0 & 0 & 0 \end{pmatrix} \tag{2.198}$$

We can combine (2.145a) and (2.146) with (2.198) to derive the following results:

$$\langle l\|C^{(0)}\|l'\rangle = \sqrt{(2l+1)}\delta_{ll'} \tag{2.199}$$

and

$$\langle l\|C^{(k)}\|0\rangle = (-1)^l\langle 0\|C^{(k)}\|l\rangle = \delta_{lk} \tag{2.200}$$

Since this is fun, we can keep going for a while longer.

Example 2.21 Consider the matrix elements of the unit tensor operator. By the W-E theorem,

$$\langle \alpha jm|I|\alpha'j'm'\rangle = (-1)^{j-m}\begin{pmatrix} j & 0 & j' \\ -m & 0 & m' \end{pmatrix}\langle \alpha j\|1\|\alpha'j'\rangle \tag{2.201}$$

Equation (2.145e) implies that, inside the 3-j symbol, we must have $j = j'$,

$$\Rightarrow \langle \alpha j m | I | \alpha' j' m' \rangle = (-1)^{j-m} \begin{pmatrix} j & 0 & j \\ -m & 0 & m' \end{pmatrix} \langle \alpha j \| 1 \| \alpha' j' \rangle \qquad (2.202)$$

Using (2.145a) and (2.146),

$$\underbrace{\langle \alpha j m | I | \alpha' j' m' \rangle}_{= \delta_{\alpha \alpha'} \delta_{j j'} \delta_{m m'}} = (-1)^{j-m} \begin{pmatrix} j & 0 & j \\ -m & 0 & \underbrace{m'}_{=m} \end{pmatrix} \langle \alpha j \| 1 \| \alpha' j' \rangle$$

$$= (-1)^{j-m} \begin{pmatrix} j & j & 0 \\ m & -m & 0 \end{pmatrix} \langle \alpha j \| 1 \| \alpha' j' \rangle$$

$$= \underbrace{(-1)^{2(j-m)}}_{=1} (2j+1)^{-1/2} \langle \alpha j \| 1 \| \alpha' j' \rangle$$

$$\Rightarrow \langle \alpha j \| 1 \| \alpha' j' \rangle = \delta_{\alpha \alpha'} \delta_{j j'} \sqrt{2j+1} \qquad (2.203)$$

∎

Example 2.22 Consider

$$\langle \alpha j j | J_z | \alpha' j' j' \rangle = j \delta_{\alpha \alpha'} \delta_{j j'} \qquad (2.204)$$

But by the W-E theorem [and anticipating the delta functions in (2.204)],

$$\langle \alpha j j | J_z | \alpha' j' j' \rangle = \langle \alpha j j | J_0^{(1)} | \alpha' j' j' \rangle = (-1)^{j-j} \underbrace{\begin{pmatrix} j & 1 & j' \\ -j & 0 & j' \end{pmatrix}}_{= \begin{pmatrix} j & 1 & j \\ -j & 0 & j \end{pmatrix}} \langle \alpha j \| J^{(1)} \| \alpha' j' \rangle \qquad (2.205)$$

We use the relation [3] (which can be proved using (2.138), see Problem 4.2)

$$\begin{pmatrix} j & 1 & j \\ -j & 0 & j \end{pmatrix} = \sqrt{\frac{j}{(j+1)(2j+1)}} \qquad (2.206)$$

Thus,

$$\langle \alpha j j | J_z | \alpha' j' j' \rangle = \sqrt{\frac{j}{(j+1)(2j+1)}} \langle \alpha j \| J^{(1)} \| \alpha' j' \rangle$$

$$\Rightarrow j \delta_{\alpha \alpha'} \delta_{j j'} = \sqrt{\frac{j}{(j+1)(2j+1)}} \langle \alpha j \| J^{(1)} \| \alpha' j' \rangle$$

$$\Rightarrow \langle \alpha j \| J^{(1)} \| \alpha' j' \rangle = \sqrt{j(j+1)(2j+1)} \delta_{\alpha \alpha'} \delta_{j j'} \qquad (2.207)$$

∎

Example 2.23 Extending our thoughts from the previous example, let us consider

$$\langle \alpha j m | J_z | \alpha' j' m' \rangle = m \delta_{\alpha \alpha'} \delta_{jj'} \delta_{mm'} \tag{2.208}$$

But, by the W-E theorem,

$$\langle \alpha j m | J_z | \alpha' j' m' \rangle = \langle \alpha j m | J_0^{(1)} | \alpha' j' m' \rangle$$

$$= (-1)^{j-m} \underbrace{\begin{pmatrix} j & 1 & j' \\ -m & 0 & m' \end{pmatrix}}_{= \begin{pmatrix} j & 1 & j \\ -m & 0 & m \end{pmatrix}} \langle \alpha j \| J^{(1)} \| \alpha' j' \rangle \tag{2.209}$$

Combining (2.207, 2.208 and 2.209),

$$m \delta_{\alpha\alpha'} \delta_{jj'} = (-1)^{j-m} \begin{pmatrix} j & 1 & j \\ -m & 0 & m \end{pmatrix} \langle \alpha j \| J^{(1)} \| \alpha' j' \rangle$$

$$= (-1)^{j-m} \begin{pmatrix} j & 1 & j \\ -m & 0 & m \end{pmatrix} \sqrt{j(j+1)(2j+1)} \delta_{\alpha\alpha'} \delta_{jj'}$$

$$\Rightarrow \begin{pmatrix} j & 1 & j \\ -m & 0 & m \end{pmatrix} = (-1)^{j-m} \frac{m}{\sqrt{j(j+1)(2j+1)}} \tag{2.210}$$

So, the W-E theorem is also useful for finding the value of 3-j symbols. ∎

We can use similar techniques to find the reduced matrix elements of the Pauli spin matrices of (6.7) and (6.8). First, we consider the spin operator of (6.9). We could write its components in a spherical basis much as we did for the orbital angular momentum operators in (2.185). The spin operators, so written, would then obey the relations of (2.186). We then argue as we did above [3]

$$\langle s \| S^{(1)} \| s' \rangle = \sqrt{s(s+1)(2s+1)} \delta_{ss'} \tag{2.211}$$

For spin 1/2,

$$\left\langle \frac{1}{2} \| \sigma^{(1)} \| \frac{1}{2} \right\rangle = \frac{2}{\hbar} \left\langle \frac{1}{2} \| S^{(1)} \| \frac{1}{2} \right\rangle = \frac{2}{\hbar} \sqrt{\frac{3}{2}} = \sqrt{6} \quad \text{in units of } \hbar = 1. \tag{2.212}$$

Problems

2.1. Angular momentum $j = 1$.

(a) Using the expressions for J_x and J_y from example 2.2, find, by matrix multiplication, the matrices $J_x J_y$ and $J_y J_x$ and thus find the commutator $[J_x, J_y]$. Ensure your result satisfies (2.1).

(b) Find the matrix representations for the operators $J_\pm = J_x \pm iJ_y$, $J_z^3, J_\pm^3$. Which of these operators, if any, are Hermitian?

(c) Find the eigenvalues and (normalized) eigenvectors for J_x in the given basis.

(d) Suppose you have a particle of spin 1. What are the probabilities that, having measured the angular momentum of the particle in the x-direction to be $\hbar$, a subsequent measurement in the z-direction yields the values $\hbar$, 0, or $-\hbar$?

(e) When a beam of spin $= 1$ particles is analyzed in the z-direction, it is found that the probability of measuring $J_z = \hbar$ is 50% and the probability of measuring $J_z = 0$ is 50%. Furthermore, when the beam is analyzed in the x-direction, it is found that $\langle J_x \rangle = \frac{\hbar}{\sqrt{2}} g$. In terms of g, what is the value, or possible range of values, of $\langle J_y \rangle$?

HINT: Expand the state in a linear superposition of eigenvectors of the J_z operator in the most general form possible consistent with the given information.

2.2. Starting with (2.36a, 2.36b and 2.36c) and (2.37a, 2.37b and 2.37c), verify (2.41a, 2.41b, and 2.41c) and (2.42a and 2.42b).

2.3. In example 2.5, we found the general form for $Y_{ll}(\theta, \phi)$ to be.

$$Y_{ll}(\theta, \phi) = \frac{(-1)^l}{2^l l!} \sqrt{\frac{(2l+1)!}{4\pi}} e^{il\phi} (\sin \theta)^l$$

(a) Operate on this form with the lowering operator of (2.34c), (2.42a), and (2.53) to verify (2.54).

(b) Use this form to construct $Y_{22}(\theta, \phi)$. Then use (2.53) and (2.55) to form $Y_{21}(\theta, \phi)$, $Y_{20}(\theta, \phi)$, $Y_{2,-1}(\theta, \phi)$, and $Y_{2,-2}(\theta, \phi)$, thus verifying (2.67d, 2.67e and 2.67f). Use your results to verify (2.80) for the family $Y_{2m}(\theta, \phi)$.

2.4. Under a rotation of coordinates through an angle $d\alpha$ about an axis along $\hat{u}$, the unit vectors $\hat{e}^{(l)}$ along the rectangular axes transform as $\hat{e}'^{(l)} = \hat{e}^{(l)} + d\alpha \hat{u} \times \hat{e}^{(l)}$. Use this information to show that any component of a vector observable $\vec{V}$ has a commutator with the components of the angular momentum operator $\vec{J}$ given by,

$$[V_i, J_j] = i\hbar \sum_k \varepsilon_{ijk} V_k.$$

2.5. Verify (2.141) and (2.142).

2.6. Prove the following recursion relation for the C-G coefficients:

$$
\sqrt{\frac{(j+m)(j-m)(j-j_1+j_2)(j+j_1-j_2)(j+j_1+j_2+1)(j_1-j+j_2+1)}{4j^2(2j-1)(2j+1)}} C^{j-1,m}_{j_1,m_1;j_2,m_2}
$$

$$
= \left[m_1 - m\frac{j_1(j_1+1)-j_2(j_2+1)+j(j+1)}{2j(j+1)} \right] C^{j,m}_{j_1,m_1;j_2,m_2}
$$

$$
- \sqrt{\frac{\left[(j+1)^2-m^2 \right](j-j_1+j_2+1)(j+j_1-j_2+1)(j+j_1+j_2+2)(j_1-j+j_2)}{4(j+1)^2(2j+1)(2j+3)}} C^{j+1,m}_{j_1,m_1;j_2,m_2}
$$

2.7. Suppose we have the total angular momentum operator $\vec{J} = \vec{J}_1 + \vec{J}_2$. Prove the following result for the matrix elements of the operator J_{1z} in the basis $|j, m\rangle$ of the total angular momentum operator $\vec{J}$.

$$
\langle j, m | J_{1z} | j, m \rangle = \frac{[j(j+1) + j_1(j_1+1) - j_2(j_2+1)]}{2j(j+1)} \hbar m
$$

2.8. Find the 3-j version of (2.125).

2.9. Prove the following:

$$
\sum_{m_1,m_2} \begin{pmatrix} l_1 & l_2 & l \\ m_1 & m_2 & m \end{pmatrix} Y_{l_1 m_1}(\theta, \phi) Y_{l_2 m_2}(\theta, \phi)
$$

$$
= \sqrt{\frac{(2l_1+1)(2l_2+1)}{4\pi(2l+1)}} \begin{pmatrix} l_1 & l_2 & l \\ 0 & 0 & 0 \end{pmatrix} Y_{lm}(\theta, \phi)
$$

HINT: Form the expansion $Y_{l_1 m_1} Y_{l_2 m_2} = \sum_{l,m} a_{lm} Y_{lm}$, and solve for the expansion coefficients.

2.10. Find the reduced matrix elements for the solid spherical harmonic operators.

Note: The definition of the solid harmonic operator can be found in Problem 4.3.

2.11. A particle of spin $\vec{S}$ moves in a central potential in a state with orbital angular momentum $\vec{L}$ and total angular momentum $\vec{J}$. A uniform magnetic field $\vec{B} = B\hat{z}$ is applied to the system. Calculate the matrix elements $\langle j, m | \vec{S} \cdot \vec{B} | j, m' \rangle$.

2.12. Prove the commutation relation:

$$\sum_i \left[J_i, \left[J_i, T_q^{(k)}\right]\right] = \hbar^2 k(k+1) T_q^{(k)}$$

2.13. A particle moves in a central potential in a state of orbital angular momentum $\vec{L}$. For purposes of calculating dipole interactions, one may need to know the matrix elements:

$$\langle l, m | \left(n_i n_j - \frac{1}{3}\delta_{ij}\right) |l, m'\rangle$$

The W-E theorem guarantees that within the $|l, m\rangle$ subspace, any two traceless symmetric rank-2 tensor operators are proportional. Therefore, one can write

$$\langle l, m | \left(n_i n_j - \frac{1}{3}\delta_{ij}\right) |l, m'\rangle = c\langle l, m | \left(L_i L_j + L_j L_i - \frac{2}{3}L^2\delta_{ij}\right) |l, m'\rangle$$

Find the coefficient c.

Note: In this problem, the components n_i, n_j and L_i, L_j are the spherical components of the vector operators $\hat{r}$ and $\vec{L}$, respectively. The unit vector $\hat{r}$ is defined in Problem 3.8.

Note: The Kronecker delta δ_{ij} is defined in Problem 3.8.

2.14. The quantum mechanical momentum operator in the position representation is given in (2.32).

(a) show that,

$$p^2 = -\hbar^2 \frac{1}{r^2}\frac{\partial}{\partial r}r^2\frac{\partial}{\partial r} + \frac{\hbar^2}{r^2}L^2.$$

(b) Find the reduced matrix elements for the momentum operator.

2.15. Prove the projection theorem,

$$\langle j, m' | A_q | j, m\rangle = \frac{\langle j, m | \vec{J}\cdot\vec{A} | j, m\rangle}{\hbar^2 j(j+1)}\langle j, m' | J_q | j, m\rangle$$

where A_q is the qth spherical component of the vector operator $\vec{A}$ and J_q is the qth spherical component of $\vec{J}$.

Chapter 3
Classical Model of Photoelectron Angular Distributions

To begin our discussion of the classical theory of atom/molecule-photon interactions, we start with Maxwell's equations (here presented in Gaussian units) [11]:

$$\vec{\nabla} \cdot \vec{B}\left(\vec{r},t\right) = 0 \tag{3.1a}$$

$$\vec{\nabla} \times \vec{E}\left(\vec{r},t\right) + \frac{1}{c}\frac{\partial \vec{B}\left(\vec{r},t\right)}{\partial t} = 0 \tag{3.1b}$$

$$\vec{\nabla} \cdot \vec{E}\left(\vec{r},t\right) = 4\pi\rho\left(\vec{r},t\right) \tag{3.1c}$$

$$\vec{\nabla} \times \vec{B}\left(\vec{r},t\right) - \frac{1}{c}\frac{\partial \vec{E}\left(\vec{r},t\right)}{\partial t} = \frac{4\pi}{c}\vec{J}\left(\vec{r},t\right) \tag{3.1d}$$

where $\rho\left(\vec{r},t\right)$ and $\vec{J}\left(\vec{r},t\right)$ are the charge and current densities, respectively, that are the sources of the fields. The fields can be written in terms of scalar and vector potentials as follows:

$$\vec{B}\left(\vec{r},t\right) = \vec{\nabla} \times \vec{A}\left(\vec{r},t\right) \tag{3.2a}$$

$$\vec{E}\left(\vec{r},t\right) = -\frac{1}{c}\frac{\partial \vec{A}\left(\vec{r},t\right)}{\partial t} - \vec{\nabla}\Phi\left(\vec{r},t\right) \tag{3.2b}$$

A major portion of this chapter follows closely the development presented in chapter 3, application 6 of [3].

© The Author(s), under exclusive license to Springer Nature Switzerland AG 2022
V. T. Davis, *Introduction to Photoelectron Angular Distributions*, Springer Tracts in Modern Physics 286, https://doi.org/10.1007/978-3-031-08027-2_3

We can manipulate the mathematical forms of the potentials in certain ways as long as the fields remain unchanged by these manipulations. These manipulations are called gauge transformations. One such gauge transformation

$$\vec{\nabla} \cdot \vec{A}\left(\vec{r}, t\right) = 0 \tag{3.3}$$

is called the Coulomb, or transverse, gauge transformation, and is convenient under certain circumstances [11, 12]. With this condition, the source-dependent Maxwell equations can be written as

$$-\nabla^2 \Phi\left(\vec{r}, t\right) = 4\pi \rho\left(\vec{r}, t\right) \tag{3.4}$$

$$\nabla^2 \vec{A}\left(\vec{r}, t\right) - \frac{1}{c^2} \frac{\partial^2 \vec{A}\left(\vec{r}, t\right)}{\partial t^2} = \frac{1}{c} \vec{\nabla} \frac{\partial \Phi\left(\vec{r}, t\right)}{\partial t} - \frac{4\pi}{c} \vec{J}\left(\vec{r}, t\right) \tag{3.5}$$

Example 3.1 Verify (3.4) and (3.5).

Start with (3.1c):

$$\vec{\nabla} \cdot \vec{E} = 4\pi\rho \Rightarrow -\frac{1}{c} \frac{\partial}{\partial t} \underbrace{\left(\vec{\nabla} \cdot \vec{A}\right)}_{=0} - \vec{\nabla}^2 \Phi = 4\pi\rho \Rightarrow -\vec{\nabla}^2 \Phi = 4\pi\rho$$

where we also used (3.2b). Now from (3.1d),

$$\vec{\nabla} \times \vec{B} - \frac{1}{c} \frac{\partial \vec{E}}{\partial t} = \frac{4\pi}{c} \vec{J} \Rightarrow \vec{\nabla} \times \left(\vec{\nabla} \times \vec{A}\right) - \frac{1}{c} \frac{\partial}{\partial t} \left(-\frac{1}{c} \frac{\partial \vec{A}}{\partial t} - \vec{\nabla}\Phi\right) = \frac{4\pi}{c} \vec{J}$$

$$\Rightarrow \underbrace{\vec{\nabla}\left(\vec{\nabla} \cdot \vec{A}\right)}_{=0} - \nabla^2 \vec{A} + \frac{1}{c^2} \frac{\partial^2 \vec{A}}{\partial t^2} + \frac{1}{c} \vec{\nabla} \frac{\partial \Phi}{\partial t} = \frac{4\pi}{c} \vec{J}$$

$$\Rightarrow \nabla^2 \vec{A} - \frac{1}{c^2} \frac{\partial^2 \vec{A}}{\partial t^2} = \frac{1}{c} \vec{\nabla} \frac{\partial \Phi}{\partial t} - \frac{4\pi}{c} \vec{J}$$

where we used (3.2a) and (3.2b), and the vector identity,

$$\vec{\nabla} \times \left(\vec{\nabla} \times \vec{A}\right) = \vec{\nabla}\left(\vec{\nabla} \cdot \vec{A}\right) - \nabla^2 \vec{A}$$

∎

Under the Coulomb gauge and in regions far from any sources, ($\rho = J = 0$), (3.2a and 3.2b) give

$$\vec{B}\left(\vec{r},t\right) = \vec{\nabla} \times \vec{A}\left(\vec{r},t\right) \tag{3.6a}$$

$$\vec{E}\left(\vec{r},t\right) = -\frac{1}{c}\frac{\partial \vec{A}\left(\vec{r},t\right)}{\partial t} \tag{3.6b}$$

where we also noted that $\Phi = 0$ from (3.4). Thus, under these conditions, finding $\vec{A}\left(\vec{r},t\right)$ is equivalent to finding $\vec{E}\left(\vec{r},t\right)$ and $\vec{B}\left(\vec{r},t\right)$. So, how do we find $\vec{A}\left(\vec{r},t\right)$? Under the Coulomb gauge and in regions far from any sources, ($\rho = J = 0$), the vector potential solves the wave equation [c.f. (3.5)]:

$$\vec{\nabla}^2\vec{A}\left(\vec{r},t\right) - \frac{1}{c^2}\frac{\partial^2\vec{A}\left(\vec{r},t\right)}{\partial^2 t} = 0 \tag{3.7}$$

The solution to this equation is

$$\vec{A}\left(\vec{r},t\right) = \vec{A}_0 e^{i\left(\vec{k}\cdot\vec{r}-\omega t\right)} + \text{c.c.};\ k^2 = \omega^2/c^2;\ \vec{A}_0 = A_0\hat{e} \tag{3.8}$$

The source-free Maxwell equations can also be manipulated to show that $\vec{E}$ and $\vec{B}$ solve similar wave equations:

$$\nabla^2\vec{E} = \frac{1}{c^2}\frac{\partial^2\vec{E}}{\partial t^2} \tag{3.9a}$$

$$\nabla^2\vec{B} = \frac{1}{c^2}\frac{\partial^2\vec{B}}{\partial t^2} \tag{3.9b}$$

with solutions

$$\vec{E} = \vec{E}_0 e^{i\left(\vec{k}\cdot\vec{r}-\omega t\right)} + \text{c.c.} \tag{3.10a}$$

$$\vec{B} = \vec{B}_0 e^{i\left(\vec{k}\cdot\vec{r}-\omega t\right)} + \text{c.c.} \tag{3.10b}$$

Example 3.2 Verify (3.9a) and (3.9b).

We start with the source-free Maxwell equations:

$$\Rightarrow \vec{\nabla} \times \left(\vec{\nabla} \times \vec{E}\right) = \vec{\nabla} \times \left(-\frac{1}{c}\frac{\partial \vec{B}}{\partial t}\right) \Rightarrow \vec{\nabla}\underbrace{\left(\vec{\nabla} \cdot \vec{E}\right)}_{=0} - \nabla^2 \vec{E} = \vec{\nabla} \times \left(-\frac{1}{c}\frac{\partial \vec{B}}{\partial t}\right)$$

$$= -\frac{1}{c}\frac{\partial}{\partial t}\underbrace{\left(\vec{\nabla} \times \vec{B}\right)}_{\frac{1}{c}\frac{\partial \vec{E}}{\partial t}} = -\frac{1}{c^2}\frac{\partial^2 \vec{E}}{\partial t^2} \Rightarrow \nabla^2 \vec{E} = \frac{1}{c^2}\frac{\partial^2 \vec{E}}{\partial t^2}$$

Similarly,

$$\vec{\nabla} \times \left(\vec{\nabla} \times \vec{B}\right) = \vec{\nabla} \times \left(\frac{1}{c}\frac{\partial \vec{E}}{\partial t}\right) \Rightarrow \vec{\nabla}\underbrace{\left(\vec{\nabla} \cdot \vec{B}\right)}_{=0} - \nabla^2 \vec{B} = \frac{1}{c}\frac{\partial}{\partial t}\underbrace{\left(\vec{\nabla} \times \vec{E}\right)}_{-\frac{1}{c}\frac{\partial \vec{B}}{\partial t}}$$

$$\Rightarrow \nabla^2 \vec{B} = \frac{1}{c^2}\frac{\partial^2 \vec{B}}{\partial t^2}$$

∎

Our choice of gauge implies $\vec{k} \cdot \vec{A}_0 = 0$, so that the fields described by (3.2a and 3.2b) [or equivalently, by (3.6a and 3.6b)] can be written as

$$\vec{E} = ik\vec{A}_0\left[e^{i\left(\vec{k}\cdot\vec{r}-\omega t\right)} + \text{c.c.}\right] \tag{3.11a}$$

$$\vec{B} = i\vec{k} \times \vec{A}_0\left[e^{i\left(\vec{k}\cdot\vec{r}-\omega t\right)} + \text{c.c.}\right] \tag{3.11b}$$

By convention, the monochromatic plane electromagnetic wave is said to be polarized in the direction the electric field vector points. For a monochromatic plane electromagnetic wave propagating in an arbitrary direction $\vec{k} = k\hat{k} = (\omega/c)\hat{k} = (2\pi/\lambda)\hat{k}$ with polarization direction $\hat{\varepsilon}$, we write

$$\vec{E}\left(\vec{r},t\right) = E_0\hat{\varepsilon}e^{i\left(\vec{k}\cdot\vec{r}-\omega t\right)} + \text{c.c.} \tag{3.12a}$$

$$\vec{B}\left(\vec{r},t\right) = \hat{k} \times \vec{E} \tag{3.12b}$$

Notice that our choice of gauge implies that $\hat{\varepsilon} \cdot \hat{k} = 0$.

A useful approximation is realized when we consider interactions of atomic (or molecular) electrons with photons of wavelengths in the visible range $\left(\lambda = 4000 \ \overset{0}{A} - 7000 \ \overset{0}{A} \right)$. The size of $\vec{r}$ in (3.8) and (3.12) is on the order of the size of the atom (or molecule), that is, on the order of a_0, the Bohr radius $\left(a_0 \sim 0.5 \ \overset{0}{A} \right)$. Under these circumstances, we have

$$\vec{k} \cdot \vec{r} = \left(\frac{2\pi}{\lambda} \right) \hat{k} \cdot \vec{r} << 1 \tag{3.13}$$

and the exponential appearing in (3.12a) can be approximated as

$$e^{i\vec{k} \cdot \vec{r}} = 1 + i\vec{k} \cdot \vec{r} + \frac{\left(i\vec{k} \cdot \vec{r} \right)^2}{2!} + \cdots \approx 1 \tag{3.14}$$

This approximation is called the dipole approximation. In the dipole approximation, the photon wavelength is so large that it is essentially constant over the extent of the target atom or molecule. This approximation begins to fail only when we approach the region of the electromagnetic spectrum occupied by x-rays, whose wavelengths are smaller than 1 nm.

The classical form of the expected angular distribution for pd. reaction products (reaction products that include photoelectrons[1][13, 14]) in the dipole approximation is derived as follows. First, we assume that the physical form of the atom or molecule in question is that of an electron-ion pair, both members of which oscillate about their equilibrium positions (with very small amplitudes, to be sure!). This charge arrangement forms a classical electric dipole which can then couple to an external field through its dipole moment. This is a very crude model, but it will serve for the moment. To begin a quantitative description, we let $(\alpha = \phi, \beta = \theta, \chi)$ be the Euler angles measured from the lab (space-fixed) frame (X, Y, Z) to the atom/molecule-fixed frame (x, y, z) with the same origin O in the usual way (see Fig. 2.2). We let the Z-axis lie along the direction of the linearly polarized photon electric field vector $\vec{E}$. Now let the z-axis lie along the electric dipole moment $\vec{\mu}_e$ of the atom/molecule. In the dipole approximation, the probability P_{pd} of a photodissociation event is proportional to $\left| \vec{\mu}_e \cdot \vec{E} \right|^2$ (we say that the photodissociation transition amplitude is proportional to the projection of the electric field, $\vec{E}$, onto the electric dipole moment direction of the atom/molecule $\hat{\mu}_e$, and that the photodissociation transition probability is proportional to the square of the transition amplitude). In this section, we ignore the radial dependance of the transition probability, leaving those issues to

---

[1] Although as subsequent chapters will show, PADs are best described by a quantum treatment.

later portions of the book. With that in mind, the probability of an electron photodissociating in the direction (ϕ, θ, γ) is

$$P_{pd}(\phi, \theta, \gamma) \propto \left|\vec{\mu}_e \cdot \vec{E}\right|^2 \propto |\hat{\mu}_e \cdot \hat{\varepsilon}|^2 \propto \left|\hat{z} \cdot \hat{Z}\right|^2 \propto \cos^2\theta = \frac{2P_2(\cos\theta) + 1}{3} \quad (3.15)$$

We account for our lack of a proportionality constant by normalizing P_{pd} as follows:

$$\int_0^{2\pi} d\phi \int_0^{\pi} \sin\theta\, d\theta \int_0^{2\pi} d\gamma\, P_{pd}(\phi, \theta, \gamma) = 1 \quad (3.16)$$

To find the normalization constant, substitute (3.15) into (3.16):

$$\frac{C}{3}\left[\int_0^{2\pi} d\phi \int_0^{\pi} \sin\theta\, d\theta \int_0^{2\pi} d\gamma + \int_0^{2\pi} d\phi \int_0^{\pi} 2P_2(\cos\theta)\sin\theta\, d\theta \int_0^{2\pi} d\gamma\right] = 1 \quad (3.17)$$

Using (2.71), we note that the last term integrates out to zero, leaving us with the result

$$C = \frac{3}{8\pi^2} \quad (3.18)$$

We now have

$$P_{pd}(\phi, \theta, \gamma) = \frac{2P_2(\cos\theta) + 1}{8\pi^2} = \frac{D_{00}^0(\phi, \theta, \gamma) + 2D_{00}^2(\phi, \theta, \gamma)}{8\pi^2} \quad (3.19)$$

where we also used (2.110). Let $f(\theta_m, \phi_m)$ be the final angular distribution of the photoelectrons in the atom/molecule frame, where (θ_m, ϕ_m) are the polar and azimuthal angles about the z-axis.

Expand the final angular distribution in a series of spherical harmonics as follows:

$$f(\theta_m, \phi_m) = \sum_{k,q} b_{kq} Y_{kq}(\theta_m, \phi_m) \quad (3.20)$$

where the expansion coefficients b_{kq} are given by

$$b_{kq} = \int_0^{2\pi} d\phi_m \int_0^{\pi} \sin\theta_m\, d\theta_m\, Y_{kq}^*(\theta_m, \phi_m) f(\theta_m, \phi_m) \quad (3.21)$$

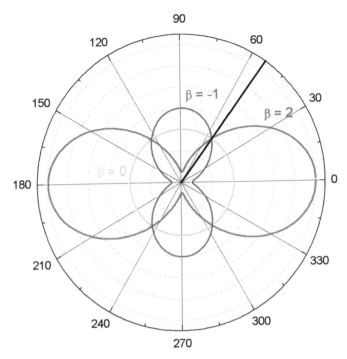

Fig. 3.1 Photoelectron angular distribution patterns for $\beta = -1, 0, 2$. The angles are measured from the (horizontal) polarization direction of the linearly polarized incident photon. In each of the three plots, the value of the radial coordinate is proportional to the relative intensity of photoelectrons ejected in that direction. Note also the so-called magic angle (54.74^0) at which the photoelectron angular distribution is independent of the asymmetry parameter β [c. f. (3.29)]

and where we have used the orthogonality condition of the spherical harmonics (2.64) to isolate the b_{kq}. Let $I(\theta_s, \phi_s)$ be the photoelectron angular distribution pattern in the stationary/lab frame, where (θ_s, ϕ_s) are the polar and azimuthal angles about the Z-axis (direction of $\vec{E}$ - see Fig. 3.1). A particular orientation (ϕ, θ, χ) contributes $P_{pd}(\phi, \theta, \chi) f(\theta_m, \phi_m)$ to $I(\theta_s, \phi_s)$, and the complete photoelectron distribution pattern is found by integrating over all possible (ϕ, θ, χ) orientations:

$$I(\theta_s, \phi_s) = \int\limits_0^{2\pi} d\phi \int\limits_0^{\pi} \sin\theta \, d\theta \int\limits_0^{2\pi} d\gamma P_{pd}(\phi, \theta, \gamma) f(\theta_m, \phi_m) \qquad (3.22)$$

Substituting the form of $P_{pd}(\phi, \theta, \gamma)$ given in (3.19) and the form of $f(\theta_m, \phi_m)$ given in (3.20) into (3.22) gives

$$I(\theta_s, \phi_s) = \iiint \left[\frac{D_{00}^0(\phi, \theta, \gamma) + 2D_{00}^2(\phi, \theta, \gamma)}{8\pi^2} \right] \sum_{k,q} b_{kq} Y_{kq}(\theta_m, \phi_m) \, d\phi \sin\theta \, d\theta \, d\gamma .$$

$$(3.23)$$

Note that $Y_{kq}(\theta_m, \phi_m)$ can be expressed in the coordinates of the lab frame by means of the rotation:

$$Y_{kq}(\theta_m, \phi_m) = \sum_m D_{mq}^k Y_{km}(\theta_s, \phi_s) \tag{3.24}$$

which when substituted into (3.23) gives

$$I(\theta_s, \phi_s) = \sum_{kqm} b_{kq} Y_{km}(\theta_s, \phi_s) \iiint \left[\frac{D_{00}^0 D_{mq}^k + 2D_{00}^2 D_{mq}^k}{8\pi^2} \right] d\Omega \tag{3.25}$$

Evaluate (3.25) using (2.110), (2.102), and (2.153):

$$I(\theta_s, \phi_s) = b_{00} Y_{00}(\theta_s, \phi_s) + \frac{2}{5} b_{20} Y_{20}(\theta_s, \phi_s)$$

$$= \frac{b_{00}}{\sqrt{4\pi}} \left[1 + \frac{2b_{20}}{\sqrt{5}\, b_{00}} P_2(\cos\theta_s) \right] \tag{3.26}$$

We define

$$\sigma_{pd} \equiv \sqrt{4\pi}\, b_{00} \tag{3.27}$$

and

$$\beta \equiv \frac{2b_{20}}{\sqrt{5}\, b_{00}} \tag{3.28}$$

to get the final form for the photoelectron angular distribution pattern,

$$I(\theta_s) = \frac{\sigma_{pd}}{4\pi} [1 + \beta P_2(\cos\theta_s)] \tag{3.29}$$

where we write $I(\theta_s)$ instead of $I(\theta_s, \phi_s)$ because the angular distribution pattern has azimuthal symmetry.

It is interesting to note the physical interpretation of σ_{pd} and β in this context. σ_{pd} represents the total cross-section for production of photoelectrons from an unpolarized target (in the dipole approximation) by 100% linearly polarized light, to wit,

$$\sigma_{pd} \equiv \sqrt{4\pi}\, b_{00} = \int\limits_{0}^{2\pi} d\phi_m \int\limits_{0}^{\pi} \sin\theta_m\, d\theta_m f(\theta_m, \phi_m) \tag{3.30}$$

where we used (3.21). We find the physical interpretation of β in the next example.

Example 3.3 Show that β represents an "angular" average over the distribution pattern $f(\theta_m, \phi_m)$.

This can be seen as follows:

$$\beta = \frac{2b_{20}}{\sqrt{5}\, b_{00}} = \frac{2 \int\limits_{0}^{2\pi} d\phi_m \int\limits_{0}^{\pi} \sin\theta_m\, d\theta_m\, Y_{20}(\theta_m, \phi_m) f(\theta_m, \phi_m)}{\sqrt{\frac{5}{4\pi}} \int\limits_{0}^{2\pi} d\phi_m \int\limits_{0}^{\pi} \sin\theta_m\, d\theta_m f(\theta_m, \phi_m)}$$

$$\equiv 2\langle P_2(\cos\theta_m)\rangle. \tag{3.31}$$

where we also used (2.67d).   ∎

Equation (3.31) tells us that all the information about the photoelectron angular distribution pattern (in the dipole approximation) is contained in β. For this reason, β is known as the asymmetry parameter for the pd. reaction. For $I(\theta_s)$ to be positive, we must have $-1 \le \beta \le 2$. Although not accounted for in this section, we will see that, in general, β is a function of the incident photon energy (see Chap. 4 and subsequent chapters). Finally, we must remember that θ_s in (3.29) is the angle between the polarization vector of the linearly polarized incident photon and the momentum vector of the photoelectron (photoelectron collection direction) as seen in the lab frame.

In many cases, we can make the approximation that, after photon absorption, the ejected photoelectron has very little interaction with the residual core. This means that the direction in which the photoelectron is ejected is unchanged during and right after the absorption-ejection process. With that in mind and given that (3.29) shows that the angular distribution pattern of the ejected photoelectrons is a function of the polar angle only, one may write the distribution of a photoelectron in the atom frame as

$$f(\theta_m, \phi_m) = \frac{\delta(\theta_m - \theta_m^0)}{2\pi \sin\theta_m} \tag{3.32}$$

where θ_m^0 is the polar angle of the photoelectron's velocity vector.

Using the form (3.32) for $f(\theta_m, \phi_m)$, we can calculate as follows:

$$b_{00} = \int_0^{2\pi} d\phi_m \int_0^{\pi} \sin\theta_m \, d\theta_m \, Y_{00}^* \frac{\delta(\theta_m - \theta_m^0)}{2\pi \sin\theta_m} = \frac{1}{\sqrt{4\pi}} \qquad (3.33)$$

and

$$b_{20} = \int_0^{2\pi} d\phi_m \int_0^{\pi} \sin\theta_m \, d\theta_m \, Y_{20}^* \frac{\delta(\theta_m - \theta_m^0)}{2\pi \sin\theta_m} = \sqrt{\frac{5}{16\pi}} \left(3\cos^2\theta_m^0 - 1\right) \qquad (3.34)$$

For $\theta_m^0 = 0$, from (3.28) and (3.29) we find

$$\beta = \frac{2\sqrt{\frac{5}{16\pi}}(3\cos^2 0 - 1)}{\sqrt{\frac{5}{4\pi}}} = 2 \qquad (3.35)$$

and

$$I(\theta_s) = \frac{\sigma_{pd}}{4\pi}[1 + 2P_2(\cos\theta_s)] = \frac{3\sigma_{pd}}{4\pi}\cos^2\theta_s \qquad (3.36)$$

For $\theta_m^0 = \frac{\pi}{2}$, we find

$$\beta = \frac{2\sqrt{\frac{5}{16\pi}}(3\cos^2\frac{\pi}{2} - 1)}{\sqrt{\frac{5}{4\pi}}} = -1 \qquad (3.37)$$

and

$$I(\theta_s) = \frac{\sigma_{pd}}{4\pi}[1 + (-1)P_2(\cos\theta_s)] = \frac{3\sigma_{pd}}{8\pi}\sin^2\theta_s \qquad (3.38)$$

For an asymmetry parameter of $\beta = 2$, the ejected photoelectrons will have a $\cos^2$ distribution (photoelectrons ejected preferentially in the direction- and anti-direction of the photon polarization vector). For an asymmetry parameter of $\beta = -1$, the ejected photoelectrons will have a $\sin^2$ distribution (photoelectrons ejected preferentially in the directions perpendicular to the photon polarization vector). For an asymmetry parameter of $\beta = 0$, the ejected photoelectrons will have an isotropic (spherical) distribution (see Fig. 3.1).

Problems

3.1. Gauge Transformations in Electrodynamics.

(a) The electromagnetic vector and scalar potentials are introduced in (3.2a) and (3.2b), respectively. Show that the forms of the electric and magnetic fields given in (3.2a and 3.2b) are unchanged if the gradient of a scalar function is added to the vector potential $\vec{A}$ and the time derivative of the same scalar function is subtracted from the scalar potential Φ.

(b) The Lorentz gauge is defined by the condition [11, 12]

$$\vec{\nabla} \cdot \vec{A} + \frac{1}{c}\frac{\partial \Phi}{\partial t} = 0$$

Find the forms of the inhomogeneous wave equations in $\vec{A}$ and Φ under the Lorentz gauge.

(c) Apply the gauge condition from part (a) to the Lorentz gauge to find under which classes of gauge the Lorentz gauge is preserved.

(d) Find expressions for A and Φ under the Coulomb gauge. Assume static conditions.

(e) Suppose $\Phi = 0$ and $\vec{A} = \vec{A}(t)$. Consider the gauge transformation generated by $\Lambda = -\vec{x} \cdot \vec{A}(t)$. Find the new potentials. Find the value of the magnetic field under both potentials.

(f) Same as part (e), but with the gauge transformation $\Lambda = \exp\left[\int_{-\infty}^{t} A^2(\tau)d\tau\right]$.

3.2. Show that $\vec{A}\left(\vec{x}\right) = -\int_{0}^{1} t dt \left(\vec{x} \times \vec{B}\left(t\vec{x}\right)\right)$, given that $\vec{B} = \vec{\nabla} \times \vec{A}$ and $\vec{\nabla} \cdot \vec{B} = 0$.

3.3. Legendre's equation can be written as.

$$\left(1 - x^2\right)\frac{d^2y}{dx^2} - 2x\frac{dy}{dx} + l(l+1)y = 0; \quad l = 1, 2, 3, \cdots$$

(a) Show that Rodrigues' formula (A.3) solves the Legendre equation. You may have to apply the boundary condition $P_l(1) = 1$.

(b) Show that the functions generated by (A.4) solve Legendre's equation.

(c) Prove (2.71) by using Rodrigues' formula (A.3).

(d) Show that

$$P_l(0) = \begin{cases} (-1)^{l/2} \dfrac{(l-1)!!}{(l)!!}; l = even \\ 0; \quad l = odd \end{cases}$$

3.4. Explain why a spin-zero particle cannot have a permanent dipole moment.

3.5. The classical energy density in an electromagnetic field is given by [11, 12].

$$u = \frac{1}{8\pi}\left(\left|\vec{E}\right|^2 + \left|\vec{B}\right|^2\right)$$

Averaging over time and imagining the energy of the field as being carried by N photons (each of energy $\hbar\omega$) of various frequencies $\omega_{\vec{k}}$ confined to a volume V, show that we can write the general solution to (3.7) as

$$\vec{A}\left(\vec{r},t\right) = \sum_{\vec{k},\lambda}\left(\frac{2\pi N\hbar c^2}{V\omega_{\vec{k}}}\right)^{1/2}\left[c\left(\vec{k},\lambda\right)\hat{\varepsilon}_\lambda e^{i\left(\vec{k}\cdot\vec{r}-\omega t\right)} + c^\dagger\left(\vec{k},\lambda\right)\hat{\varepsilon}_\lambda e^{-i\left(\vec{k}\cdot\vec{r}-\omega t\right)}\right]$$

Here the vector nature of the vector potential is contained in the polarization vector $\hat{\varepsilon}_\lambda$.

3.6. Find the angle θ at which the photoelectron angular distribution is independent of the asymmetry parameter β. Also prove that $-1 \leq \beta \leq 2$.

3.7. Atomic Anions.

The formation of a negative ion from a neutral atom involves the subtle interplay of attractive Coulombic nuclear-centered forces and the interactions (Coulombic and otherwise) of the atomic electrons with one another. The simplest atomic negative ion is H⁻. We construct this system as follows. First, the neutral H atom is modeled as a nuclear point charge of $+e$ surrounded by an atomic electron in the form of a spherically symmetric "cloud" of negative charge, $-e$, extending to infinity. The electron of the neutral H atom as a charge distribution is given by

$$\rho\left(\vec{r}_2\right) = e\left|\psi_{100}\left(\vec{r}_2\right)\right|^2$$

where ψ_{100} is the ground state wave function of the electron

$$\psi_{100}\left(\vec{r}\right) = \left(\pi a_0^3\right)^{-1/2}\exp\left(-r/a_0\right)$$

We then create the H⁻ anion by adding to the neutral H atom, an "extra" electron of a point charge e, placed at $\vec{r}_1$. Show that the potential in which this extra electron finds itself is an exponentially decaying one, which is why atomic anions are generally weakly bound systems compared to neutral atoms or atomic cations. Do not forget to include the influence of the nucleus.

3.8. The Magnetic Dipole [11, 12].

The Biot-Savart law states that the magnetic field due to a steady current is given by (in SI units)

$$\vec{B}(\vec{r}) = \frac{\mu_0}{4\pi} \int \left[\vec{J}(\vec{r}') \times \frac{\hat{\mathcal{R}}}{\mathcal{R}^2} \right] d^3x'; \mu_0 = 4\pi \times 10^{-7} N/A^2; \vec{\mathcal{R}} = \vec{r} - \vec{r}'$$

where $\vec{J}$ is the volume current density that is responsible for the magnetic field and $\hat{\mathcal{R}}$ is a unit vector that points from the differential element of current (located with respect to some origin by the position vector $\vec{r}'$) to the point in space at which the field is to be calculated (located with respect to the same origin by the position vector $\vec{r}$). Specifically,

$$\vec{r} \equiv x\hat{x} + y\hat{y} + z\hat{z} = x_1\hat{e}_1 + x_2\hat{e}_2 + x_3\hat{e}_3 = \sum_{i=1}^{3} x_i\hat{e}_i$$

is a position vector of some point with respect to some origin, with magnitude $r = |\vec{r}|$, and $\hat{r} = \vec{r}/r$ is a unit vector in the direction of $\vec{r}$, with (Cartesian) components

$$n_i = \frac{x_i}{r} = x_i / \sqrt{\sum_{m=1}^{3} x_m^2}$$

Similar relations hold for the position vector $\vec{r}'$.

In classical electrodynamical theory, the magnetic field is known to have a zero divergence:

$$\vec{\nabla} \cdot \vec{B} = 0$$

Since the divergence of the curl is always zero, the above relation invites a vector potential $\vec{A}$ of the form

$$\vec{B} = \vec{\nabla} \times \vec{A}$$

with the following caveats:

- $\vec{A}$ is specified only to within an additive term, since we can add to $\vec{A}$ any quantity whose curl is zero (i.e., the gradient of some scalar). This ambiguity is the source of the gauge freedoms mentioned in this chapter.
- $\vec{B}$ depends on the space derivatives of $\vec{A}$, and not on $\vec{A}$ itself. The value of $\vec{B}$ at a given point can thus be calculated from $\vec{A}$ only if $\vec{A}$ is known in the region around the point in question.

(a) Use the Biot-Savart law to derive the general form of the vector potential $\vec{A}$ in terms of the volume current density. Show that this form of the vector potential obeys the Coulomb gauge in magnetostatics. Assume the current is bounded.

Potentially useful formulae:

$$\vec{\nabla} \times \left(\varphi \vec{v} \right) = -\vec{v} \times \left(\vec{\nabla} \varphi \right) + \varphi \vec{\nabla} \times \vec{v}$$

$$\vec{\nabla}_r \frac{1}{\left| \vec{r} - \vec{r}' \right|} = -\frac{\left(\vec{r} - \vec{r}' \right)}{\left| \vec{r} - \vec{r}' \right|^3}$$

$$\vec{\nabla} \cdot \left(\varphi \vec{v} \right) = \vec{v} \cdot \vec{\nabla} \varphi + \varphi \left(\vec{\nabla} \cdot \vec{v} \right)$$

(b) Using the formula,

$$\frac{1}{\left| \vec{r} - \vec{r}' \right|} = \frac{1}{\left| \vec{r} \right|} \left[1 + \frac{\hat{r} \cdot \vec{r}'}{\left| \vec{r} \right|} + \frac{1}{2} \left(3 \frac{\left(\hat{r} \cdot \vec{r}' \right)^2}{\left| \vec{r} \right|^2} - \frac{\left| \vec{r}' \right|^2}{\left| \vec{r} \right|^2} \right) - \cdots \right]$$

express your result from part (a) in terms of a magnetic multipole expansion. Show that the magnetic monopole term is always zero. That is, show that,

$$\int d^3 x J_i = 0$$

where J_i is the ith Cartesian component of the volume current density. Thus, ultimately show that

$$A_i \left(\vec{r} \right) \approx \frac{\mu_0}{4\pi} \sum_{j=1}^{3} \frac{x_j}{\left| \vec{r} \right|^3} \int x_j' J_i \left(\vec{r}' \right) d^3 x'$$

where $A_i \left(\vec{r} \right)$ the ith Cartesian component of the vector potential.

Hint: Remember that $\vec{\nabla} \cdot \vec{J} = 0$ in magnetostatics. Also, the divergence theorem may come in handy.

(c) Show that the integral $\int x'_j J_i\left(\vec{r}'\right) d^3x'$ is antisymmetric in i and j. Use this result to show that your result from part (b) can be written as.

$$\vec{A}\left(\vec{r}\right) \approx \frac{\mu_0}{4\pi} \frac{\vec{m} \times \hat{r}}{|\vec{r}|^2}$$

where: $\vec{m} \equiv \left(\frac{1}{2}\right) \int \left[\left(\vec{r}' \times \vec{J}\left(\vec{r}'\right)\right)\right] d^3x'$

is the magnetic dipole moment of a general volume current density $\vec{J}$.

Potentially useful formulae:

1.
$$\vec{a} \times \vec{b} = \sum_{i=1}^{3} \sum_{j=1}^{3} \sum_{k=1}^{3} \varepsilon_{ijk} a_i b_j \hat{e}_k$$

where:

$$\varepsilon_{ijk} = \left\{ \begin{array}{l} 0 \text{ if } i = j, j = k, \text{ or } i = k \\ 1 \text{ if } i, j, k, \text{ are a cyclic (even) permutation of } 1, 2, 3 \\ \text{-}1 \text{ if } i, j, k, \text{ are an anti-cyclic (odd) permutation of } 1, 2, 3 \end{array} \right\}$$

2. $\sum_{k=1}^{3} \varepsilon_{ijk} \varepsilon_{lmk} = \delta_{il}\delta_{jm} - \delta_{im}\delta_{jl}; \quad i, j, l, m = 1, 2, 3$ (prove this)

where the Kronecker delta δ_{ij} is defined as

$$\delta_{ij} \equiv \left\{ \begin{array}{l} 1 \text{ if } i = j \\ 0 \text{ if } i \neq j \end{array} \right\}; \quad i, j = 1, 2, 3$$

3. $\sum_{k=1}^{3} \varepsilon_{ijk} \left(\vec{a} \times \vec{b}\right)_k = a_i b_j - a_j b_i; \quad i, j = 1, 2, 3$ (prove this).

(d) Use the result from part (c) to show that the magnetic field of a point magnetic dipole of constant moment $\vec{m}$ located at the origin can be written as.

$$\vec{B}_{dip}\left(\vec{r}\right) = \frac{\mu_0}{4\pi} \left[\frac{8\pi}{3} \vec{m}\delta^3\left(\vec{r}\right) + \frac{3\left(\vec{m} \cdot \vec{r}\right)\vec{r}}{r^5} - \frac{\vec{m}}{r^3} \right]$$

This expression is central to the Hamiltonian responsible for hyperfine structure in hydrogen (see Problem 6.12).

Potentially useful formula:

$$\partial_j \partial_i \frac{1}{r} = -\frac{\delta_{ij}}{r^3} + \frac{3x_i x_j}{r^5} - \frac{4\pi}{3} \delta_{ij} \delta^3 \left(\vec{r} \right)$$

where $\partial_i \equiv \partial/\partial x_i$

3.9. Atomic Polarization [12].[2]

A simple model of an atom consists of a nucleus (point charge) of positive charge $q = N|e|$ surrounded by a spherical electron "cloud" of the same total negative charge (N is the atomic number of the atom and e is the fundamental electronic charge). A constant external field $\vec{E}_0$ will cause a relative displacement of the nucleus until it is a distance r_0 from the center of the cloud (which maintains its spherical shape) thus polarizing the atom. Assuming a uniform volume charge distribution within the electronic cloud of radius b, find r_0 in terms of the given parameters/constants.

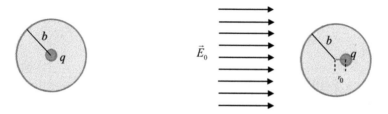

Hint: The nucleus is in equilibrium at r_0.

Chapter 4
Quantum Treatment of Photoelectron Angular Distributions (Dipole Approximation)

We first recall the classical Hamiltonian for the interaction of an electron of mass m and charge $(-e)$ with the field of a photon [c.f. (E.13)]:

$$H = \frac{\left[\vec{p} + \frac{e}{c}\vec{A}\left(\vec{r},t\right)\right]^2}{2m} - e\Phi \tag{4.1}$$

where $e \approx 4.8 \times 10^{-10}$ esu and $m \approx 9.1 \times 10^{-28}$ g. The quantum treatment starts by converting the classical dynamical quantities into quantum operators and then expanding

$$H = \frac{e}{2mc}\vec{A} \cdot \frac{\hbar}{i}\vec{\nabla} + \frac{e}{2mc}\frac{\hbar}{i}\vec{\nabla} \cdot \vec{A} + \frac{e^2}{2mc^2}A^2 - e\Phi \tag{4.2}$$

where we have used (2.32) and noted that, since $\vec{A} = \vec{A}\left(\vec{r},t\right)$, we must account for (2.31a) as well. We operate in the source-free condition $(\rho = J = 0)$ and under the Coulomb gauge [c.f. (3.3)] which means we can set the last term on the RHS of (4.2) equal to zero. In the Coulomb gauge, $\vec{A}$ and $\vec{p}$ commute

$$\vec{\nabla} \cdot \vec{A}\Psi = \left(\vec{\nabla} \cdot \vec{A}\right)\Psi + \vec{A} \cdot \vec{\nabla}\Psi = \vec{A} \cdot \vec{\nabla}\Psi \tag{4.3}$$

As a result, under these conditions, the interaction Hamiltonian is

Portions of this chapter follow closely the development presented in [13].

The original version of this chapter was revised. The correction to this chapter is available at https://doi.org/10.1007/978-3-031-08027-2_11

V. T. Davis, *Introduction to Photoelectron Angular Distributions*, Springer Tracts in Modern Physics 286, https://doi.org/10.1007/978-3-031-08027-2_4

$$H = \frac{e}{mc} \vec{A} \cdot \frac{\hbar}{i} \vec{\nabla} + \frac{e^2}{2mc^2} A^2 \qquad (4.4a)$$

Because it is multiplied by the (relatively small) term e^2/c^2, we neglect the term quadratic in $\vec{A}\left(\vec{r}, t\right)$, noting that it does not contribute to any bound-free transitions involving single photons (we confine our discussions in this document to single-photon transitions). Strictly speaking, to justify that last statement, and to show that the vector potential $\vec{A}\left(\vec{r}, t\right)$ is associated with the absorption (or emission) of a single photon, we would have to treat the electromagnetic field as a quantum field. However, as will be seen, that is not essential, so we will avoid that unnecessary complication. In other words, the approach we use here is a semiclassical one in which we let quantum objects (atoms and molecules) interact with an external classical field, while at the same time also (incongruously) speaking of photons (electromagnetic field quanta). We must also realize that by invoking photons in this scheme, the dipole approximation we are now set to apply is equivalent to a first-order perturbation theory. On a final note, we remark that the Hamiltonian of (4.2) does not account for the interaction of the intrinsic electron spin with the magnetic field of the photon. Given all the above caveats, we write our interaction Hamiltonian (in the dipole approximation) as follows:

$$H = -\vec{\mu}_e \cdot \vec{E} \qquad (4.4b)$$

The details of how this form of the interaction Hamiltonian came about are presented in the following example.

Example 4.1 Justify the form of the interaction Hamiltonian in (4.4b).

Consider the following Hamiltonian;

$$H_0 = \frac{p^2}{2m} + V(r) \qquad (4.5)$$

We calculate the following commutator:

$$[H_0, r_i] = \frac{1}{2m}\left[p^2, r_i\right] = \frac{1}{2m}\sum_{j=x,y,z}\left[p_j p_j, r_i\right] = \frac{1}{2m}\sum_{j=x,y,z}\left\{p_j\left[p_j, r_i\right] + \left[p_j, r_i\right]p_j\right\}$$

$$= -\frac{1}{m}\sum_{j=x,y,z}p_j i\hbar\delta_{ij} = -\frac{i\hbar}{m}p_i$$

$$(4.6)$$

where we used (2.31a) and (2.4). Therefore, for $|i\rangle$, $|f\rangle$ eigenstates of H_0, we have:

$$\langle f|p_i|i\rangle = \frac{im}{\hbar}\langle f|[H_0, r_i]|i\rangle = \frac{im}{\hbar}(E_f - E_i)\langle f|r_i|i\rangle = -im\omega\langle f|r_i|i\rangle$$

$$\Rightarrow \langle f|\vec{p}|i\rangle = -im\omega\langle f|\vec{r}|i\rangle \qquad (4.7)$$

where $\hbar\omega = E_i - E_f$ represents (in what is to follow) the energy carried by a single photon. We use this result to rewrite our interaction Hamiltonian (in the dipole approximation) as follows:

$$H = \frac{e}{mc}\vec{A}\cdot\frac{\hbar}{i}\vec{\nabla} \rightarrow \frac{e}{mc}\vec{A}\cdot\left(-im\omega\vec{r}\right) = \frac{e}{c}\frac{\partial\vec{A}}{\partial t}\cdot\vec{r} = -e\vec{E}\cdot\vec{r} = -\vec{\mu}_e\cdot\vec{E} \quad (4.8)$$

where $\vec{\mu}_e = e\vec{r}$ is the dipole moment of the atom and we used (3.8) and (3.6b). ∎

In the dipole approximation, the differential cross-section for photodissociation $d\sigma_{pd}/d\Omega$ for the single-photon, bound-free photodissociation transition is given by [3, 15, 16]

$$\frac{d\sigma_{pd}}{d\Omega} \propto |\langle f|O|i\rangle|^2 \quad (4.9)$$

where the operator O, defined as

$$O = \vec{r}\cdot\hat{Z} = r\cos\theta_s = r\sqrt{\frac{4\pi}{3}}Y_{10}(\hat{r}) = rC_0^{(1)}(\hat{r}) \quad (4.10)$$

is the dipole operator of a linearly polarized photon polarized along the Z-direction (in the lab frame) [c.f. (3.15)], the $C_q^{(k)}(\hat{r})$ are the "normalized" spherical harmonic tensor operators [3, 15, 16]:

$$C_q^{(k)}(\hat{r}) \equiv \sqrt{\frac{4\pi}{2k+1}}Y_{kq}(\hat{r}) \quad (4.11)$$

and $\vec{r}$ gives the position of the targeted electron. The polar angles are measured from the Z-axis. Furthermore, if we assume that the target atoms are isotropic in their angular momentum states (i.e., all orbital angular momentum magnetic sublevels are equally occupied), then (4.9) will have to be averaged over all possible initial orbital angular momentum states [3, 13].

The initial (bound) state of the electron in the atom (or molecule) is a state of definite angular momentum, and can be described by

$$|i\rangle = R_{nl'}(r)|l', m'\rangle \quad (4.12)$$

where $R_{nl'}(r)$ is the radial portion of the wave function that describes the electron in its initial bound state. One way to model the final state of the photoelectron in the continuum is as a plane wave plus an incoming spherical wave [3, 6, 8]:

$$|f\rangle \propto \left[e^{ikZ} + \frac{f(\theta_s)}{r}e^{-ikr}\right] \quad (4.13)$$

where k is the wave number of the photoelectron and $f(\theta_s)$ is a form factor that represents the effect of the atomic/molecular potential on the outgoing photoelectron (assuming azimuthal symmetry-a consequence of an assumed spherically symmetric potential). The asymptotic form of the final state is [3, 6, 13, 15]

$$|f\rangle \underset{r \to \infty}{\propto} \sum_{l=0}^{\infty} (2l+1) i^l e^{-i\delta_l} P_l(\cos\theta_s) \left[\frac{e^{i\delta_l} \sin\left(kr - \frac{l\pi}{2}\right) + e^{-i\left(kr + \frac{l\pi}{2}\right)} e^{-2i\delta_l} \sin\delta_l}{kr} \right]$$

(4.14)

where δ_l is the phase shift of the lth partial wave, and we recall that θ_s is the polar angle in the stationary/lab frame measured from the Z-axis. We now invoke (2.79), which gives

$$|f\rangle \propto 4\pi \sum_{l,m} (i)^l e^{-i\delta_l} Y_{lm}^*\left(\widehat{k}\right) Y_{lm}(\widehat{r}) G_{kl}(r)$$

(4.15)

where $\widehat{k}$ is a unit vector along the direction of the photoelectron momentum vector and

$$G_{kl}(r) \equiv \frac{e^{i\delta_l} \sin\left(kr - \frac{l\pi}{2}\right) + e^{-i\left(kr + \frac{l\pi}{2}\right)} e^{-2i\delta_l} \sin\delta_l}{kr}$$

(4.16)

is the radial portion of the final-state wave function divided by kr [16]. The matrix element $\langle f|rC_0^{(1)}(\widehat{r})|i\rangle$ is now

$$\langle f|rC_0^{(1)}(\widehat{r})|i\rangle = 4\pi \sum_{l,m} (i)^l e^{-i\delta_l} Y_{lm}^*\left(\widehat{k}\right) \langle l,m|C_0^{(1)}(\widehat{r})|l',m'\rangle \int_0^\infty r^3 R_{nl'}(r) G_{kl}(r) dr$$

(4.17)

Note that in this model, reflected in (4.17) above, we are implicitly assuming LS-coupling and ignoring any lingering interactions (other than with the centrifugal barrier of the core potential) between the escaping photoelectron and the residual core [c.f. (9.2)].

To proceed, we make use of (2.158) [or equivalently, (2.197) for $q = 0$]:

$$\langle l,m|C_0^{(1)}(\widehat{r})|l',m'\rangle = (-1)^m [(2l'+1)(2l+1)]^{\frac{1}{2}} \begin{pmatrix} l & 1 & l' \\ -m & 0 & m' \end{pmatrix} \begin{pmatrix} l & 1 & l' \\ 0 & 0 & 0 \end{pmatrix}$$

(4.18)

to get

$$\langle f|rC_0^{(1)}(\hat{r})|i\rangle = 4\pi \sum_{l,m} (i)^l e^{-i\delta_l} Y_{lm}^*\left(\hat{k}\right)(-1)^m[(2l'+1)(2l+1)]^{\frac{1}{2}}$$

$$\times \begin{pmatrix} l & 1 & l' \\ -m & 0 & m' \end{pmatrix} \begin{pmatrix} l & 1 & l' \\ 0 & 0 & 0 \end{pmatrix} \mathfrak{R}_l \tag{4.19}$$

where the radial dipole integral is

$$\mathfrak{R}_l \equiv \int_0^\infty r^3 R_{nl'}(r)\, G_{kl}(r)\, dr \tag{4.20}$$

all the while remembering that the $R_{nl'}(r)$ are normalized as

$$\int_0^\infty [R_{nl'}(r)]^2\, r^2\, dr = 1 \tag{4.21}$$

Example 4.2 The reduced matrix elements of the normalized spherical harmonic tensor operators were important in the derivation of (4.18). As an important aside, let us derive some additional properties of these reduced matrix elements. Specifically, we show that.

$$\langle l\|C^{(k)}\|l'\rangle = (-1)^l \langle l'\|C^{(k)}\|l\rangle \text{ and } \langle l\|C^{(k)}\|l'\rangle = 0 \text{ if } l+k+l' = \text{odd number.}$$

Using (2.145b),

$$\langle l\|C^{(k)}\|l'\rangle = (-1)^l \sqrt{(2l+1)(2l'+1)} \begin{pmatrix} l & k & l' \\ 0 & 0 & 0 \end{pmatrix}$$

$$= (-1)^{2l+k+l'} \sqrt{(2l+1)(2l'+1)} \begin{pmatrix} l' & k & l \\ 0 & 0 & 0 \end{pmatrix}$$

$$= (-1)^k(-1)^{l'} \sqrt{(2l+1)(2l'+1)} \begin{pmatrix} l' & k & l \\ 0 & 0 & 0 \end{pmatrix} = (-1)^k \langle l'\|C^{(k)}\|l\rangle$$

Using (2.145c),

$$\begin{pmatrix} l & k & l' \\ 0 & 0 & 0 \end{pmatrix} = (-1)^{l+k+l'} \begin{pmatrix} l & k & l' \\ 0 & 0 & 0 \end{pmatrix}$$

Since $\langle l\|C^{(k)}\|l'\rangle$ is proportional to this 3-j symbol, $\langle l\|C^{(k)}\|l'\rangle$ is zero if $l+k+l'$=odd number. ∎

The differential cross-section is found by squaring (4.19) and then averaging over all possible initial states m':

$$\frac{d\sigma_{pd}}{d\Omega} = \frac{16\pi^2}{(2l'+1)} \sum_{m'} \sum_{l_1,m_1 l_2,m_2} (i)^{l_1-l_2} e^{-i(\delta_{l_1}-\delta_{l_2})}$$

$$\times Y^*_{l_1 m_1}(k) Y_{l_2 m_2}(k)(-1)^{m_1+m_2}(2l'+1)[(2l_1+1)(2l_2+1)]^{\frac{1}{2}}$$

$$\times \begin{pmatrix} l_1 & 1 & l' \\ -m_1 & 0 & m' \end{pmatrix} \begin{pmatrix} l_1 & 1 & l' \\ 0 & 0 & 0 \end{pmatrix} \begin{pmatrix} l_2 & 1 & l' \\ -m_2 & 0 & m' \end{pmatrix} \begin{pmatrix} l_2 & 1 & l' \\ 0 & 0 & 0 \end{pmatrix} \mathfrak{R}_{l_1} \mathfrak{R}_{l_2}$$

$$(4.22)$$

The triangle selection rules inherent in the 3-j symbols collapse the sums over m_1, m_2 to the value $m_1 = m_2 = m'$. Dropping the primes (which now means that the label l refers to the photoelectron's orbital angular momentum while still bound, i.e., prior to the pd. event) leaves us with

$$\frac{d\sigma_{pd}}{d\Omega} = \frac{16\pi^2}{(2l+1)} \sum_{m} \sum_{l_1,l_2} (i)^{l_1-l_2} e^{-i(\delta_{l_1}-\delta_{l_2})} Y^*_{l_1 m_1}(\hat{k}) Y_{l_2 m_2}(\hat{k})(-1)^{2m}(2l+1)$$

$$\times [(2l_1+1)(2l_2+1)]^{\frac{1}{2}} \begin{pmatrix} l_1 & 1 & l \\ -m & 0 & m \end{pmatrix} \begin{pmatrix} l_1 & 1 & l \\ 0 & 0 & 0 \end{pmatrix} \begin{pmatrix} l_2 & 1 & l \\ -m & 0 & m \end{pmatrix} \begin{pmatrix} l_2 & 1 & l \\ 0 & 0 & 0 \end{pmatrix} \mathfrak{R}_{l_1} \mathfrak{R}_{l_2}$$

$$(4.23)$$

Again, we invoke triangle selection rules and symmetries of the 3-j symbols to realize that we must have $|l-1| \le l_1, l_2 \le |l+1|$ while simultaneously (l_1+1+l) and (l_2+1+l) must be even. As a result, the sums over l_1, l_2 collapse to just four terms: $l_1 = l \pm 1$; $l_2 = l \pm 1$. These four terms can be evaluated with the help of the following algebraic expressions for the 3-j symbols we will need:

$$\begin{pmatrix} l+1 & 1 & l \\ -m & 0 & m \end{pmatrix} = (-1)^{l-m-1} \left[\frac{(l+1)^2 - m^2}{(2l+3)(2l+1)(l+1)} \right]^{\frac{1}{2}} \Rightarrow \begin{pmatrix} l+1 & 1 & l \\ 0 & 0 & 0 \end{pmatrix}$$

$$= (-1)^{l-1} \left[\frac{(l+1)^2}{(2l+3)(2l+1)(l+1)} \right]^{\frac{1}{2}} \qquad (4.24a)$$

$$\begin{pmatrix} l-1 & 1 & l \\ -m & 0 & m \end{pmatrix} = (-1)^{l-m} \left[\frac{l^2 - m^2}{(2l+1)(2l-1)(l)} \right]^{\frac{1}{2}} \Rightarrow \begin{pmatrix} l-1 & 1 & l \\ 0 & 0 & 0 \end{pmatrix}$$

$$= (-1)^{l} \left[\frac{l^2}{(2l+1)(2l-1)(l)} \right]^{\frac{1}{2}} \qquad (4.24b)$$

Example 4.3 Determine the possible values of $(i)^{l_1-l_2}$ for $l_1 = l \pm 1$; $l_2 = l \pm 1$.

$$l_1 = l+1; l_2 = l+1 \Rightarrow (i)^{l_1-l_2} = 1$$

$$l_1 = l + 1; l_2 = l - 1 \Rightarrow (i)^{l_1 - l_2} = -1$$

$$l_1 = l - 1; l_2 = l + 1 \Rightarrow (i)^{l_1 - l_2} = -1$$

$$l_1 = l - 1; l_2 = l - 1 \Rightarrow (i)^{l_1 - l_2} = 1$$

∎

A quick remark on algebraic expressions for individual 3-j symbols found here and throughout this document. These expressions could (in principle) be found via (2.143) and (2.144) and other expressions (e.g., see Problem 4.2). Usually, however, it is easier to use tables such as those found in [3, 5, 10, 16].

In any case, invoking these identities gives

$$
\frac{d\sigma_{pd}}{d\Omega} = \frac{16\pi^2}{(2l+1)} \sum_m \left\{ (-1)^{4l}(2l+1)[2(l-1)+1] \left[\frac{l^2 - m^2}{(2l+1)(2l-1)(l)} \right] \right.
$$
$$
\times \left[\frac{l^2}{(2l+1)(2l-1)(l)} \right] \left| Y_{l-1,m}\left(\hat{k}\right) \right|^2 \mathfrak{R}_{l-1}^2 + (-1)^{4l}(2l+1)[2(l+1)+1]
$$
$$
\times \left[\frac{(l+1)^2 - m^2}{(2l+3)(2l+1)(l+1)} \right] \left[\frac{(l+1)^2}{(2l+3)(2l+1)(l+1)} \right] \left| Y_{l+1,m}\left(\hat{k}\right) \right|^2 \mathfrak{R}_{l+1}^2
$$
$$
+ (-1)^{4l}(2l+1)[2(l-1)+1]^{\frac{1}{2}}[2(l+1)+1]^{\frac{1}{2}} \left[\frac{l^2 - m^2}{(2l+1)(2l-1)(l)} \right]^{\frac{1}{2}}
$$
$$
\times \left[\frac{l^2}{(2l+1)(2l-1)(l)} \right]^{\frac{1}{2}} \left[\frac{(l+1)^2 - m^2}{(2l+3)(2l+1)(l+1)} \right]^{\frac{1}{2}} \left[\frac{(l+1)^2}{(2l+3)(2l+1)(l+1)} \right]^{\frac{1}{2}}
$$
$$
\times \left[Y_{l+1,m}^*\left(\hat{k}\right) Y_{l-1,m}\left(\hat{k}\right) e^{-i(\delta_{l+1}-\delta_{l-1})} + Y_{l-1,m}^*\left(\hat{k}\right) Y_{l+1,m}\left(\hat{k}\right) e^{i(\delta_{l+1}-\delta_{l-1})} \right]
$$
$$
\left. \times \mathfrak{R}_{l+1}\mathfrak{R}_{l-1} \right\}
$$

(4.25)

which, after a little algebra (okay-a lot of algebra!), reduces to

$$
\frac{d\sigma_{pd}}{d\Omega} = \frac{16\pi^2}{(2l+1)}
$$
$$
\times \sum_m \left\{ \begin{array}{l} \left[\dfrac{l^2 - m^2}{(2l+1)(2l-1)} \right] \left| Y_{l-1,m}\left(\hat{k}\right) \right|^2 \mathfrak{R}_{l-1}^2 + \left[\dfrac{(l+1)^2 - m^2}{(2l+3)(2l+1)} \right] \left| Y_{l+1,m}\left(\hat{k}\right) \right|^2 \mathfrak{R}_{l+1}^2 \\[2ex] + \left[\dfrac{l^2 - m^2}{(2l+1)(2l-1)} \right]^{\frac{1}{2}} \left[\dfrac{(l+1)^2 - m^2}{(2l+3)(2l+1)} \right]^{\frac{1}{2}} \\[2ex] \times \left[Y_{l+1,m}^*\left(\hat{k}\right) Y_{l-1,m}\left(\hat{k}\right) e^{-i(\delta_{l+1}-\delta_{l-1})} + Y_{l-1,m}^*\left(\hat{k}\right) Y_{l+1,m}\left(\hat{k}\right) e^{i(\delta_{l+1}-\delta_{l-1})} \right] \mathfrak{R}_{l+1}\mathfrak{R}_{l-1} \end{array} \right\}
$$

(4.26)

We now call upon (2.68), which we reproduce here in a slightly different form,

$$\cos\theta Y_{lm} = \left[\frac{(l+1)^2 - m^2}{(2l+3)(2l+1)}\right]^{\frac{1}{2}} Y_{l+1,m} + \left[\frac{l^2 - m^2}{(2l+1)(2l-1)}\right]^{\frac{1}{2}} Y_{l-1,m} \qquad (4.27)$$

First, we square both sides of (4.27) and then sum over m,

$$\sum_m \cos^2\theta |Y_{lm}|^2$$

$$= \sum_m \left\{ \begin{array}{l} \left[\dfrac{(l+1)^2 - m^2}{(2l+3)(2l+1)}\right]|Y_{l+1,m}|^2 + \left[\dfrac{l^2 - m^2}{(2l+1)(2l-1)}\right]|Y_{l-1,m}|^2 \\[3mm] + \left[\dfrac{(l+1)^2 - m^2}{(2l+3)(2l+1)}\right]^{\frac{1}{2}}\left[\dfrac{l^2 - m^2}{(2l+1)(2l-1)}\right]^{\frac{1}{2}} \left(Y_{l+1,m}^* Y_{l-1,m} + Y_{l+1,m}Y_{l-1,m}^*\right) \end{array} \right\}$$

$$= \frac{1}{(2l+3)(2l+1)}\left[(l+1)^2\sum_m |Y_{l+1,m}|^2 - \sum_m m^2 |Y_{l+1,m}|^2\right]$$

$$+ \frac{1}{(2l+1)(2l-1)}\left[l^2\sum_m |Y_{l-1,m}|^2 - \sum_m m^2 |Y_{l-1,m}|^2\right]$$

$$+ \sum_m \left[\frac{(l+1)^2 - m^2}{(2l+3)(2l+1)}\right]^{\frac{1}{2}}\left[\frac{l^2 - m^2}{(2l+1)(2l-1)}\right]^{\frac{1}{2}} \left(Y_{l+1,m}^* Y_{l-1,m} + Y_{l+1,m}Y_{l-1,m}^*\right)$$

$$(4.28)$$

Rearranging,

$$\sum_m \left[\frac{(l+1)^2 - m^2}{(2l+3)(2l+1)}\right]^{\frac{1}{2}}\left[\frac{l^2 - m^2}{(2l+1)(2l-1)}\right]^{\frac{1}{2}} \left(Y_{l+1,m}^* Y_{l-1,m} + Y_{l+1,m}Y_{l-1,m}^*\right)$$

$$= \sum_m \cos^2\theta |Y_{lm}|^2 - \frac{1}{(2l+3)(2l+1)}\left[(l+1)^2\sum_m |Y_{l+1,m}|^2 - \sum_m m^2 |Y_{l+1,m}|^2\right]$$

$$- \frac{1}{(2l+1)(2l-1)}\left[l^2\sum_m |Y_{l-1,m}|^2 - \sum_m m^2 |Y_{l-1,m}|^2\right]$$

$$(4.29)$$

If you look closely, you will notice that (4.29) is symmetric in the quantity:

$$\sum_m \left[\frac{(l+1)^2 - m^2}{(2l+3)(2l+1)}\right]^{\frac{1}{2}}\left[\frac{l^2 - m^2}{(2l+1)(2l-1)}\right]^{\frac{1}{2}} \left(Y_{l+1,m}^* Y_{l-1,m} + Y_{l+1,m}Y_{l-1,m}^*\right)$$

which can be seen by taking the complex conjugate of (4.27) before squaring and summing over m. This means that [13]

$$\sum_m \left[\frac{(l+1)^2 - m^2}{(2l+3)(2l+1)}\right]^{\frac{1}{2}} \left[\frac{l^2 - m^2}{(2l+1)(2l-1)}\right]^{\frac{1}{2}} \left(Y^*_{l+1,m} Y_{l-1,m}\right)$$

$$= \sum_m \left[\frac{(l+1)^2 - m^2}{(2l+3)(2l+1)}\right]^{\frac{1}{2}} \left[\frac{l^2 - m^2}{(2l+1)(2l-1)}\right]^{\frac{1}{2}} \left(Y_{l+1,m} Y^*_{l-1,m}\right)$$

$$= \frac{1}{2} \left\{ \begin{array}{l} \cos^2\theta \sum_m |Y_{lm}|^2 - \dfrac{1}{(2l+3)(2l+1)} \left[(l+1)^2 \sum_m |Y_{l+1,m}|^2 - \sum_m m^2 |Y_{l+1,m}|^2\right] \\[2ex] - \dfrac{1}{(2l+1)(2l-1)} \left[l^2 \sum_m |Y_{l-1,m}|^2 - \sum_m m^2 |Y_{l-1,m}|^2\right] \end{array} \right\}$$

$$(4.30)$$

Applying (2.80) and (2.195) to the RHS of (4.30) gives

$$\sum_m \left[\frac{(l+1)^2 - m^2}{(2l+3)(2l+1)}\right]^{\frac{1}{2}} \left[\frac{l^2 - m^2}{(2l+1)(2l-1)}\right]^{\frac{1}{2}} \left(Y^*_{l+1,m} Y_{l-1,m}\right)$$

$$= \sum_m \left[\frac{(l+1)^2 - m^2}{(2l+3)(2l+1)}\right]^{\frac{1}{2}} \left[\frac{l^2 - m^2}{(2l+1)(2l-1)}\right]^{\frac{1}{2}} \left(Y_{l+1,m} Y^*_{l-1,m}\right)$$

$$= \frac{1}{2} \left\{ \begin{array}{l} \cos^2\theta_s \dfrac{(2l+1)}{4\pi} - \dfrac{1}{(2l+3)(2l+1)} \\[2ex] \times \left[(l+1)^2 \dfrac{(2l+3)}{4\pi} - \dfrac{(l+1)(l+2)(2l+3)}{8\pi} \sin^2\theta_s\right] \\[2ex] - \dfrac{1}{(2l+1)(2l-1)} \left[l^2 \dfrac{(2l-1)}{4\pi} - \dfrac{(l-1)(l)(2l-1)}{8\pi} \sin^2\theta_s\right] \end{array} \right\}$$

$$(4.31)$$

After some additional algebra, and noting that $\cos^2\theta = 1 - \sin^2\theta$, we get

$$\sum_m \left[\frac{(l+1)^2 - m^2}{(2l+3)(2l+1)}\right]^{\frac{1}{2}} \left[\frac{l^2 - m^2}{(2l+1)(2l-1)}\right]^{\frac{1}{2}} \left(Y^*_{l+1,m} Y_{l-1,m}\right)$$

$$= \sum_m \left[\frac{(l+1)^2 - m^2}{(2l+3)(2l+1)}\right]^{\frac{1}{2}} \left[\frac{l^2 - m^2}{(2l+1)(2l-1)}\right]^{\frac{1}{2}} \left(Y_{l+1,m} Y^*_{l-1,m}\right)$$

$$= \frac{l(l+1)(2 - 3\sin^2\theta_s)}{8\pi(2l+1)}$$

$$(4.32)$$

We now apply the results of (4.32) to (4.26):

$$\frac{d\sigma_{pd}}{d\Omega} = \frac{16\pi^2}{(2l+1)} \sum_m \left\{ \left[\frac{l^2 - m^2}{(2l+1)(2l-1)} \right] \left| Y_{l-1,m}\left(\hat{k}\right) \right|^2 \mathfrak{R}_{l-1}^2 \right.$$

$$+ \left[\frac{(l+1)^2 - m^2}{(2l+3)(2l+1)} \right] \left| Y_{l+1,m}\left(\hat{k}\right) \right|^2 \mathfrak{R}_{l+1}^2 \right\}$$

$$+ \frac{16\pi^2}{(2l+1)} \frac{l(l+1)\left(2 - 3\sin^2\theta_s\right)}{8\pi(2l+1)} \underbrace{\left(e^{-i(\delta_{l+1}-\delta_{l-1})} + e^{i(\delta_{l+1}-\delta_{l-1})} \right)}_{2\cos(\delta_{l+1}-\delta_{l-1})} \mathfrak{R}_{l+1}\mathfrak{R}_{l-1}$$

(4.33)

Rearranging again,

$$\frac{d\sigma_{pd}}{d\Omega} = \frac{16\pi^2}{(2l+1)}$$

$$\times \left\{ \begin{array}{l} \left[\dfrac{\mathfrak{R}_{l-1}^2}{(2l+1)(2l-1)} \right] \left[l^2 \sum_m \left| Y_{l-1,m}\left(\hat{k}\right) \right|^2 - \sum_m m^2 \left| Y_{l-1,m}\left(\hat{k}\right) \right|^2 \right] \\[2ex] + \left[\dfrac{\mathfrak{R}_{l+1}^2}{(2l+3)(2l+1)} \right] \left[(l+1)^2 \sum_m \left| Y_{l+1,m}\left(\hat{k}\right) \right|^2 - \sum_m m^2 \left| Y_{l+1,m}\left(\hat{k}\right) \right|^2 \right] \\[2ex] + \dfrac{l(l+1)\left(2 - 3\sin^2\theta_s\right)}{8\pi(2l+1)} 2\cos(\delta_{l+1} - \delta_{l-1})\mathfrak{R}_{l+1}\mathfrak{R}_{l-1} \end{array} \right\}$$

(4.34)

Equation (4.34) is in a form against which we can again apply (2.80) and (2.195):

$$\frac{d\sigma_{pd}}{d\Omega} = \frac{16\pi^2}{(2l+1)}$$

$$\times \left\{ \begin{array}{l} \left[\dfrac{\mathfrak{R}_{l-1}^2}{(2l+1)(2l-1)} \right] \left[l^2 \dfrac{(2l-1)}{4\pi} - \dfrac{(l-1)(l)(2l-1)}{8\pi} \sin^2\theta_s \right] \\[2ex] + \left[\dfrac{\mathfrak{R}_{l+1}^2}{(2l+3)(2l+1)} \right] \\[2ex] \times \left[(l+1)^2 \dfrac{(2l+3)}{4\pi} - \dfrac{(l+1)(l+2)(2l+3)}{8\pi} \sin^2\theta_s \right] \\[2ex] + \dfrac{l(l+1)\left(2 - 3\sin^2\theta_s\right)}{8\pi(2l+1)} 2\cos(\delta_{l+1} - \delta_{l-1})\mathfrak{R}_{l+1}\mathfrak{R}_{l-1} \end{array} \right\}$$

(4.35)

After some more algebra, and again noting that $\sin^2\theta = 1 - \cos^2\theta$, and after collecting like terms, we get

$$I(\theta_s) = \frac{d\bar{\sigma}_{pd}}{d\Omega}$$

$$= \frac{2\pi}{(2l+1)^2} \left\{ \begin{array}{l} l(l+1)\left[\mathfrak{R}_{l-1}^2 + \mathfrak{R}_{l+1}^2 + 2\mathfrak{R}_{l-1}\mathfrak{R}_{l+1}\cos(\delta_{l+1} - \delta_{l+1})\right] \\ + \begin{bmatrix} l(l-1)\mathfrak{R}_{l-1}^2 + (l+1)(l+2)\mathfrak{R}_{l+1}^2 \\ -6l(l+1)\mathfrak{R}_{l-1}\mathfrak{R}_{l+1}\cos(\delta_{l+1} - \delta_{l-1}) \end{bmatrix} \cos^2\theta_s \end{array} \right\}$$

$$(4.36)$$

Multiply the term in the last bracket on the RHS by $\left(\frac{2}{3}\right)\left(\frac{3}{2}\right)$ and make the following identifications:

$$\alpha \equiv l(l+1)\left[\mathfrak{R}_{l-1}^2 + \mathfrak{R}_{l+1}^2 + 2\mathfrak{R}_{l-1}\mathfrak{R}_{l+1}\cos(\delta_{l+1} - \delta_{l+1})\right]$$
$$\chi \equiv \left[l(l-1)\mathfrak{R}_{l-1}^2 + (l+1)(l+2)\mathfrak{R}_{l+1}^2 - 6l(l+1)\mathfrak{R}_{l-1}\mathfrak{R}_{l+1}\cos(\delta_{l+1} - \delta_{l-1})\right]$$
$$\xi \equiv (2l+1)\left[l\mathfrak{R}_{l-1}^2 + (l+1)\mathfrak{R}_{l+1}^2\right]$$

$$(4.37)$$

which gives [13]

$$I(\theta_s) = \frac{d\bar{\sigma}_{pd}}{d\Omega} = \frac{2\pi}{(2l+1)^2}\left[\alpha + \frac{2}{3}\chi\frac{3}{2}\cos^2\theta_s\right]$$

$$= \frac{2\pi}{(2l+1)^2}\left[\alpha + \frac{2\chi}{3}\frac{\xi}{\xi}\frac{3}{2}\cos^2\theta_s + \left(-\frac{1}{2}\frac{2\chi}{3}\frac{\xi}{\xi} + \frac{1}{3}\chi\right)\right]$$

$$(4.38)$$

$$= \frac{2\pi}{(2l+1)^2}\frac{2}{3}\xi\left[\underbrace{\frac{3}{2}\frac{\alpha}{\xi} + \frac{1}{2}\frac{\chi}{\xi}}_{=1} + \frac{\chi}{\xi}\left(\frac{3}{2}\cos^2\theta_s - \frac{1}{2}\right)\right]$$

$$= \frac{\sigma_{pd}}{4\pi}\left[1 + \beta P_2(\cos\theta_s)\right]$$

where

$$\sigma_{pd} = \frac{16\pi^2}{3(2l_0+1)}\left[l_0\mathfrak{R}_{l_0-1}^2 + (l_0+1)\mathfrak{R}_{l_0+1}^2\right] \qquad (4.39a)$$

$$\beta = \frac{\chi}{\xi} = \frac{\begin{bmatrix} l_0(l_0-1)\mathfrak{R}_{l_0-1}^2 + (l_0+1)(l_0+2)\mathfrak{R}_{l_0+1}^2 \\ -6l_0(l_0+1)\mathfrak{R}_{l_0-1}\mathfrak{R}_{l_0+1}\cos(\delta_{l_0+1} - \delta_{l_0-1}) \end{bmatrix}}{(2l_0+1)\left[l_0\mathfrak{R}_{l_0-1}^2 + (l_0+1)\mathfrak{R}_{l_0+1}^2\right]} \qquad (4.39b)$$

$$P_2(\cos\theta_s) = \frac{1}{2}\left(3\cos^2\theta_s - 1\right) \qquad (4.39c)$$

$$\mathfrak{R}_{l_0\pm1} = \int_0^\infty r^3 R_{nl_0}(r)\, G_{k,l_0\pm1}(r)\, dr \qquad (4.39d)$$

Here we have relabeled slightly (for the sake of clarity in what is to follow), so that the label l_0 refers to the orbital angular momentum of the electron prior to photodetachment.

Equations (4.38) and (4.39) are the main results of what is known as the Cooper-Zare (C-Z) theory of photoelectron angular distributions. Eq. (4.38) has the same form as (3.29), and, as before, for $I(\theta_s)$ to be positive, we must have $-1 \leq \beta \leq 2$. The key feature of the C-Z theory is that the anisotropy of the photoelectron angular distribution is determined entirely by the asymmetry parameter β. For example, for s-electron photodissociation, $\beta(l_0 = 0) = 2$, which implies a $\cos^2$ distribution [c.f. (3.36)]. A $\cos^2$ distribution is consistent with what we might expect from the dipole selection rules (c.f. Problem 4.4), which, for an s-electron pd. event, predict p-waves emitted in the same direction (and anti-direction) as the photon polarization vector. Equation (4.39a), contains the sum of the two radial dipole integrals $\mathfrak{R}^2_{l_0-1}$ and $\mathfrak{R}^2_{l_0+1}$. We conclude that the total pd. cross-section is the sum of contributions from the transitions $l_0 \to l_0 - 1$ and $l_0 \to l_0 + 1$. Since there are no other terms on the RHS of (4.39a), we also conclude that these transitions do not interfere in the total pd. cross-section.

Example 4.4 Find the expression for the asymmetry parameter for p-electron photodissociation.

For p-electron detachment, we have $l_0 = 1$:

$$\beta(l_0 = 1) = \frac{2\mathfrak{R}^2_2 - 4\mathfrak{R}_0\mathfrak{R}_2 \cos(\delta_2 - \delta_0)}{\left(\mathfrak{R}^2_0 + 2\mathfrak{R}^2_2\right)} \tag{4.40}$$

∎

The dependence of β on the kinetic energy of the photoelectrons is realized through the radial dipole integrals and can be made explicit by rearranging (4.39b) in terms of their ratios [(4.39b) also indicates that β depends on phase differences between continuum partial waves]. In anion photodetachment, for example, these ratios are often found to vary linearly with the photoelectron kinetic energy, a consequence of the Wigner threshold law, demonstrated as follows. First, we recall from the Wigner threshold law that near threshold [17, 18],

$$\sigma_{pd} \propto k^{2l_0+1} \propto E_c^{\frac{2l_0+1}{2}} \tag{4.41}$$

where k is the wavenumber and E_c is the (center-of-mass) kinetic energy of the continuum photoelectron, respectively. It is also found that (see Problem 7.3) [17, 19]

$$\sigma_{pd} \propto k|\mathfrak{R}_{l_0}|^2 \tag{4.42}$$

So, it must be that [20]

$$|\mathfrak{R}_{l_0}| \propto k^{l_0} \Rightarrow |\mathfrak{R}_{l_0 \pm 1}| \propto k^{l_0 \pm 1} \qquad (4.43)$$

If the final state wavelength is large compared to the size of the initial state, and if we neglect any interactions between the photoelectron and the residual neutral, we can then say [20, 21]

$$\mathfrak{R}_{l_0+1}/\mathfrak{R}_{l_0-1} \propto k^2 \Rightarrow \mathfrak{R}_{l_0+1}/\mathfrak{R}_{l_0-1} = AE_c \qquad (4.44)$$

where A gives the relative size of the of the two radial matrix elements.

Example 4.5 Find an expression for the asymmetry parameter for the photodetachment of an atomic anion near threshold [17].

Divide the numerator and denominator of (4.39b) by the factor $\mathfrak{R}_{l_0-1}^2$ and apply (4.44)

$$\Rightarrow \beta = \frac{l_0(l_0 - 1) + (l_0 + 1)(l_0 + 2)A^2 E_c^2 - 6l_0(l_0 + 1)AE_c \cos\left(\delta_{l_0+1} - \delta_{l_0-1}\right)}{(2l_0 + 1)\left[l_0 + (l_0 + 1)A^2 E_c^2\right]}$$

$$\qquad (4.45)$$

Although the Wigner threshold law is strictly valid only near threshold, partial wave cross-section ratios have been found to be valid even at energies several electron volts above the threshold regime [18, 20, 22]. ∎

Now, for p-electron detachment ($l_0 = 1$) near threshold, from (4.44),

$$\mathfrak{R}_2/\mathfrak{R}_0 = A_{20}E_c \qquad (4.46)$$

where A_{20} gives the relative size of the two radial dipole integrals $\mathfrak{R}_2$ & $\mathfrak{R}_0$. Combining (4.40) or (4.45) with (4.46), we have (after a little algebra) the following expression for p-electron detachment near threshold [20]:

$$\beta = \frac{2(A_{20}E_c)^2 - 4A_{20}E_c c}{1 + 2A_{20}^2 E_c^2} \qquad (4.47)$$

where $c = \cos(\delta_2 - \delta_0)$. Eq. (4.47) is known as the Hanstorp model. A typical plot of the Hanstorp model for p-electron detachment can be found in Fig. 4.1.

The plot in Fig. 4.1 shows the spectral dependence of the asymmetry parameter for p-electron photodetachment. According to the dipole selection rules (see Problem 4.4), the photodetachment of a p-electron gives rise to s- and d-waves. In accordance with the Wigner threshold law, in a central potential, the centrifugal barrier suppresses the d-wave near threshold, leaving an isotropic

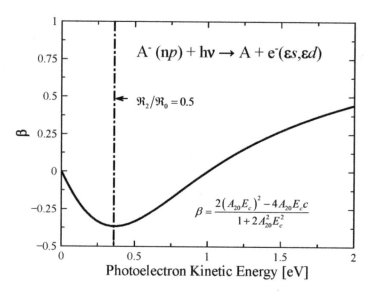

Fig. 4.1 Plot of the Hanstorp model for p-electron photodetachment. The plot shown is for the value $A_{20} = 1$. Note the location of the minimum at the ratio $\Re_2/\Re_0 = 0.5$ [23]. Reproduced from [23] with permission from the author

s-wave distribution with the concomitant asymmetry parameter $\beta = 0$. Away from threshold, the two waves interfere destructively, driving the asymmetry parameter into negative territory and the plot reaches a minimum when $\Re_2/\Re_0 = 0.5$. For full destructive interference ($\delta_2 - \delta_0 = 0 \Rightarrow c = 1$), the asymmetry parameter will reach its minimum possible value $\beta = -1$. At higher photoelectron kinetic energies, the d-wave becomes increasingly dominant, and β becomes positive, eventually approaching the asymptotic limit $\beta = 1$. Note that for the photoionization of a p-electron from a neutral or positive atom, β will not display the same behavior, and the plot of Fig. 4.1 will not apply [20, 24, 25].

For p-electron photodetachment near threshold, we have [17, 20–22]

$$\frac{|\Re_2|^2}{|\Re_0|^2} << 1 \Rightarrow \lim_{E_c \to 0} \beta(l_0 = 1) \approx -4A_{20}E_c c \qquad (4.48a)$$

And from (4.45), for orbital angular momentum $l \geq 2$,

$$\lim_{E_c \to 0} \beta \to \frac{(l_0 - 1)}{(2l_0 + 1)} \qquad (4.48b)$$

Remembering that $\beta(l_0 = 0) = 2$, and by comparing (4.48a) and (4.48b), we see that, except for the case $l_0 = 1$, all photodetachment asymmetry parameters are nonzero at photodetachment thresholds.

The C-Z theory can be applied even if the electron orbital must be described as a mixture of states (as is usually the case for molecular valence electrons, for example). Suppose the electron is removed from an orbital which may be expanded as a linear combination of an s-type and a p-type state. The initial state could be written as [21]

$$|i\rangle = \sqrt{1-f}|s\rangle + \sqrt{f}|p\rangle \qquad (4.49)$$

where f is the fraction of the state that can be described as a p state. As usual in this type of approximation, we describe the photoelectron as a superposition of well-defined orbital angular momentum states in the angular momentum quantum number l. Assuming LS-coupling and using the dipole-selection rules for linearly polarized photons (see Problem 4.4), the s term will give rise to p partial waves, and the p term will produce s-type and d-type partial waves. This state may then lead to an angular distribution described by [21]

$$I(\theta_s) = \left| \sqrt{f}e^{i\delta_0}R_0Y_{00} + \sqrt{f}e^{i(\delta_2+\pi)}R_2Y_{20} \right|^2 + 3\left| \sqrt{1-f}e^{i\left(\delta_1+\frac{\pi}{2}\right)}R_1Y_{10} \right|^2$$
$$+ Y_{21} \& Y_{2-1} \text{ terms} \qquad (4.50)$$

Specifically highlighted are the phase shifts associated with each partial wave as it interacts with the central potential of the residual neutral, as well as phase shifts resulting from partial wave interactions with the centrifugal barrier [c.f. (4.14)]. The coefficients R_l are proportional to the radial matrix elements, $\mathfrak{R}_{l_0 \pm 1}$, as appropriate for each term. The "3" in front of the second term on the RHS accounts for all possible orientations of the post-pd p states with respect to the spherically symmetric initial s state.

We will use (9.23) to construct β for this state. We will need the following quantities [c.f. (2.67a), (2.67d), and (2.67b)]:

$$I(0) = \left| \sqrt{f}e^{i\delta_0}R_0\underbrace{Y_{00}(0)}_{\propto 1} + \sqrt{f}e^{i(\delta_2+\pi)}R_2\underbrace{Y_{20}(0)}_{\propto 2} \right|^2 + 3\left| \sqrt{1-f}e^{i\left(\delta_1+\frac{\pi}{2}\right)}R_1\underbrace{Y_{10}(0)}_{\propto 1} \right|^2$$
$$= fR_0^2 + 3(1-f)R_1^2 + 4fR_2^2 - 4fR_0R_2 \cos(\delta_2 - \delta_0) \qquad (4.51)$$

and

$$I(\pi/2) = \left| \sqrt{f}e^{i\delta_0}R_0\underbrace{Y_{00}(\pi)}_{\propto 1} + \sqrt{f}e^{i(\delta_2+\pi)}R_2\underbrace{Y_{20}(\pi)}_{\propto -1} \right|^2 \qquad (4.52)$$
$$= fR_0^2 + fR_2^2 + 2fR_0R_2 \cos(\delta_2 - \delta_0)$$

The spherical harmonics here are un-normalized. We are incorporating all necessary normalization constants into the coefficient R_l. The Y_{21} and Y_{2-1} terms did not enter into consideration because they both have nodes at $\theta = 0$, $\pi/2$. Substituting (4.51) and (4.52) into (9.23) gives [21]

$$\beta = \frac{2(1-f)\left(\frac{R_1}{R_0}\right)^2 + 2f\left(\frac{R_2}{R_0}\right)^2 - 4f\left(\frac{R_2}{R_0}\right)\cos(\delta_2 - \delta_0)}{f + 2f\left(\frac{R_2}{R_0}\right)^2 + (1-f)\left(\frac{R_1}{R_0}\right)^2} \tag{4.53}$$

To check, note that in the limit $f \to 0$, (4.52) reduces to $\beta = 2$, as expected for the photodetachment from a pure s state. Also, in the limit $f \to 1$ (4.53) reduces to (4.40) for photodetachment of a p-electron, since $R_2/R_0 = \mathfrak{R}_2/\mathfrak{R}_0$.

At this point, it becomes necessary to clarify the conditions under which the C-Z theory can be considered valid. Obviously, we are operating under the dipole approximation as mentioned above. We have also been assuming LS-coupling, in which the orbital and spin angular momenta separately are good quantum numbers and under which we can assign to the electron (before it is ejected) a definite orbital angular momentum l_0. We have also neglected spin-orbit interactions and other relativistic phenomena. But, in addition, we have implicitly used a central-field, independent-particle approximation in which the effect of electron-electron correlations has been neglected. Now, as it turns out, when calculations similar to the one just performed for a single-electron atom are performed for multi-electron atoms, we find that, as long as we properly antisymmetrize the many-electron wave function, the form of the photo-electron angular distribution is the same as that of (4.38).[1] Therefore, under this approximation, when as s-electron is removed from an atom heavier than hydrogen, a pure $\cos^2$ distribution is still the expected result. When any other type of electron is removed, the resulting distribution will depend (in the dipole approximation) on the interference between the $l_0 \pm 1$ partial waves and the relative magnitudes of the one-electron radial dipole integrals $\mathfrak{R}_{l_0 \pm 1}$ (in which rest, e.g., the important relation of the incident photon energy to the pd. energy threshold of the atom) [13, 26, 27]. We should note that, in evaluating the radial integrals $\mathfrak{R}_{l_0 \pm 1}$, single-electron, hydrogen-like orbitals are often used as approximations for the radial portion $R_{nl'}(r)$ of the initial atomic state in (4.12), but there are other types of orbitals that can also be used [28–30], some of which can even account for electron correlations [31–34].

Finally, it should be remembered that, except for the effects of the central potential, we have ignored all interactions between the photoelectron and the residual core left behind. This may sound like a very restrictive approximation, but the C-Z theory describes photoelectron angular distributions remarkably well (for low photon energies), particularly for anion photodetachment [20, 35–44] although not for every anionic species [45–47]. This is because, unlike the photoionization of neutrals or positive ions which leaves the photoelectron subject to a long-range Coulombic force, the photodetachment of negative ions leaves behind a neutral core, considerably reducing residual interactions with the ejected photoelectron.

---

[1] If we include the effects of configuration mixing, however, the picture becomes more complicated [13, 26].

Problems

4.1. By taking the asymptotic limit of (4.13), derive (4.14). Also show that, in the asymptotic limit

$$f(\theta_s) = \frac{1}{k} \sum_{l=0}^{\infty} (2l+1) P_l(\cos\theta_s) e^{-i\delta_l} \sin\delta_l$$

4.2. 3-j symbols

(a) Using (2.138), prove (2.206). That is, show that

$$\begin{pmatrix} j & 1 & j \\ -j & 0 & j \end{pmatrix} = \sqrt{\frac{j}{(j+1)(2j+1)}}$$

(b) Using (B.82) and your results from part (a), verify (4.24a).
(c) Show that (2.210), (4.24a), and (4.24b) together satisfy (B.82).

4.3. In Cartesian form, the position vector is $\vec{r} = x\hat{x} + y\hat{y} + z\hat{z}$. Starting from the definition of an irreducible tensor operator (2.175), find the spherical forms $r_{\pm1}^{(1)}, r_0^{(1)}$.

Note: The operator $r_q^{(1)}$ is a special case of the solid harmonic operator $R_q^{(k)}$ defined as $R_q^{(k)} = r^k C_q^{(k)}(\hat{r})$. The operator O of (4.10) would then be the solid harmonic operator $R_0^{(1)}$.

4.4. Dipole selection rules.

(a) Find the dipole selection rules for single-particle transitions between states of definite angular momentum. Ignore spin. Differentiate between selection rules for linearly polarized and circularly polarized light.
(b) Use these rules to explain the form of (4.50).

4.5. Find an expression for the asymmetry parameter for photodetachment near threshold due to linearly polarized photons for initial states of the form:

$$|i\rangle = \sqrt{1-f}|s\rangle + \sqrt{f}|d\rangle$$
$$|i\rangle = \sqrt{1-f}|p\rangle + \sqrt{f}|d\rangle$$

Ensure your expressions reduce to the correct results in the appropriate limits.

4.6. Prove the following:

(a)

$$Y^*_{l_1,m+1}(\theta,\phi)Y_{l_2,m+1}(\theta,\phi) = \sqrt{(2l_1+1)(2l_2+1)}\sum_L \frac{2L+1}{4\pi}\begin{pmatrix} l_1 & l_2 & L \\ 0 & 0 & 0 \end{pmatrix}$$

$$\times \begin{pmatrix} l_1 & l_2 & L \\ -m-1 & m+1 & 0 \end{pmatrix}P_L(\cos\theta)$$

(b)

$$D^j_{m'm} = \sum_{m_1}\sum_{m_1'}\langle j_1,j_2;m_1,m_2|j,m\rangle\langle j_1,m_1';j_2,m_2'|j,m'\rangle D^{j_1}_{m_1'm_1}D^{j_2}_{m_2'm_2}$$

(c)

$$\sum_m (-1)^m \begin{pmatrix} l+1 & 1 & l \\ -m & 0 & m \end{pmatrix}\begin{pmatrix} l-1 & 1 & l \\ -m & 0 & m \end{pmatrix}Y_{l+1,m}Y_{l-1,-m}$$

$$= \frac{\sqrt{l(l+1)}}{4\pi(2l+1)}P_2(\cos\theta)$$

(d)

$$\frac{2l+1}{2}\int_{-1}^{1} d(\cos\theta)P_l(\cos\theta)P_{l_1}(\cos\theta)P_{l_2}(\cos\theta) = |\langle l_1,l_2;0,0|l,0\rangle|^2$$

4.7. Following the same basic analysis as presented in this chapter, find an expression for $I(\theta_s)$ (in the dipole approximation) for photodissociation caused by circularly polarized light [see also (9.27)]. Note that, in this case, the z-axis is taken as the photon beam propagation direction.

4.8. Matrix elements for electric dipole transitions.

Given an electric field $\vec{E} = E_0\hat{\varepsilon}_\lambda$ where

$$\hat{\varepsilon}_\lambda = \varepsilon_{\lambda x}\hat{e}_x + \varepsilon_{\lambda y}\hat{e}_y + \varepsilon_{\lambda z}\hat{e}_z$$

is the polarization direction.

(a) Show that we can write.

$$\vec{E}\cdot\vec{r} = E_0\sum_{\xi=-1}^{1}(-1)^\xi\varepsilon_{\lambda\xi}r^{(1)}_{-\xi}; \quad \xi = -1,0,1$$

where $\varepsilon_{\lambda\xi}$ are the components of $\hat{\varepsilon}_\lambda$ in a spherical tensor basis and $r_\xi^{(1)}$ are the components of $\vec{r} = x\hat{e}_x + y\hat{e}_y + z\hat{e}_z$ in a spherical tensor basis [c.f. Problem 4.3].

(b) Use your result from part (a) to show that, if we characterize $|i\rangle \rightarrow |n'l'm'\rangle$ and $|f\rangle \rightarrow |nlm\rangle$, then we can write,

$$\langle f|\vec{E}\cdot\vec{r}|i\rangle = E_0 \mathfrak{R}_l(l - l')\sqrt{l_{\max}} \sum_{\xi=-1}^{1} (-1)^{\xi-m}\varepsilon_{\lambda\xi} \begin{pmatrix} l & 1 & l' \\ -m & -\xi & m' \end{pmatrix}$$

where $l_{\max} = \max(l, l')$ and $\mathfrak{R}_l$ is defined as in (4.20). Show also that the matrix elements for electric dipole transitions can be written as

$$\left|\langle f|\vec{E}\cdot\vec{r}|i\rangle\right|^2 = |E_0|^2 \mathfrak{R}_l^2(l_{\max})(-1)^{-m-m'}\varepsilon_{\lambda(m'-m)}\varepsilon_{\lambda(m-m')}^* \begin{pmatrix} l & 1 & l' \\ -m & m-m' & m' \end{pmatrix}^2$$

To achieve the transition $|11\rangle \rightarrow |00\rangle$, would you use right circularly polarized light, left circularly polarized light, or linearly polarized light?

4.9. Starting with (2.149), derive (A.25), (A.26a), and (A.26b).

4.10. Atomic polarizability (revisited).

Due to its spherical symmetry, an isolated, neutral atom has no permanent electric dipole moment. In the presence of an external field, however, a dipole moment can be induced in an atom (see Problem 3.9).

(a) The dipole moment for a hydrogen atom (in the ground state) is given by the following:

$$\langle 100|e\vec{r}|100\rangle$$

where the ket $|100\rangle$ represents the ground-state wave function of the hydrogen atom. Show that this quantity is zero, confirming that this atom has no permanent electric dipole moment.

(b) Use first-order perturbation theory to find the electric dipole moment induced in a hydrogen atom (in the ground state) by a constant external electric field $\vec{E} = E_0\hat{z}$. The perturbation Hamiltonian takes the form $H' = -eE_0z$.

Chapter 5
Higher-Order Multipole Terms in Photoelectron Angular Distributions

For most valence-shell pd. experiments, in which typical photon energies are given by $E_\gamma = \hbar\omega < 100\text{eV}$, deviations (in photoelectron angular distributions) from the dipole approximation models presented earlier are expected to be negligible [48, 49]. But deviations due to higher-order multipole terms may become important for inner-shell photodetachment experiments (for example) in which photon energies can be higher than 100 eV. Therefore, it is prudent to examine how (4.38) will change when higher-order multipole terms are included in the expansion of the interaction Hamiltonian.

The interaction Hamiltonian is given in (4.4). Once again, if we neglect the quadratic term, we have

$$H = \frac{e}{mc}\vec{A}\cdot\vec{p} \tag{5.1}$$

We now include the next term in the expansion of the vector potential [cf. (3.8), (3.11a), (3.12a), and (3.14)]:

$$H \propto e^{i\vec{k}\cdot\vec{r}}\left(\hat{\varepsilon}\cdot\vec{p}\right) \approx \left(1 + i\vec{k}\cdot\vec{r}\right)\left(\hat{\varepsilon}\cdot\vec{p}\right) \tag{5.2}$$

The differential cross-section for photodissociation $d\sigma_{pd}/d\Omega$ for the single-photon, bound-free pd. transition in this approximation is given by

$$\frac{d\sigma_{pd}}{d\Omega} \propto |\langle f|O|i\rangle|^2 = \left|\langle f|\left(1 + i\vec{k}\cdot\vec{r}\right)\left(\hat{\varepsilon}\cdot\vec{p}\right)|i\rangle\right|^2 \tag{5.3}$$

Including all proportionality constants,

V. T. Davis, *Introduction to Photoelectron Angular Distributions*, Springer Tracts in Modern Physics 286, https://doi.org/10.1007/978-3-031-08027-2_5

$$\frac{d\sigma_{pd}}{d\Omega} = \frac{e^2 |A_0|^2}{m^2 c^2} \left| \langle f | \left(1 + i\vec{k} \cdot \vec{r} \right) \left(\hat{\varepsilon} \cdot \vec{p} \right) | i \rangle \right|^2$$

$$= \frac{2\pi \hbar^2 c \alpha}{m^2 \omega} \left| \langle f | \left(1 + i\vec{k} \cdot \vec{r} \right) \left(\hat{\varepsilon} \cdot \vec{p} \right) | i \rangle \right|^2 \tag{5.4}$$

where α is the fine-structure constant and the (magnitude of) the vector potential A_0 has been normalized to a unit volume such that the total energy carried by the electromagnetic field (i.e., the photon) is [15]

$$\frac{\omega^2 V}{2\pi c^2} |A_0|^2 = N\hbar\omega \tag{5.5}$$

and where V is the unit volume and $N = 1$ for a single photon (see problem 3.5).
    Start by examining the matrix element:

$$\left| \langle f | \left(1 + i\vec{k} \cdot \vec{r} \right) \left(\hat{\varepsilon} \cdot \vec{p} \right) | i \rangle \right| = \left| \langle f | \left(\hat{\varepsilon} \cdot \vec{p} \right) | i \rangle \right| + \left| \langle f | \left(i\vec{k} \cdot \vec{r} \right) \left(\hat{\varepsilon} \cdot \vec{p} \right) | i \rangle \right| \tag{5.6}$$

Before we proceed, we must first re-specify our choice of gauge (implicit up until this point):

$$\vec{k} \cdot \vec{A}_0 = \vec{k} \cdot \hat{\varepsilon} = 0 \tag{5.7}$$

The operators $\vec{k} \cdot \vec{r}$ and $\hat{\varepsilon} \cdot \vec{p}$ are thus orthogonal position and momentum operators, respectively, which means [c.f. (2.31a)]

$$\left[\left(\vec{k} \cdot \vec{r} \right), \left(\hat{\varepsilon} \cdot \vec{p} \right) \right] = 0 \Rightarrow \left(\vec{k} \cdot \vec{r} \right) \left(\hat{\varepsilon} \cdot \vec{p} \right) = \left(\hat{\varepsilon} \cdot \vec{p} \right) \left(\vec{k} \cdot \vec{r} \right) \tag{5.8}$$

Now,

$$i\left(\vec{k} \cdot \vec{r} \right) \left(\hat{\varepsilon} \cdot \vec{p} \right) = \frac{i}{2} \left[\left(\hat{\varepsilon} \cdot \vec{p} \right) \left(\vec{k} \cdot \vec{r} \right) + \left(\hat{\varepsilon} \cdot \vec{r} \right) \left(\vec{p} \cdot \vec{k} \right) \right] + \frac{i}{2}$$

$$\times \left[\left(\hat{\varepsilon} \cdot \vec{p} \right) \left(\vec{k} \cdot \vec{r} \right) - \left(\hat{\varepsilon} \cdot \vec{r} \right) \underbrace{\left(\vec{p} \cdot \vec{k} \right)}_{=\left(\vec{k} \cdot \vec{p} \right)} \right] \tag{5.9}$$

Using the vector identity

$$\left(\vec{A} \cdot \vec{C} \right) \left(\vec{B} \cdot \vec{D} \right) - \left(\vec{A} \cdot \vec{D} \right) \left(\vec{B} \cdot \vec{C} \right) = \left(\vec{A} \times \vec{B} \right) \cdot \left(\vec{C} \times \vec{D} \right) \tag{5.10}$$

we can manipulate the second term on the RHS of (5.9) as follows [15]:

$$\left(\hat{\varepsilon} \cdot \vec{p}\right)\left(\vec{k} \cdot \vec{r}\right) - \left(\hat{\varepsilon} \cdot \vec{r}\right)\left(\vec{k} \cdot \vec{p}\right) = \left(\hat{\varepsilon} \times \vec{k}\right) \cdot \left(\vec{p} \times \vec{r}\right)$$
$$= \left(\vec{k} \times \hat{\varepsilon}\right) \cdot \left(\vec{r} \times \vec{p}\right) \qquad (5.11)$$

If we designate the polarization direction as $\hat{\varepsilon} = \hat{z}$, and the photon propagation direction as the $\hat{x}$-direction, then [1]

$$|\langle f|(i\vec{k} \cdot \vec{r})(\hat{\varepsilon} \cdot \vec{p})|i\rangle|$$
$$= |\langle f|\frac{i}{2}\left[\underbrace{\left(\hat{\varepsilon} \cdot \vec{p}\right)}_{p_z}\underbrace{\left(\vec{k} \cdot \vec{r}\right)}_{\frac{\omega}{c}x} + \underbrace{\left(\hat{\varepsilon} \cdot \vec{r}\right)}_{z}\underbrace{\left(\vec{p} \cdot \vec{k}\right)}_{\frac{\omega}{c}p_x}\right]|i\rangle|$$
$$+ |\langle f|\frac{i}{2}\underbrace{\left(\vec{k} \times \hat{\varepsilon}\right)}_{\frac{\omega}{c}(x \times z)} \cdot \underbrace{\left(\vec{r} \times \vec{p}\right)}_{\vec{L}}|i\rangle| \qquad (5.12)$$
$$= |\langle f|\frac{i\omega}{2c}\left(p_z x + z p_x\right)|i\rangle| - |\langle f|\frac{i\omega}{2c}\vec{L}_y|i\rangle|$$

where we have also noted that $k = \omega/c$. We now have

$$\frac{d\sigma_{pd}}{d\Omega} = \frac{2\pi\hbar^2 c\alpha}{m^2\omega}\left|\langle f|\overbrace{\left(\hat{\varepsilon} \cdot \vec{p}\right)}^{p_z}|i\rangle + \langle f|\frac{i\omega}{2c}\left(p_z x + z p_x\right)|i\rangle - \langle f|\frac{i\omega}{2c}\vec{L}_y|i\rangle\right|^2 \qquad (5.13)$$

The first term on the RHS of (5.13) is the familiar electric dipole term, the second is the electric quadrupole term, and the third term is the magnetic dipole term [13, 48, 49]. Note that the electric quadrupole term and the magnetic dipole term are a factor of $1/c$ smaller than the electric dipole term, and so these higher-order multipole transition probabilities will be smaller by a factor of $1/c^2$ (as expected). However, we generally expect the magnetic dipole term to be even smaller since it leads to the selection rule $\Delta l = 0$ (see Problem 5.6), and the radial portion of the associated matrix element is usually zero in the single-particle, central-field (nonrelativistic) approximation [49]. Therefore, we will neglect the magnetic dipole term for the remainder of this analysis.

Example 5.1 Show that the first significant correction term to the dipole approximation is the interference term between the electric dipole and electric quadrupole terms:

---

[1] For the sake of clarity, we are no longer using capital letters to indicate the Cartesian directions in the stationary/lab frame.

Consider the following for $H_0 = \frac{p^2}{2m} + V(r)$ [c.f. (4.6)]:

$$
\begin{aligned}
[xz, H_0] &= -[H_0, xz] = -[H_0, x]z - x[H_0, z] = [x, H_0]z + x[z, H_0] \\
&= \frac{1}{2m}[x, p^2]z + \frac{1}{2m}x[z, p^2] = \frac{i\hbar}{m}(p_x z + xp_z) = \frac{i\hbar}{m}(zp_x + p_z x) \quad (5.14) \\
\Rightarrow (p_z x + zp_x) &= \frac{m}{i\hbar}[xz, H_0]
\end{aligned}
$$

where we also used (2.31a). By the same token [c.f. (4.6)],

$$
\begin{aligned}
\left[\hat{\varepsilon}\cdot\vec{r}, H_0\right] &= [z, H_0] = \frac{1}{2m}[z, p^2] = \frac{i\hbar}{m}p_z \\
\Rightarrow p_z &= \frac{m}{i\hbar}[z, H_0]
\end{aligned} \quad (5.15)
$$

Therefore, for $|i\rangle$, $|f\rangle$ eigenstates of H_0, we have [c.f. (4.7)]:

$$
\langle f|\frac{i\omega}{2c}(p_z x + zp_x)|i\rangle = \langle f|\frac{m\omega}{2\hbar c}[xz, H_0]|i\rangle = \frac{m\omega^2}{2c}\langle f|xz|i\rangle \quad (5.16)
$$

and

$$
\langle f|p_z|i\rangle = \frac{m}{i\hbar}\langle f|[z, H_0]|i\rangle = im\omega\langle f|z|i\rangle \quad (5.17)
$$

Now we have [48]:

$$
\begin{aligned}
\frac{d\sigma_{pd}}{d\Omega} &= 2\pi\hbar^2 ca\omega\left|i\langle f|z|i\rangle + \frac{\omega}{2c}\langle f|xz|i\rangle\right|^2 \\
&= 2\pi\hbar^2 ca\omega\left|\langle z\rangle\langle z\rangle^* + \frac{i\omega}{2c}(\langle xz\rangle\langle z\rangle^* - \langle z\rangle\langle xz\rangle^*) + O\left(\frac{1}{c^2}\right)\right| \quad (5.18) \\
&\approx 2\pi\hbar^2 ca\omega[\langle z\rangle\langle z\rangle^* + \frac{\omega}{c}\text{Im}\{\langle z\rangle^*\langle xz\rangle\}]
\end{aligned}
$$

The second term on the RHS of (5.18) shows us that the first significant correction term to the dipole approximation is the interference term between the electric dipole and electric quadrupole terms. Although this interference term is smaller than the dipole term by a factor of $1/c$, it is also multiplied by the photon frequency ω, and if the photon frequency is large enough, will produce observable effects on the photoelectron angular distribution, and so cannot be neglected at higher photon frequencies [48, 49].[2] ∎

---

[2]There is also an interference term between the electric dipole and magnetic dipole terms, but as previously stated, the magnetic dipole term vanishes in a nonrelativistic calculation (in the current approximation) and, in any case, is negligible below energies of ~5 keV [50].

The forms we will use for $\langle i|$, $\langle f|$ are found in (4.12) and (4.14), respectively:

$$|i\rangle = R_{nl_0}(r)|l_0, m\rangle = R_{nl_0}(r)Y_{l_0m}(\theta_m, \phi_m) \tag{5.19}$$

$$|f\rangle = \sum_{l'=0}^{\infty} (2l' + 1)i^{l'}e^{-i\delta_{l'}}P_{l'}(\cos\theta_m)G_{kl'}(r)$$

$$= \sum_{l'=0}^{\infty} \sqrt{4\pi(2l' + 1)}i^{l'}e^{-i\delta_{l'}}Y_{l'0}(\theta_m, \phi_m)G_{kl'}(r) \tag{5.20}$$

We also note from (2.67b):

$$z = r\cos\theta_s = r\sqrt{\frac{4\pi}{3}}Y_{10}(\theta_s, \phi_s) \tag{5.21}$$

And from (2.36a), (2.36c), and (2.67e),

$$xz = r^2\sin\theta_s\cos\theta_s\cos\phi_s = r^2\sqrt{\frac{2\pi}{15}}[Y_{2,-1}(\theta_s, \phi_s) - Y_{21}(\theta_s, \phi_s)] \tag{5.22}$$

Combining (5.19), (5.20), (5.21), and (5.22) gives

$$\langle z\rangle^* = \frac{4\pi}{\sqrt{3}}\sum_{l'=0}^{\infty}$$

$$\times \sqrt{(2l' + 1)}\, i^{l'}e^{-i\delta_{l'}}\left\langle Y^*_{l_0m}(\theta_m, \phi_m)Y_{l'0}(\theta_m, \phi_m)Y_{10}(\theta_s, \phi_s)\right\rangle\Re_{l'} \tag{5.23}$$

where $\Re_{l'}$ is defined in (4.20). We also get,

$$\langle xz\rangle = 2\pi\sqrt{\frac{2}{15}}\sum_{l''=0}^{\infty}\sqrt{(2l'' + 1)}\, i^{-l''}e^{i\delta_{l''}}\left\langle R_{nl_0}\big|r^2\big|G_{kl''}\right\rangle$$

$$\times\left\langle Y_{l_0m}(\theta_m, \phi_m)[Y_{2,-1}(\theta_s, \phi_s) - Y_{21}(\theta_s, \phi_s)]\underbrace{Y^*_{l''0}(\theta_m, \phi_m)}_{Y_{l''0}(\theta_m, \phi_m)}\right\rangle \tag{5.24}$$

Example 5.2 Put the angular quantities from (5.23) and (5.24) into the center-of-mass (CM) atom/molecule reference frame.

From (2.95) we get,

$$Y_{10}(\theta_s, \phi_s) = \sum_{\overline{m}} D^1_{\overline{m}0}(\alpha, \beta, \gamma) \, Y_{1\overline{m}}(\theta_m, \phi_m) \tag{5.25}$$

and

$$Y_{2,-1}(\theta_s, \phi_s) - Y_{21}(\theta_s, \phi_s) = \sum_M \left[D^2_{M,-1}(\alpha, \beta, \gamma) - D^2_{M1}(\alpha, \beta, \gamma) \right] Y_{2M}(\theta_m, \phi_m) \tag{5.26}$$

which leaves:

$$\langle z \rangle^* = \frac{4\pi}{\sqrt{3}} \sum_{l'\overline{m}} \sqrt{(2l'+1)} i^{l'} e^{-i\delta_{l'}} \Big\langle \underbrace{Y^*_{l_0 m}}_{(-1)^m Y_{l_0,-m}} \, Y_{l'0} Y_{1\overline{m}} \Big\rangle D^1_{\overline{m}0} \, \mathfrak{R}_{l'} \tag{5.27}$$

and

$$\langle xz \rangle = 2\pi \sqrt{\frac{2}{15}} \sum_{l''M} \sqrt{(2l''+1)} i^{-l''} e^{i\delta_{l''}} \langle R_{nl_0} | r^2 | G_{kl''} \rangle \langle Y_{l_0 m} Y_{2M} Y_{l''0} \rangle$$
$$\times \left[D^2_{M,-1} - D^2_{M1} \right] \tag{5.28}$$

∎

We use the following formulas gleaned from (2.158):

$$\langle Y_{l_0,-m} Y_{l'0} Y_{1\overline{m}} \rangle = \sqrt{\frac{3(2l_0+1)(2l'+1)}{4\pi}} \begin{pmatrix} l_0 & l' & 1 \\ 0 & 0 & 0 \end{pmatrix} \begin{pmatrix} l_0 & l' & 1 \\ -m & 0 & \overline{m} \end{pmatrix} \tag{5.29a}$$

$$\langle Y_{l_0 m} Y_{2M} Y_{l''0} \rangle = \sqrt{\frac{5(2l_0+1)(2l''+1)}{4\pi}} \begin{pmatrix} l_0 & 2 & l'' \\ 0 & 0 & 0 \end{pmatrix} \begin{pmatrix} l_0 & 2 & l'' \\ m & M & 0 \end{pmatrix} \tag{5.29b}$$

Combining (5.27), (5.28), and (5.29a and 5.29b), and realizing that we must again average over the initial (unobserved) states m,

$$\langle z \rangle^* \langle xz \rangle = \frac{2\pi\sqrt{6}}{3(2l_0 + 1)} \sum_{m\bar{m}M} \sum_{l'l''} (-1)^m (2l_0 + 1)(2l' + 1)$$

$$\times (2l'' + 1) i^{l'-l''} e^{-i(\delta_{l'} - \delta_{l''})} \Re_{l'} \langle R_{nl_0} | r^2 | G_{kl''} \rangle$$

$$\times \begin{pmatrix} l_0 & l' & 1 \\ 0 & 0 & 0 \end{pmatrix} \begin{pmatrix} l_0 & l' & 1 \\ -m & 0 & \bar{m} \end{pmatrix} \begin{pmatrix} l_0 & 2 & l'' \\ 0 & 0 & 0 \end{pmatrix} \begin{pmatrix} l_0 & 2 & l'' \\ m & M & 0 \end{pmatrix}$$

$$\times D^1_{\bar{m}0} [D^2_{M,-1} - D^2_{M1}]$$

$$(5.30)$$

The triangle relations in the 3-j symbols in (5.30) collapse the summations over $\bar{m}, M$ to $\bar{m} = m, M = -m$:

$$\langle z \rangle^* \langle xz \rangle = \frac{2\pi\sqrt{6}}{3(2l_0 + 1)} \sum_m \sum_{l'l''} (-1)^m (2l_0 + 1)(2l' + 1)$$

$$\times (2l'' + 1) i^{l'-l''} e^{-i(\delta_{l'} - \delta_{l''})} \underbrace{\Re_{l'} \langle R_{nl_0} | r^2 | G_{kl''} \rangle}_{\equiv Q_{l''}} \begin{pmatrix} l_0 & l' & 1 \\ 0 & 0 & 0 \end{pmatrix}$$

$$\times \begin{pmatrix} l_0 & l' & 1 \\ -m & 0 & m \end{pmatrix} \begin{pmatrix} l_0 & 2 & l'' \\ 0 & 0 & 0 \end{pmatrix} \begin{pmatrix} l_0 & 2 & l'' \\ m & -m & 0 \end{pmatrix} D^1_{m0} [D^2_{-m,-1} - D^2_{-m1}]$$

$$(5.31)$$

where $Q_{l''}$ is the quadrupole radial matrix element. We now take advantage of (2.149) [3]:

$$D^1_{m0} D^2_{-m,-1} - D^1_{m0} D^2_{-m1} = \sum_L \left\{ \begin{array}{l} \langle 1, 2; 0, -1 | L, M' \rangle \langle 1, 2; m, -m | L, M'' \rangle D^L_{M'',-1} \\ -\langle 1, 2; 0, 1 | L, M' \rangle \langle 1, 2; m, -m | L, M'' \rangle D^L_{M''1} \end{array} \right\}$$

$$= \sum_L \left\{ \begin{array}{l} \langle 1, 2; 0, -1 | L, -1 \rangle \langle 1, 2; m, -m | L, 0 \rangle D^L_{0,-1} \\ -\langle 1, 2; 0, 1 | L, 1 \rangle \langle 1, 2; m, -m | L, 0 \rangle D^L_{01} \end{array} \right\}$$

$$\Rightarrow D^1_{m0} D^2_{-m,-1} - D^1_{m0} D^2_{-m1}$$

$$= \sum_L \left\{ \begin{array}{l} \langle 1, 2; 0, -1 | L, -1 \rangle \langle 1, 2; m, -m | L, 0 \rangle D^L_{0,-1} \\ -(-1)^{3-L} \langle 1, 2; 0, -1 | L, -1 \rangle \langle 1, 2; m, -m | L, 0 \rangle D^L_{01} \end{array} \right\} \quad (5.32)$$

where we have used (2.141) and the triangle conditions inherent in the C-G coefficients to specify some of the magnetic quantum numbers.

We convert the C-G coefficients to 3-j symbols using (2.144b):

$$D^1_{m0}D^2_{-m,-1} - D^1_{m0}D^2_{-m1} = \begin{pmatrix} 1 & 2 & L \\ m & -m & 0 \end{pmatrix}\begin{pmatrix} 1 & 2 & L \\ 0 & -1 & 1 \end{pmatrix}$$

$$\times \left[(-1)D^L_{01} + \underbrace{(-1)^{-L}(-1)}_{(-1)^L} D^L_{0,-1} \right] \qquad (5.33)$$

Here we must have $L=$ odd (see remarks below):

$$\Rightarrow D^1_{m0}D^2_{-m,-1} - D^1_{m0}D^2_{-m1} = \begin{pmatrix} 1 & 2 & L \\ m & -m & 0 \end{pmatrix}\begin{pmatrix} 1 & 2 & L \\ 0 & -1 & 1 \end{pmatrix}[D^L_{0,-1} - D^L_{01}]$$

$$(5.34)$$

From (2.96) and example 2.10, we have [10, 43].

$$D^L_{01} = -D^{L*}_{0,-1} \text{ and } D^{j*}_{m'm} = D^j_{mm'} \qquad (5.35)$$

Putting everything together and noting that (2.145c) applied to two of the 3-j symbols in (5.31) demands that $l_0 + l' + 1=$even and $l_0 + l'' + 2=$even

$$\langle z \rangle^* \langle xz \rangle = \frac{2\pi\sqrt{6}}{3(2l_0 + 1)} \sum_m \sum_{l'l''L} i^{l'-l''} e^{-i(\delta_{l'} - \delta_{l''})} \Re_{l'} Q_{l''} (-1)^m (2l_0 + 1)$$

$$\times (2l' + 1)(2l'' + 1)(2L + 1)\begin{pmatrix} l_0 & 1 & l' \\ 0 & 0 & 0 \end{pmatrix}\begin{pmatrix} 1 & l_0 & l' \\ m & -m & 0 \end{pmatrix}$$

$$\times \begin{pmatrix} l_0 & 2 & l'' \\ 0 & 0 & 0 \end{pmatrix}\begin{pmatrix} l_0 & 2 & l'' \\ m & -m & 0 \end{pmatrix}\begin{pmatrix} 1 & 2 & L \\ m & -m & 0 \end{pmatrix}$$

$$\times \begin{pmatrix} 1 & 2 & L \\ 0 & -1 & 1 \end{pmatrix}[D^L_{-10} + D^{L*}_{-10}] \qquad (5.36)$$

From (2.109) [49],

$$[D^L_{-10} + D^{L*}_{-10}] = \sqrt{\frac{4\pi}{(2L + 1)}}[Y_{L,-1} + Y^*_{L,-1}] \qquad (5.37)$$

which gives

$$\langle z \rangle^* \langle xz \rangle = \frac{4\pi\sqrt{6\pi}}{3(2l_0+1)} \sum_m \sum_{l'l''L} i^{l'-l''} e^{-i(\delta_{l'}-\delta_{l''})} \mathfrak{R}_{l'} Q_{l''} (-1)^m (2l_0+1)$$

$$\times (2l'+1)(2l''+1)\sqrt{(2L+1)} \begin{pmatrix} l_0 & 1 & l' \\ 0 & 0 & 0 \end{pmatrix} \begin{pmatrix} 1 & l_0 & l' \\ m & -m & 0 \end{pmatrix}$$

$$\times \begin{pmatrix} l_0 & 2 & l'' \\ 0 & 0 & 0 \end{pmatrix} \begin{pmatrix} l_0 & 2 & l'' \\ m & -m & 0 \end{pmatrix} \begin{pmatrix} 1 & 2 & L \\ m & -m & 0 \end{pmatrix}$$

$$\times \begin{pmatrix} 1 & 2 & L \\ 0 & -1 & 1 \end{pmatrix} [Y_{L,-1} + Y^*_{L,-1}] \tag{5.38}$$

We are now faced with the task of evaluating the summations. Fortunately, we can once again appeal to the properties of the 3-j symbols to simplify the situation. We first see that the last 3-j symbol on the RHS of (5.38) restricts the terms in the sum over L to $L = 1, 2, 3$. Additional symmetry considerations in the 3-j symbols present here require that $L + 1 + 2 =$ even, thus eliminating the possibility $L = 2$ [49]. Evaluating the sum over L,

$$\langle z \rangle^* \langle xz \rangle = \frac{4\pi\sqrt{6\pi}}{3(2l_0+1)} \sum_{l'l''} i^{l'-l''} e^{-i(\delta_{l'}-\delta_{l''})} \mathfrak{R}_{l'} Q_{l''} (2l_0+1)(2l'+1)(2l''+1)$$

$$\times \begin{pmatrix} l_0 & 1 & l' \\ 0 & 0 & 0 \end{pmatrix} \begin{pmatrix} l_0 & 2 & l'' \\ 0 & 0 & 0 \end{pmatrix} \sum_m (-1)^m \begin{pmatrix} 1 & l_0 & l' \\ m & -m & 0 \end{pmatrix} \begin{pmatrix} l_0 & 2 & l'' \\ m & -m & 0 \end{pmatrix}$$

$$\times \left\{ \begin{array}{l} \sqrt{3} \begin{pmatrix} 1 & 2 & 1 \\ m & -m & 0 \end{pmatrix} \begin{pmatrix} 1 & 2 & 1 \\ 0 & -1 & 1 \end{pmatrix} [Y_{1,-1} + Y^*_{1,-1}] \\[4mm] + \sqrt{7} \begin{pmatrix} 1 & 2 & 3 \\ m & -m & 0 \end{pmatrix} \begin{pmatrix} 1 & 2 & 3 \\ 0 & -1 & 1 \end{pmatrix} [Y_{3,-1} + Y^*_{3,-1}] \end{array} \right\}$$

$$\tag{5.39}$$

Tabulated values of 3-j symbols allow us to evaluate the following 3-j symbols [3, 7, 10]:

$$\begin{pmatrix} 1 & 2 & 1 \\ m & -m & 0 \end{pmatrix} = (-1)^m \sqrt{\frac{(2-m)(2+m)}{30}} \tag{5.40a}$$

$$\begin{pmatrix} 1 & 2 & 1 \\ 0 & -1 & 1 \end{pmatrix} = -\sqrt{\frac{1}{10}} \tag{5.40b}$$

$$\begin{pmatrix} 1 & 2 & 3 \\ m & -m & 0 \end{pmatrix} = (-1)^{m+1}(3-5m^2)\sqrt{\frac{(2-m)(2+m)}{7\cdot5\cdot4\cdot3}} \tag{5.40c}$$

$$\begin{pmatrix} 1 & 2 & 3 \\ 0 & -1 & 1 \end{pmatrix} = 4\sqrt{\frac{1}{7\cdot5\cdot3\cdot2}} \tag{5.40d}$$

Substituting,

$$\langle z \rangle^* \langle xz \rangle = \frac{4\pi\sqrt{6\pi}}{3(2l_0+1)} \sum_{l'l''} i^{l'-l''} e^{-i(\delta_{l'}-\delta_{l''})} \mathfrak{R}_{l'} Q_{l''} (2l_0+1)(2l'+1)(2l''+1)$$

$$\times \begin{pmatrix} l_0 & 1 & l' \\ 0 & 0 & 0 \end{pmatrix} \begin{pmatrix} l_0 & 2 & l'' \\ 0 & 0 & 0 \end{pmatrix} \sum_m (-1)^m \begin{pmatrix} 1 & l_0 & l' \\ m & -m & 0 \end{pmatrix}$$

$$\times \begin{pmatrix} l_0 & 2 & l'' \\ m & -m & 0 \end{pmatrix}$$

$$\times \left\{ \begin{array}{l} (-1)^{m+1} \left(\frac{1}{10}\right) \sqrt{(2-m)(2+m)} \left[Y_{1,-1} + Y_{1,-1}^* \right] \\[2mm] +(-1)^{m+1} \left(\frac{2}{7\cdot 5\cdot 3}\right)(3-5m^2) \sqrt{\frac{7(2-m)(2+m)}{2}} \left[Y_{3,-1} + Y_{3,-1}^* \right] \end{array} \right\}$$

$$(5.41)$$

The triangle conditions of the 3-j symbols limit the sum over m to the terms $m = 0, \pm 1$:

$$\langle z \rangle^* \langle xz \rangle = \frac{4\pi\sqrt{6\pi}}{3(2l_0+1)} \sum_{l'l''} i^{l'-l''} \underbrace{e^{-i(\delta_{l'}-\delta_{l''})}}_{2\cos(\delta_{l''}-\delta_{l'})} \mathfrak{R}_{l'} Q_{l''} (2l_0+1)(2l'+1)(2l''+1)$$

$$\times \begin{pmatrix} l_0 & 1 & l' \\ 0 & 0 & 0 \end{pmatrix} \begin{pmatrix} l_0 & 2 & l'' \\ 0 & 0 & 0 \end{pmatrix}$$

$$\times (-1) \left\{ \begin{array}{l} \left[\begin{array}{l} \left(\frac{2}{10}\right) \begin{pmatrix} 1 & l_0 & l' \\ 0 & 0 & 0 \end{pmatrix} \begin{pmatrix} l_0 & 2 & l'' \\ 0 & 0 & 0 \end{pmatrix} \\[2mm] + \left(\frac{\sqrt{3}}{10}\right) \begin{pmatrix} 1 & l_0 & l' \\ -1 & 1 & 0 \end{pmatrix} \begin{pmatrix} l_0 & 2 & l'' \\ -1 & 1 & 0 \end{pmatrix} \\[2mm] + \left(\frac{\sqrt{3}}{10}\right) \begin{pmatrix} 1 & l_0 & l' \\ 1 & -1 & 0 \end{pmatrix} \begin{pmatrix} l_0 & 2 & l'' \\ 1 & -1 & 0 \end{pmatrix} \end{array} \right] \left[Y_{1,-1} + Y_{1,-1}^* \right] \\[10mm] + \left[\begin{array}{l} \left(\frac{2\sqrt{7\cdot 2}}{7\cdot 5}\right) \begin{pmatrix} 1 & l_0 & l' \\ 0 & 0 & 0 \end{pmatrix} \begin{pmatrix} l_0 & 2 & l'' \\ 0 & 0 & 0 \end{pmatrix} \\[2mm] + \left(\frac{-4}{7\cdot 5\cdot 3}\right) \sqrt{\frac{7\cdot 3}{2}} \begin{pmatrix} 1 & l_0 & l' \\ -1 & 1 & 0 \end{pmatrix} \begin{pmatrix} l_0 & 2 & l'' \\ -1 & 1 & 0 \end{pmatrix} \\[2mm] + \left(\frac{-4}{7\cdot 5\cdot 3}\right) \sqrt{\frac{7\cdot 3}{2}} \begin{pmatrix} 1 & l_0 & l' \\ 1 & -1 & 0 \end{pmatrix} \begin{pmatrix} l_0 & 2 & l'' \\ 1 & -1 & 0 \end{pmatrix} \end{array} \right] \left[Y_{3,-1} + Y_{3,-1}^* \right] \end{array} \right\}$$

$$(5.42)$$

Recalling that the symmetry properties of the 3-j symbols appearing in (5.42) demand that $l_0 + l' + 1 =$even and $l_0 + l'' + 2 =$even, we can say,

$$
\begin{pmatrix} 1 & l_0 & l' \\ -1 & 1 & 0 \end{pmatrix} \begin{pmatrix} l_0 & 2 & l'' \\ -1 & 1 & 0 \end{pmatrix} = \begin{pmatrix} 1 & l_0 & l' \\ 1 & -1 & 0 \end{pmatrix} \begin{pmatrix} l_0 & 2 & l'' \\ 1 & -1 & 0 \end{pmatrix}
\tag{5.43}
$$

Combining (5.42) and (5.43),

$$
\langle z \rangle^* \langle xz \rangle = \frac{4\pi\sqrt{6\pi}}{3(2l_0 + 1)} \sum_{l'l''} i^{l'-l''} e^{-i(\delta_{l'} - \delta_{l''})} \, \Re_{l'} Q_{l''} (2l_0 + 1)(2l' + 1)(2l'' + 1)
$$

$$
\times \begin{pmatrix} l_0 & 1 & l' \\ 0 & 0 & 0 \end{pmatrix} \begin{pmatrix} l_0 & 2 & l'' \\ 0 & 0 & 0 \end{pmatrix}
$$

$$
\times (-1) \left\{
\begin{aligned}
&\left[\left(\tfrac{1}{5}\right) \begin{pmatrix} 1 & l_0 & l' \\ 0 & 0 & 0 \end{pmatrix} \begin{pmatrix} l_0 & 2 & l'' \\ 0 & 0 & 0 \end{pmatrix} \right. \\
&\left. + \left(\tfrac{\sqrt{3}}{5}\right) \begin{pmatrix} 1 & l_0 & l' \\ -1 & 1 & 0 \end{pmatrix} \begin{pmatrix} l_0 & 2 & l'' \\ -1 & 1 & 0 \end{pmatrix} \right] \bigl[Y_{1,-1} + Y^*_{1,-1} \bigr] \\[2ex]
&+ \left[\left(\tfrac{2\sqrt{7\cdot 2}}{7\cdot 5}\right) \begin{pmatrix} 1 & l_0 & l' \\ 0 & 0 & 0 \end{pmatrix} \begin{pmatrix} l_0 & 2 & l'' \\ 0 & 0 & 0 \end{pmatrix} \right. \\
&\left. + \left(\tfrac{-8}{7\cdot 5\cdot 3}\right)\sqrt{\tfrac{7\cdot 3}{2}} \begin{pmatrix} 1 & l_0 & l' \\ -1 & 1 & 0 \end{pmatrix} \begin{pmatrix} l_0 & 2 & l'' \\ -1 & 1 & 0 \end{pmatrix} \right] \bigl[Y_{3,-1} + Y^*_{3,-1} \bigr]
\end{aligned}
\right\}
\tag{5.44}
$$

Now we make explicit all the angular factors:

$$
\bigl[Y_{1,-1} + Y^*_{1,-1} \bigr] = \sqrt{\frac{3}{2\pi}} \sin\theta \cos\phi
\tag{5.45a}
$$

$$
\bigl[Y_{3,-1} + Y^*_{3,-1} \bigr] = \frac{1}{4} \sqrt{\frac{21}{\pi}} \sin\theta \left(5\cos^2\theta - 1\right) \cos\phi
\tag{5.45b}
$$

Combining (5.44) and (5.45) and rearranging,

$$\langle z \rangle^* \langle xz \rangle = \frac{4\pi}{3(2l_0+1)} \sum_{l'l''} i^{l'-l''} \cos(\delta_{l''} - \delta_{l'}) \Re_{l'} Q_{l''} (2l_0+1)(2l'+1)$$

$$\times (2l''+1) \begin{pmatrix} l_0 & 1 & l' \\ 0 & 0 & 0 \end{pmatrix} \begin{pmatrix} l_0 & 2 & l'' \\ 0 & 0 & 0 \end{pmatrix}$$

$$\times (-2\sqrt{3}) \left\{ \begin{array}{l} \left[\left[\begin{pmatrix} 1 & l_0 & l' \\ -1 & 1 & 0 \end{pmatrix} \begin{pmatrix} l_0 & 2 & l'' \\ -1 & 1 & 0 \end{pmatrix} \right] \right] [\sin\theta\cos\phi] \\[3em] \left[\sqrt{3} \begin{pmatrix} 1 & l_0 & l' \\ 0 & 0 & 0 \end{pmatrix} \begin{pmatrix} l_0 & 2 & l'' \\ 0 & 0 & 0 \end{pmatrix} \right] \\[2em] + \\[1em] \left. -2 \begin{pmatrix} 1 & l_0 & l' \\ -1 & 1 & 0 \end{pmatrix} \begin{pmatrix} l_0 & 2 & l'' \\ -1 & 1 & 0 \end{pmatrix} \right] [\sin\theta\cos^2\theta\cos\phi] \end{array} \right\}$$

$$(5.46)$$

Again, we invoke the triangle rules and symmetries inherent in the 3-j symbols to see that the sums over l' and l'' are limited to the terms $l' = l_0 \pm 1$ and $l'' = l_0, l_0 \pm 2$. Tabulated values of 3-j symbols allow us to calculate the following factors [3, 7, 10]:

For $l' = l_0 - 1, l'' = l_0 - 2$,

$$i^{l'-l''} \frac{-2}{\sqrt{3}} (2l_0+1)(2l'+1)(2l''+1) \begin{pmatrix} 1 & l_0 & l' \\ 0 & 0 & 0 \end{pmatrix} \begin{pmatrix} l_0 & 2 & l'' \\ 0 & 0 & 0 \end{pmatrix} \begin{pmatrix} 1 & l_0 & l' \\ -1 & 1 & 0 \end{pmatrix} \begin{pmatrix} l_0 & 2 & l'' \\ -1 & 1 & 0 \end{pmatrix}$$

$$= \frac{-il_0(l_0+1)(l_0-1)}{(2l_0-1)(2l_0+1)}$$

$$(5.47a)$$

$$i^{l'-l''} \frac{-2}{\sqrt{3}} (2l_0+1)(2l'+1)(2l''+1) \begin{pmatrix} 1 & l_0 & l' \\ 0 & 0 & 0 \end{pmatrix}^2 \begin{pmatrix} l_0 & 2 & l'' \\ 0 & 0 & 0 \end{pmatrix}^2 = \frac{-\sqrt{3}il_0^2(l_0-1)}{(2l_0-1)(2l_0+1)}$$

$$(5.47b)$$

For $l' = l_0 - 1, l'' = l_0$,

$$i^{l'-l''} \frac{-2}{\sqrt{3}} (2l_0+1)(2l'+1)(2l''+1) \begin{pmatrix} 1 & l_0 & l' \\ 0 & 0 & 0 \end{pmatrix} \begin{pmatrix} l_0 & 2 & l'' \\ 0 & 0 & 0 \end{pmatrix} \begin{pmatrix} 1 & l_0 & l' \\ -1 & 1 & 0 \end{pmatrix} \begin{pmatrix} l_0 & 2 & l'' \\ -1 & 1 & 0 \end{pmatrix}$$

$$= \frac{il_0(l_0+1)}{(2l_0+3)(2l_0-1)}$$

$$(5.47c)$$

$$i^{l'-l''}\frac{-2}{\sqrt{3}}(2l_0+1)(2l'+1)(2l''+1)\begin{pmatrix}1 & l_0 & l' \\ 0 & 0 & 0\end{pmatrix}^2\begin{pmatrix}l_0 & 2 & l'' \\ 0 & 0 & 0\end{pmatrix}^2=\frac{2il_0^2(l_0+1)}{\sqrt{3}(2l_0+3)(2l_0-1)}$$

$$(5.47\text{d})$$

For $l'=l_0-1$, $l''=l_0+2$,

$$i^{l'-l''}\frac{-2}{\sqrt{3}}(2l_0+1)(2l'+1)(2l''+1)\begin{pmatrix}1 & l_0 & l' \\ 0 & 0 & 0\end{pmatrix}\begin{pmatrix}l_0 & 2 & l'' \\ 0 & 0 & 0\end{pmatrix}\begin{pmatrix}1 & l_0 & l' \\ -1 & 1 & 0\end{pmatrix}\begin{pmatrix}l_0 & 2 & l'' \\ -1 & 1 & 0\end{pmatrix}$$

$$=\frac{il_0(l_0+1)(l_0+2)}{(2l_0+3)(2l_0+1)}$$

$$(5.47\text{e})$$

$$i^{l'-l''}\frac{-2}{\sqrt{3}}(2l_0+1)(2l'+1)(2l''+1)\begin{pmatrix}1 & l_0 & l' \\ 0 & 0 & 0\end{pmatrix}^2\begin{pmatrix}l_0 & 2 & l'' \\ 0 & 0 & 0\end{pmatrix}^2=\frac{-\sqrt{3}il_0(l_0+1)(l_0+2)}{(2l_0+3)(2l_0+1)}$$

$$(5.47\text{f})$$

For $l'=l_0+1$, $l''=l_0+2$,

$$i^{l'-l''}\frac{-2}{\sqrt{3}}(2l_0+1)(2l'+1)(2l''+1)\begin{pmatrix}1 & l_0 & l' \\ 0 & 0 & 0\end{pmatrix}\begin{pmatrix}l_0 & 2 & l'' \\ 0 & 0 & 0\end{pmatrix}\begin{pmatrix}1 & l_0 & l' \\ -1 & 1 & 0\end{pmatrix}\begin{pmatrix}l_0 & 2 & l'' \\ -1 & 1 & 0\end{pmatrix}$$

$$=\frac{il_0(l_0+1)(l_0+2)}{(2l_0+3)(2l_0+1)}$$

$$(5.47\text{g})$$

$$i^{l'-l''}\frac{-2}{\sqrt{3}}(2l_0+1)(2l'+1)(2l''+1)\begin{pmatrix}1 & l_0 & l' \\ 0 & 0 & 0\end{pmatrix}^2\begin{pmatrix}l_0 & 2 & l'' \\ 0 & 0 & 0\end{pmatrix}^2=\frac{\sqrt{3}il_0(l_0+1)(l_0+2)}{(2l_0+3)(2l_0+1)}$$

$$(5.47\text{h})$$

For $l'=l_0+1$, $l''=l_0$,

$$i^{l'-l''}\frac{-2}{\sqrt{3}}(2l_0+1)(2l'+1)(2l''+1)\begin{pmatrix}1 & l_0 & l' \\ 0 & 0 & 0\end{pmatrix}\begin{pmatrix}l_0 & 2 & l'' \\ 0 & 0 & 0\end{pmatrix}\begin{pmatrix}1 & l_0 & l' \\ -1 & 1 & 0\end{pmatrix}\begin{pmatrix}l_0 & 2 & l'' \\ -1 & 1 & 0\end{pmatrix}$$

$$=\frac{il_0(l_0+1)}{(2l_0+3)(2l_0-1)}$$

$$(5.47\text{i})$$

$$i^{l'-l''}\frac{-2}{\sqrt{3}}(2l_0+1)(2l'+1)(2l''+1)\begin{pmatrix} 1 & l_0 & l' \\ 0 & 0 & 0 \end{pmatrix}^2\begin{pmatrix} l_0 & 2 & l'' \\ 0 & 0 & 0 \end{pmatrix}^2 = \frac{-\sqrt{3}il_0(l_0+2)^2}{(2l_0+3)(2l_0-1)}$$

$$(5.47\text{j})$$

For $l' = l_0 + 1$, $l'' = l_0 - 2$,

$$i^{l'-l''}\frac{-2}{\sqrt{3}}(2l_0+1)(2l'+1)(2l''+1)\begin{pmatrix} 1 & l_0 & l' \\ 0 & 0 & 0 \end{pmatrix}\begin{pmatrix} l_0 & 2 & l'' \\ 0 & 0 & 0 \end{pmatrix}\begin{pmatrix} 1 & l_0 & l' \\ -1 & 1 & 0 \end{pmatrix}\begin{pmatrix} l_0 & 2 & l'' \\ -1 & 1 & 0 \end{pmatrix}$$

$$= \frac{-il_0(l_0+1)(l_0-1)}{(2l_0-1)(2l_0+1)}$$

$$(5.47\text{k})$$

$$i^{l'-l''}\frac{-2}{\sqrt{3}}(2l_0+1)(2l'+1)(2l''+1)\begin{pmatrix} 1 & l_0 & l' \\ 0 & 0 & 0 \end{pmatrix}^2\begin{pmatrix} l_0 & 2 & l'' \\ 0 & 0 & 0 \end{pmatrix}^2$$

$$= \frac{\sqrt{3}il_0(l_0+1)(l_0-1)}{(2l_0-1)(2l_0+1)}$$

$$(5.47\text{l})$$

We can now combine (5.18), (5.46), and all the iterations of (5.47). The result is usually presented in the form [40, 41]:

$$\frac{d\sigma_{pd}}{d\Omega} = \frac{\sigma_{pd}}{4\pi}\left[1 + \beta P_2(\cos\theta) + (\delta + \gamma\cos^2\theta)\sin\theta\cos\phi\right] \qquad (5.48)$$

where σ_{pd} is given by (4.39a) and β by (4.39b). The factors δ and γ are given by [49]

$$\gamma \equiv \frac{8\pi^2\hbar^2\alpha\omega^2}{\left[l\Re_{l_0-1}^2 + (l_0+1)\Re_{l_0+1}^2\right]}\sum_{l'l''}A_{l'l''}\Re_{l'}Q_{l''}\cos(\delta_{l''}-\delta_{l'}) \qquad (5.49\text{a})$$

$$\delta \equiv \frac{8\pi^2\hbar^2\alpha\omega^2}{\left[l\Re_{l_0-1}^2 + (l_0+1)\Re_{l_0+1}^2\right]}\sum_{l'l''}B_{l'l''}\Re_{l'}Q_{l''}\cos(\delta_{l''}-\delta_{l'}) \qquad (5.49\text{b})$$

$A_{l'l''}$ and $B_{l'l''}$ are found by combining (the imaginary part of) the appropriate iterations of (5.47) with the appropriate factors inside of the curly brackets on the RHS of (5.46).

Example 5.3 Find B_{l_0-1,l_0-2} and A_{l_0-1,l_0-2}.

For $l' = l_0 - 1, l'' = l_0 - 2, B_{l'l''} = (5.47a) = \dfrac{-l_0(l_0 + 1)(l_0 - 1)}{(2l_0 - 1)(2l_0 + 1)}$, while $A_{l'l''}$

$$= \sqrt{3} \times (5.47b) - 2 \times (5.47a) = \frac{l_0(-l_0 + 2)(l_0 - 1)}{(2l_0 - 1)(2l_0 + 1)}.$$
∎

The complete list of the factors $A_{l'l''}$ and $B_{l'l''}$ is found in [43]. Notice that the interference between the electric dipole and electric quadrupole terms has introduced odd powers of $\cos\theta$ into the differential pd. cross-section as parity considerations demand [1, 51].

A few remarks are in order here. Eq. (5.48) shows us that the non-dipolar terms are greatest in the direction (and anti-direction) of photon propagation, $\phi = 0, \pi$. Also, the parameter δ multiplies the factor $\sin\theta \cos \phi$, which is the direction of the photon momentum vector along the x-axis (direction of photon propagation) [52]. It appears that the inclusion of the first higher-order multipole term in the expression for the photoelectron angular distribution reveals an asymmetry that depends on the direction of photon propagation [53]. Eq. (5.48) also shows that the term with δ is rotationally symmetric about the direction of photon travel but is forward-backward asymmetric with respect to the reversal of the photon propagation direction as is the term with γ. The term with γ also has reflection symmetry in the xy and xz planes [54].

Example 5.4 Explain under what conditions experimentalists will be able to isolate higher-order multipole corrections to photoelectron angular distributions.

First, experimentalists attempting to measure total cross-sections should note that the concept of the "magic angle" $\theta = 54.7^0$ holds only in the $\phi = \pi/2$ plane. Experimentalists should also note that measurements carried out in a plane perpendicular to the photon direction, $\theta = 0, \phi = \pi/2$, will not reveal any of the interference terms. Or, putting it another way, higher-order multipole corrections to photoelectron angular distributions can only be measured in a plane that is not the polarization plane (for linearly polarized light). On the other hand, it is possible to isolate the higher-order terms if measurements are made at the magic angle, $\theta = 54.7^0$ (and in the direction $\phi = 0$). In this case, (5.48) gives [54, 55]

$$\frac{d\sigma_{pd}}{d\Omega} = \frac{\sigma_{pd}}{4\pi}\left[1 + \sqrt{\frac{2}{27}}(3\delta + \gamma)\right] \tag{5.50}$$

although only the combined quantity $(3\delta + \gamma)$ can be extracted from such a measurement.
∎

Example 5.5 Find expressions for the higher-order asymmetry parameters for photodissociation from s- and p-subshells.

Using (5.46), all relevant iterations of (5.47), and (5.49a and 5.49b), and the dipole and quadrupole selection rules, we can determine the expressions for δ and γ for photodissociation from s- and p-subshells [49, 56, 57]:

$$\delta_s = 0 \tag{5.51a}$$

$$\gamma_s \propto \frac{Q_2}{\mathfrak{R}_1} \cos(\delta_2 - \delta_1) \tag{5.51b}$$

$$\delta_p \propto \frac{\{\mathfrak{R}_0[Q_1 \cos(\delta_1 - \delta_0) + Q_3 \cos(\delta_3 - \delta_0)] + \mathfrak{R}_2[Q_1 \cos(\delta_1 - \delta_2) + Q_3 \cos(\delta_3 - \delta_2)]\}}{[\mathfrak{R}_0^2 + 2\mathfrak{R}_2^2]} \tag{5.51c}$$

$$\gamma_p \propto \frac{\{-5\mathfrak{R}_0 Q_3 \cos(\delta_3 - \delta_0) + 2\mathfrak{R}_2[2Q_3 \cos(\delta_3 - \delta_2) - 3Q_1 \cos(\delta_1 - \delta_2)]\}}{[\mathfrak{R}_0^2 + 2\mathfrak{R}_2^2]} \tag{5.51d}$$

Equation (5.51a) can be derived most directly by considering (5.33) for the case $m = 0$.

$$D_{00}^1[D_{-10}^2 - D_{10}^2] = \sqrt{6}\cos^2\theta \sin\theta \cos\phi \tag{5.52}$$

where we also used (2.109) and (2.67e). This result shows that the $m = 0$ component of the sum over m in (5.41) contributes only to γ. Nonzero values of δ arise from the $m = \pm 1$ terms in the sum [49]. By measuring the angular distributions of electrons photodissociated from s-subshells at the magic angle $\theta = 54.7^0$ and in the direction $\phi = 0$, one could isolate the non-dipole parameter γ_s, which would then be determined by (5.50) under the condition $\delta_s = 0$. ∎

It should be remembered that the higher-order multipole correction terms in (5.48) generally will not manifest themselves for photon energies below 100 eV (although important exceptions exist for which the electric dipole-electric quadrupole interference term is comparable or larger than the dipole term [58–63]). Conversely, at photon energies greater than ~10 keV, (5.48) is no longer strictly valid since the magnetic dipole term (and additional higher-order multipole terms) will become important [48].

On a final note, expressions other than (5.48) have been derived that include higher-order multipole terms. These expressions differ in the details, but not in the physics [50, 64, 65].

Problems

5.1. Explain why a spin 1/2 particle cannot have a permanent quadrupole moment.

5.2. Fill in the following table:

l'	l''	$A_{l'l''}$	$B_{l'l''}$
$l_0 - 1$	$l_0 - 2$		
$l_0 - 1$	l		
$l_0 - 1$	$l_0 + 2$		
$l_0 + 1$	$l_0 - 2$		
$l_0 + 1$	l		
$l_0 + 1$	$l_0 + 2$		

5.3. Prove.

$$\frac{1}{|\vec{r} - \vec{r}'|} = \sum_{l=0}^{\infty} \left(\frac{r_<^l}{r_>^{l+1}} \right) P_l(\cos \alpha)$$

where $r_>/r_<$ is the greater/lesser of $\vec{r}/\vec{r}'$. Use your analysis to derive (2.71).

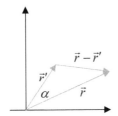

Finally, show that

$$\frac{1}{|\vec{r} - \vec{r}'|} = 4\pi \sum_{l=0}^{\infty} \sum_{m=-l}^{l} \frac{1}{2l+1} Y_{lm}^*(\Omega') Y_{lm}(\Omega) \left(\frac{r_<^l}{r_>^{l+1}} \right)$$

5.4. The potential due to a continuous volume charge density $\rho\left(\vec{r}'\right)$ can be described as.

$$\Phi\left(\vec{r}\right) = k \int \frac{\rho\left(\vec{r}'\right)}{|\vec{r} - \vec{r}'|} d^3 r'$$

For regions outside of the distribution $\left(|\vec{r}| > |\vec{r}'| \right)$, we can construct a multipole expansion of the potential in spherical coordinates using spherical harmonics. We might imagine that such an expansion would take the form

$$\Phi\left(\vec{r}\right) = k \int \frac{\rho\left(\vec{r}'\right)}{\left|\vec{r} - \vec{r}'\right|} d^3 r' = k \sum_{l=0}^{\infty} \sum_{m=-l}^{l} \frac{4\pi}{2l+1} q_{lm} \frac{Y_{lm}(\Omega)}{r^{l+1}}$$

where the q_{lm} are the multipole moments. Find an expression for the q_{lm}.

5.5. From Problem 5.4, we can infer that the electric quadrupole moment is proportional to the term $r^2 C_q^{(2)}$. Use this fact to find the electric quadrupole selection
rules for single-particle transitions between states of definite angular momentum. Ignore spin.

5.6. In (5.13), we found that the magnetic dipole operator was proportional to the
term $\langle f | \frac{i\omega}{2c} \vec{L}_y | i \rangle$. Use this term to find the magnetic dipole selection rules for
single-particle transitions between states of definite angular momentum.
Ignore spin.

5.7. Average displacement of the electron from the nucleus in hydrogenic atoms.
The radial dipole and quadrupole integrals introduced in (4.20) and (5.31),
respectively, usually cannot be evaluated analytically. One set of radial integrals that can be evaluated analytically and are of interest [e.g., see Problem
6.12(c)] is the set of integrals that give the expectation values of various powers
of the average displacement of the electron from the nucleus in hydrogenic
atoms.
When expressed in spherical coordinates, the Schrodinger equation

$$H\Psi = E\Psi$$

is separable into a radial equation and an angular equation with

$$\Psi = Y_{lm}(\theta, \phi) R_{El}(r)$$

where $R_{El}(r)$ is the radial portion of the wave function. $R_{El}(r)$ solves the radial
Schrodinger eq. [15]:

$$\left(\frac{d^2}{dr^2} + \frac{2}{r}\frac{d}{dr}\right) R_{El}(r) - \frac{2\mu}{\hbar^2}\left[V(r) + \frac{l(l+1)\hbar^2}{2\mu r^2}\right] R_{El}(r) + \frac{2\mu E}{\hbar^2} R_{El}(r) = 0$$

where μ is the reduced mass of the system. If we let $u_{nl}(r) \equiv r R_{El}(r)$, the radial
equation becomes

$$\frac{d^2}{dr^2} u_{nl}(r) + \frac{2\mu}{\hbar^2}\left[E - V(r) - \frac{l(l+1)\hbar^2}{2\mu r^2}\right] u_{nl}(r) = 0$$

For hydrogenic atoms, the potential is

$$V(r) = -\frac{Ze^2}{r}$$

where Z is the number of protons in the nucleus. This potential results in the energy eigenvalues [15]:

$$|E_n| = \frac{1}{2}\mu c^2 \frac{(Z\alpha)^2}{n^2} = \frac{e^2 Z^2}{2n^2 a_0}$$

where $\alpha = \frac{e^2}{\hbar c}$ and $a_0 = \frac{\hbar^2}{\mu e^2}$.

(a) Let,

$$\kappa^2 \equiv \frac{2\mu|E_n|}{\hbar^2}$$

$$\lambda \equiv \frac{Ze^2}{\hbar^2}\left(\frac{\mu}{2|E_n|}\right)^{1/2} = Z\alpha\left(\frac{\mu c^2}{2|E_n|}\right)^{1/2} = \left(\frac{Z}{\kappa a_0}\right)$$

$$\rho \equiv 2\kappa r = \left(\frac{8\mu|E_n|}{\hbar^2}\right)^{1/2} r = \frac{2\mu c Z\alpha}{\hbar n} r = \frac{2Z}{n a_0} r$$

Using these definitions, show that the radial equation can be written as [15]

$$\frac{d^2 u}{d\rho^2} - \frac{l(l+1)}{\rho^2} u + \left(\frac{\lambda}{\rho} - \frac{1}{4}\right) u = 0$$

Note that,

$$\lim_{\rho \to 0} u \sim \rho^{l+1} \quad \text{and} \quad \lim_{\rho \to \infty} u \to 0$$

(b) Use this form of the radial equation to prove Kramer's relation [9]:

$$\left(\frac{s+1}{n^2}\right)\langle r^s\rangle - (2s+1)\left(\frac{a_0}{Z}\right)\langle r^{s-1}\rangle + \frac{s}{4}\left[(2l+1)^2 - s^2\right]\left(\frac{a_0^2}{Z^2}\right)\langle r^{s-2}\rangle$$

$$= 0; s + 2l + 3 > 0$$

where

$$\langle r^s\rangle_{nl} = \int_0^\infty dr\, r^{s+2}[R_{nl}(r)]^2$$

Hint: Multiply the radial equation of part (a) by [9]:

$$\left[\rho^{s+1}u' - \left(\frac{s+1}{2}\right)\rho^s u\right]$$

and integrate over ρ.

(c) Use Kramer's relation to compute various radial integrals $\langle r^s\rangle_{nl}$ of interest (i.e., for all integer values $s = -6 \rightarrow 4$).

5.8. The Cartesian components of a 2nd-rank tensor can be formed from the Cartesian components of two vectors via,

$$T_{ij} = U_i V_j$$

Follow the procedure outlined in example 2.19 to compute the spherical components $T_q^{(2)}$ of this 2nd-rank tensor in terms of the product $U_{q_1}^{(1)} V_{q_2}^{(1)}$. Also, express the $T_q^{(2)}$ in terms of the T_{ij}.

Chapter 6
Relativistic Theory of Photoelectron Angular Distributions

In general, relativistic interactions have little effect on photoelectron angular distributions for atoms of low or moderate atomic weight. For most (open-shell) atoms, the asymmetry parameter β is energy-dependent due to non-relativistic (term-dependent) electron-core interactions, which permit several additional final-state channels to open (see Chap. 7). But for heavier alkali and for closed-shell atoms, the energy dependence of β has been found to be the result of relativistic (mainly spin-orbit) effects [51, 66–70]. In addition, experimental evidence has revealed different values of β for different values of $j = l \pm 1/2$, indicating that relativistic jj-coupling may influence photoelectron angular distributions in some systems [68, 71]. Discrepancies between the angular distributions predicted in non-relativistic and relativistic theories also tend to grow with increasing Z [51, 72]. What follows is a simplified, relativistic, independent-particle treatment in which core-continuum interactions are neglected.

In relativistic quantum theories, the "non-radial" part of the total wavefunction (for a particle in a central potential) is a combination of spherical harmonics and two-component spinors. These combinations are called spin-angular or spin-orbital functions. Single-particle spin-orbital functions are classified by their total angular momentum j and their projection onto the z-axis m_j in accordance with (2.119) and (2.144b) as follows [66, 73, 74],

$$\chi_\kappa^{m_j}(\hat{r}) \equiv \langle \hat{r} | \lambda j m_j \rangle =$$

$$\sum_{m_s} (-1)^{\lambda - \frac{1}{2} - m_j} \sqrt{2j+1} \begin{pmatrix} \lambda & \frac{1}{2} & j \\ m_j - m_s & m_s & -m_j \end{pmatrix} Y_{\lambda, m_j - m_s}(\hat{r}) \chi^{m_s} \tag{6.1}$$

The original version of this chapter was revised. The correction to this chapter is available at https://doi.org/10.1007/978-3-031-08027-2_11

V. T. Davis, *Introduction to Photoelectron Angular Distributions*, Springer Tracts in Modern Physics 286, https://doi.org/10.1007/978-3-031-08027-2_6

where χ^{m_s} is the two-component Pauli spinor; m_s being the projection of the spin angular momentum onto the z-axis,

$$\chi^{1/2} = \begin{pmatrix} 1 \\ 0 \end{pmatrix} \qquad \chi^{-1/2} = \begin{pmatrix} 0 \\ 1 \end{pmatrix} \tag{6.2}$$

Note that the $\chi_\kappa^{m_j}(\hat{r})$ are simultaneous eigenfunctions of L^2, S^2, J^2 and J_z [72]. The spin-orbit functions are normalized as follows [73],

$$\int [\chi_\kappa^{m_j}(\hat{r})]^\dagger \chi_{\kappa'}^{m'_j}(\hat{r}) d\Omega = \delta_{\kappa\kappa'} \delta_{m_j m'_j} \tag{6.3}$$

The label κ stands for the eigenvalue of the operator $K = \vec{L} \cdot \tilde{\sigma} + \hbar$ and takes the values

$$\kappa = -l - 1 = -\left(j + \frac{1}{2}\right) \text{ for } j = l + \frac{1}{2} \tag{6.4a}$$

$$\kappa = l = \left(j + \frac{1}{2}\right) \text{ for } j = l - \frac{1}{2} \tag{6.4b}$$

which we write as [66, 74],

$$\kappa = -\left(j + \frac{1}{2}\right)a \tag{6.5}$$

with

$$a = +1 \text{ for } l = j - \frac{1}{2} \tag{6.6a}$$

$$a = -1 \text{ for } l = j + \frac{1}{2} \tag{6.6b}$$

and

$$\tilde{\sigma} \equiv \hat{x}\sigma_x + \hat{y}\sigma_y + \hat{z}\sigma_z \tag{6.7}$$

And, of course, σ_x, σ_y, σ_z are the Pauli spin matrices,

$$\sigma_x = \begin{pmatrix} 0 & 1 \\ 1 & 0 \end{pmatrix}; \qquad \sigma_y = \begin{pmatrix} 0 & -i \\ i & 0 \end{pmatrix}; \qquad \sigma_z = \begin{pmatrix} 1 & 0 \\ 0 & -1 \end{pmatrix} \tag{6.8}$$

The spin operators are related to the Pauli spin matrices as follows,

$$S_i = \frac{\hbar}{2}\sigma_i; i = x, y, z \tag{6.9}$$

Example 6.1 Verify the relations in (6.4a and 6.4b)

$$
\begin{aligned}
K\chi_\kappa^{m_j} &= \left(\vec{L}\cdot\vec{\sigma}+\hbar\right)\chi_\kappa^{m_j} = \frac{2}{\hbar}\left(\vec{L}\cdot\vec{S}+\hbar\right)\chi_\kappa^{m_j} = \frac{2}{\hbar}\left[\frac{1}{2}\left(J^2-L^2-S^2\right)+\hbar\right]\chi_\kappa^{m_j}\\
&= \hbar\left[j(j+1)-l(l+1)-\frac{1}{2}\left(\frac{1}{2}+1\right)+1\right]\hbar\chi_\kappa^{m_j}\\
&= \hbar\left[j(j+1)-l(l+1)+\frac{1}{4}\right]\hbar\chi_\kappa^{m_j} = -\kappa\hbar\chi_\kappa^{m_j}
\end{aligned}
$$

where we used (2.111), (2.22a), (2.34a), and the fact that $s = 1/2$ for electrons.
So, we have,

$$
\kappa = -\left[j(j+1)-l(l+1)+\frac{1}{4}\right]
$$

Now, if $j = l+\frac{1}{2}$, then,

$$
\kappa = -\left[\left(l+\frac{1}{2}\right)\left(l+\frac{3}{2}\right)-l(l+1)+\frac{1}{4}\right] = -(l+1)
$$

and if $j = l-\frac{1}{2}$, then,

$$
\kappa = -\left[\left(l-\frac{1}{2}\right)\left(l+\frac{1}{2}\right)-l(l+1)+\frac{1}{4}\right] = l
$$

Thus, the relations in (6.4) are verified. ∎

Example 6.2 The conventional forms of the Pauli spin matrices are given in (6.8). What physical arbitrariness has been exploited in reducing the results from Problem 6.2(a) to these forms?

In Problem 6.2, we found the most general forms for the Hermitian matrices σ_x and σ_y that satisfy the given angular momentum commutation rules to be,

$$
\sigma_x = \begin{pmatrix} 0 & iC \\ -iC^* & 0 \end{pmatrix} \text{ and } \sigma_y = \begin{pmatrix} 0 & C \\ C^* & 0 \end{pmatrix} \text{ where } CC^* = 1.
$$

Because $CC^* = 1$ is our only constraint, we can write $C = -i$.

$$
\Rightarrow \sigma_x = e^{i\varphi}\begin{pmatrix} 0 & 1 \\ 1 & 0 \end{pmatrix} \text{ and } \sigma_y = e^{i\varphi}\begin{pmatrix} 0 & -i \\ i & 0 \end{pmatrix}
$$

where we have explicitly expressed any lingering phase factors that may remain. These phase factors are global, and therefore cannot be measured, which allows us to arbitrarily set them to unity. ∎

For a central potential $V\left(\vec{r}\right) = V(r)$, the relativistic single-particle bound orbitals are written as [6, 66, 69, 74],

$$\psi_{n\kappa m_j}\left(\vec{r}\right) = \frac{1}{r}\begin{bmatrix} P_{n\kappa}(r)\chi_\kappa^{m_j}(\hat{r}) \\ iQ_{n\kappa}(r)\chi_{-\kappa}^{m_j}(\hat{r}) \end{bmatrix} \tag{6.10}$$

where $P_{n\kappa}(r)$ and $Q_{n\kappa}(r)$ solve the coupled radial Dirac eqs. [71–74],

$$\frac{\partial P_{n\kappa}(r)}{\partial r} = -\frac{\kappa}{r}P_{n\kappa}(r) + \frac{1}{c\hbar}\left(E - V(r) + mc^2\right)Q_{n\kappa}(r) \tag{6.11a}$$

$$\frac{\partial Q_{n\kappa}(r)}{\partial r} = \frac{\kappa}{r}Q_{n\kappa}(r) - \frac{1}{c\hbar}\left(E - V(r) - mc^2\right)P_{n\kappa}(r) \tag{6.11b}$$

and E is the total (relativistic) energy of the particle. These radial wave functions are usually normalized as follows [73, 75],

$$\int \psi_{n\kappa m_j}^\dagger\left(\vec{r}\right)\psi_{n\kappa m_j}\left(\vec{r}\right)d^3r = 1 \Rightarrow \int_0^\infty \left[P_{n\kappa}^2(r) + Q_{n\kappa}^2(r)\right]dr = 1 \tag{6.12}$$

Since $j = |\kappa| - \frac{1}{2}$, we can instead use the label κ, and write the 4-component wave function for the initial bound state/orbital (in Dirac notation) as [66, 74],

$$\psi_{n\kappa m_j}(\vec{r}) = \begin{bmatrix} \langle r|n\kappa m_j, \varsigma = -1\rangle \\ \langle r|n\kappa m_j, \varsigma = +1\rangle \end{bmatrix} \tag{6.13}$$

where,

$$\langle r|n\kappa m_j, \varsigma\rangle = \langle r|n\kappa\varsigma\rangle\langle\hat{r}|\lambda j m_j\rangle\delta_{\lambda, j+\frac{1}{2}\varsigma} \tag{6.14}$$

and

$$\langle r|n\kappa\varsigma\rangle = \frac{1}{r}\begin{bmatrix} P_{n\kappa}(r), \varsigma = -1 \\ iQ_{n\kappa}(r), \varsigma = +1 \end{bmatrix} \tag{6.15}$$

The spin-angular part of $\psi_{n\kappa m_j}\left(\vec{r}\right)$, $\langle\hat{r}|\lambda j m_j\rangle$, is given in (6.1).

Example 6.3 Show that the operator K anti-commutes with the operator $\sigma_r = \hat{r} \cdot \tilde{\sigma}$ [73]. That is, show that,

$$\{K, \sigma_r\} = 0$$

$$
\begin{aligned}
\{K, \sigma_r\} &= \frac{1}{r}[K\vec{r} \cdot \tilde{\sigma} + \vec{r} \cdot \tilde{\sigma}K] \\
&= \frac{1}{r}[(\vec{L} \cdot \tilde{\sigma} + \hbar)(\vec{r} \cdot \tilde{\sigma}) + (\vec{r} \cdot \tilde{\sigma})(\vec{L} \cdot \tilde{\sigma} + \hbar)] \\
&= \frac{1}{r}[(\vec{L} \cdot \tilde{\sigma})(\vec{r} \cdot \tilde{\sigma}) + (\vec{r} \cdot \tilde{\sigma})(\vec{L} \cdot \tilde{\sigma}) + 2\hbar(\vec{r} \cdot \tilde{\sigma})] \\
&= \frac{1}{r}[\underbrace{\vec{L} \cdot \vec{r}}_{=0} + i\tilde{\sigma} \cdot (\vec{L} \times \vec{r}) + \underbrace{\vec{L} \cdot \vec{r}}_{=0}] \\
&\quad + \frac{1}{r}[i\tilde{\sigma} \cdot (\vec{r} \times \vec{L}) + 2\hbar(\vec{r} \cdot \tilde{\sigma})] \\
&= \frac{1}{r}\{i\tilde{\sigma} \cdot [(\vec{L} \times \vec{r}) + (\vec{r} \times \vec{L})] + 2\hbar(\vec{r} \cdot \tilde{\sigma})\}
\end{aligned}
$$

Examining the z-component of this equation,

$$
\begin{aligned}
\{K, \sigma_r\}_z &= \frac{1}{r}[i\sigma_z(L_x y - L_y x + x L_y - y L_x) + 2\hbar\sigma_z z] \\
&= \frac{1}{r}[i\sigma_z([L_x, y] - [L_y, x]) + 2\hbar\sigma_z z] \\
&= \frac{1}{r}[i\sigma_z(i\hbar z + i\hbar z) + 2\hbar\sigma_z z] = \frac{1}{r}(-2\hbar\sigma_z z + 2\hbar\sigma_z z) = 0
\end{aligned}
$$

where we used the results of Problem 6.2(b)(x) below. Similar results for the other components prove the desired relationship. One final remark. As a consequence of this relationship, we observe,

$$K\sigma_r \chi_\kappa^{m_j}(\hat{r}) = -\sigma_r K \chi_\kappa^{m_j}(\hat{r}) = -\sigma_r(-\hbar\kappa)\chi_\kappa^{m_j}(\hat{r}) = \hbar\kappa\sigma_r \chi_\kappa^{m_j}(\hat{r})$$

For this result to be consistent with example 6.1, we must have,

$$\sigma_r \chi_\kappa^{m_j}(\hat{r}) = -\chi_{-\kappa}^{m_j}(\hat{r})$$

∎

The final state consists of a continuum photoelectron, which can be expanded in a series of partial waves either of the form in (F.53) or in the equivalent form [66, 73],

$$\psi_{p\kappa m_j}(pr) = 4\pi \sum_{\kappa m_j} \exp\left(i\delta_\kappa\right)\chi_\kappa^{m_j}\left(\vec{p}\right)\left[\begin{array}{c}\langle r|p\kappa m_j, \varsigma = -1\rangle \\ \langle r|p\kappa m_j, \varsigma = +1\rangle\end{array}\right] \tag{6.16}$$

Here $\vec{p}$ is the momentum of the photoelectron.

The function in the brackets in (6.16) must have the correct asymptotic behavior [66, 73],

$$\frac{P_\kappa(pr)}{r} \to j_l(pr)\cos\delta_\kappa \tag{6.17}$$

The phase shift δ_κ is given by [73],

$$\tan\delta_\kappa = \frac{j_l(pr)[Q_\kappa(pr)/P_\kappa(pr)] - \frac{pca}{E+mc^2}j_{\bar{l}}(pr)}{\eta_l(pr)[Q_\kappa(pr)/P_\kappa(pr)] - \frac{pca}{E+mc^2}\eta_{\bar{l}}(pr)} \tag{6.18}$$

where [73]

$$\bar{l} = l + 1 = -\kappa \text{ for } \kappa < 0 \tag{6.19a}$$

$$\bar{l} = l - 1 = \kappa - 1 \text{ for } \kappa > 0 \tag{6.19b}$$

In the relativistic case, the matrix elements arising from multipole transitions in an external field have the form [66, 74],

$$D = \langle p\kappa'm_j'|\tilde{\alpha}\cdot\vec{A}|n\kappa m_j\rangle \tag{6.20}$$

where $\tilde{\alpha} = \hat{x}\alpha_x + \hat{y}\alpha_y + \hat{z}\alpha_z$ are the (4x4) Dirac matrices,

$$\alpha_i = \begin{pmatrix} 0 & \sigma_i \\ \sigma_i & 0 \end{pmatrix}; \ i = x, y, z \tag{6.21}$$

and $\vec{A}$ the electromagnetic vector potential. The ket $|n\kappa m_j\rangle$ represents the initial (bound) state, and $|p\kappa'm_j'\rangle$ the continuum photoelectron. The vector potential can be written as [cf. (3.8)],

$$\vec{A}\left(\vec{r}\right) \propto \hat{e}_q \exp\left(i\vec{k}\cdot\vec{r}\right) \tag{6.22}$$

where $\hat{e}_q$ is the polarization vector direction, $\vec{k}$ is the propagation vector of the photon which causes the pd. event, and $\omega = c|\vec{k}|$ is the angular frequency of the photon.

If we take the z-axis to be along $\vec{k}$ and considering (for now) circularly polarized light, we can write the polarization vector as [74],

$$\hat{e}_q = \frac{1}{\sqrt{2}}\left[\hat{e}_x + i\hat{e}_y\right] = -q\xi_q; \quad q = \pm 1 \tag{6.23}$$

This choice means we are working in the spherical vector basis $\{\xi_0, \xi_1, \zeta_{-1}\}$.

With the z-axis along $\vec{k}$, we can expand the vector potential as a sum of the normalized spherical harmonics using (F.47), (2.60), and (4.11) [66, 73, 74],

$$\exp\left(i\vec{k}\cdot\vec{r}\right) = \sum_l (i)^l (2l+1) j_l(kr) C_0^l(\hat{r}) \tag{6.24}$$

The operator $\vec{A}$ operates in coordinate space and the operator $\tilde{\alpha}$ operates in spin space. With that in mind, and using (6.22, 6.23 and 6.24), we construct the tensor product [3, 66],

$$\tilde{\alpha}\cdot\vec{A} = \alpha_q\left|\vec{A}\right| = \sum_l\sum_{q_1}\underbrace{(-1)^{1-l+q}}_{=(-1)^l}\sqrt{2L+1}\begin{pmatrix} 1 & l & L \\ q_1 & 0 & -q \end{pmatrix}\left[\tilde{\alpha}\otimes C^l\right]_q^{(L)}(i)^l(2l+1)j_l(kr) \tag{6.25}$$

where we also used (2.184) and (2.144b). Note that α_q is the component of $\tilde{\alpha}$ in the q-direction and is a rank-one tensor. Defining $X_q^{(1l,L)} \equiv \left[\tilde{\alpha}\otimes C^l\right]_q^{(L)}$, and seeing that the triangle relations inherent in the 3-j symbol in (6.25) require $q_1 = q$ (thus collapsing the sum over q_1), gives us,

$$\tilde{\alpha}\cdot\vec{A} = \sum_l(-1)^l(i)^l\sqrt{2L+1}(2l+1)\begin{pmatrix} 1 & l & L \\ q & 0 & -q \end{pmatrix}j_l(kr)X_q^{(1l,L)} \tag{6.26}$$

Example 6.4 Using what we have done so far, find expressions for the electric 2^L-pole operators and the magnetic 2^L-pole operators.

We can write the qth component of the interaction $\tilde{\alpha}\cdot\vec{A}$ as a linear combination of tensors of rank $L \geq 1$as follows [74],

$$\left(\tilde{\alpha}\cdot\vec{A}\right)_q = \sum_L(-1)^L\sum_l(-1)^l(i)^l\sqrt{2L+1}(2l+1)$$
$$\times \begin{pmatrix} 1 & l & L \\ q & 0 & -q \end{pmatrix}j_l(kr)X_q^{(1l,L)} \tag{6.27}$$

The triangle rule of the 3-j symbol in (6.27) tells us that the allowed values of l are $l = L, L \pm 1$. Thus, the tensors in the summation of (6.27) fall into two classes of opposite parity. Those for which the parity is $(-1)^l = (-1)^{L \pm 1}$ are the electric 2^L-pole operators, and those for which the parity is $(-1)^l = (-1)^L$ are the magnetic 2^L-pole operators.

We now evaluate the sum over l in (6.27), by finding the needed 3-j symbols. Realizing that $(L \pm 1) + 1 + L$ is even, and recalling that $q = \pm 1$ allows us to compute the following [7, 10],

$$
\begin{pmatrix} 1 & L-1 & L \\ q & 0 & -q \end{pmatrix} = (-1)^{1-L} \sqrt{\frac{(L+1)}{2(2L+1)(2L-1)}}
\tag{6.28}
$$

Similarly,

$$
\begin{pmatrix} 1 & L+1 & L \\ q & 0 & -q \end{pmatrix} = (-1)^{1-L} \sqrt{\frac{L}{2(2L+1)(2L+3)}}
\tag{6.29}
$$

Thus, we find the expression for the electric 2^L-pole operators to be [66, 74],

$$
\left(\tilde{\alpha} \cdot \vec{A} \right)^L_{q}\bigg|_{(e)} = (i)^{L-1} \left[\sqrt{\frac{(2L-1)(L+1)}{2}} j_{L-1}(kr) X_q^{(1L-1,L)} \right.
$$
$$
\left. - \sqrt{\frac{(2L+3)L}{2}} j_{L+1}(kr) X_q^{(1L+1,L)} \right]
\tag{6.30}
$$

We also find [7, 10],

$$
\begin{pmatrix} 1 & L & L \\ q & 0 & -q \end{pmatrix} = (-1)^{1-L} \sqrt{\frac{1}{2(2L+1)}}
\tag{6.31}
$$

so that the magnetic 2^L-pole operators are [66, 74],

$$
\left(\tilde{\alpha} \cdot \vec{A} \right)^L_{q}\bigg|_{(m)} = \frac{1}{\sqrt{2}} (2L+1)(i)^L j_L(kr) X_q^{(1L,L)}
\tag{6.32}
$$

∎

Since we are writing the interaction $\tilde{\alpha} \cdot \vec{A}$ as a linear combination of tensors of rank $L \geq 1$, we re-write (6.20) as follows [66, 74],

$$
D = \sum_L \langle p\kappa''m_j'' | \left(\tilde{\alpha} \cdot \vec{A} \right)^L_{q} | n\kappa m_j \rangle
\tag{6.33}
$$

As in earlier chapters, when computing the differential cross-section, we sum over all possible final states, square the interaction matrix element, and then average over all possible initial states [3, 66],

$$
\begin{aligned}
\frac{d\sigma_{pd}}{d\Omega} &= \frac{1}{(2j+1)} \sum_{m_j} |D|^2 = \frac{16\pi^2}{(2j+1)} \sum_{m_j} \sum_{LL''m'_jm''_jj'j''} \chi^{m''_j\,*}_{\kappa''}(p)\chi^{m'_j}_{\kappa'}(p) \\
&\quad \times \langle p\kappa''j''m''_j| \left(\vec{\alpha}\cdot\vec{A}\right)^{L''}_q |n\kappa jm_j\rangle\langle n\kappa jm_j|\left(\vec{\alpha}\cdot\vec{A}\right)^{L}_q|p\kappa'j'm'_j\rangle
\end{aligned}
\tag{6.34}
$$

The explicit dependence on j has been reinstated in the matrix elements to remind us that we need to sum over those states as well.

We begin the process of working out the terms in the expression of the differential cross section in detail.

First, we take care of the possible terms in ς by writing [74],

$$
\langle n\kappa jm_j|\left(\vec{\alpha}\cdot\vec{A}\right)^{L}_q|p\kappa'j'm'_j\rangle = \sum_\varsigma \langle n\kappa jm_j,\varsigma|\left(\vec{\alpha}\cdot\vec{A}\right)^{L}_q|p\kappa'j'm'_j,-\varsigma\rangle
\tag{6.35}
$$

If we write a typical term of the operators in (6.30) and (6.32) as $a_{lL}(r)X^{(1l,L)}_q$, we find that [73, 74],

$$
\begin{aligned}
\langle n\kappa jm_j|\left(\vec{\alpha}\cdot\vec{A}\right)^{L}_q|p\kappa'j'm'_j\rangle &= \langle n\kappa jm_j|a_{lL}(r)X^{(1l,L)}_q|p\kappa'j'm'_j\rangle \\
&= \sum_\varsigma \langle n\kappa jm_j,\varsigma|a_{lL}(r)X^{(1l,L)}_q|p\kappa'j'm'_j,-\varsigma\rangle \\
&= \sum_\varsigma \langle n\kappa j,\varsigma|a_{lL}(r)|p\kappa'j',-\varsigma\rangle\langle\lambda jm_j|X^{(1l,L)}_q|\lambda'j'm'_j\rangle\delta_{\lambda,j+\frac{1}{2}a\varsigma}\delta_{\lambda'j'-\frac{1}{2}a'\varsigma}
\end{aligned}
\tag{6.36}
$$

We operate on $\langle\lambda jm_j|X^{(1l,L)}_q|\lambda'j'm'_j\rangle$ with the Wigner-Eckhart theorem,

$$
\langle\lambda jm_j|X^{(1l,L)}_q|\lambda'j'm'_j\rangle = (-1)^{j-m_j}\begin{pmatrix} j & L & j' \\ -m_j & q & m'_j \end{pmatrix}\langle\lambda j\|X^{(1l,L)}\|\lambda'j'\rangle
\tag{6.37}
$$

As stated earlier, this product of two tensors consists of one operator which operates in spin space, and the other which operates in coordinate space, and we can form the tensor product via (D.15) [3, 74],

$$
\langle\lambda j\|X^{(1l,L)}\|\lambda'j'\rangle = \sqrt{2j+1}\sqrt{2j'+1}\sqrt{2L+1}\begin{Bmatrix} \lambda & \lambda' & l \\ \frac{1}{2} & \frac{1}{2} & 1 \\ j & j' & L \end{Bmatrix}
$$

$$
\times \langle\lambda\|C^{(l)}\|\lambda'\rangle\langle 1/2\|\widetilde{\sigma}^{(1)}\|1/2\rangle
\tag{6.38}
$$

where we have rewritten the 6-j symbol as follows,

$$
\begin{Bmatrix} \lambda & \lambda' & l \\ 1/2 & 1/2 & 1 \\ j & j' & L \end{Bmatrix} = \begin{Bmatrix} \lambda & 1/2 & j \\ \lambda' & 1/2 & j' \\ l & 1 & L \end{Bmatrix}
\tag{6.39}
$$

and realized that the summation over ς in (6.36) has reduced the Dirac matrix $\tilde{\alpha}$ to the Pauli spin matrix $\tilde{\sigma}$.

We now appeal to the identities (2.198) and (2.212) [3, 16],

$$
\langle \lambda \| C^{(l)} \| \lambda' \rangle = (-1)^\lambda \sqrt{2\lambda + 1} \sqrt{2\lambda' + 1} \begin{pmatrix} \lambda & l & \lambda' \\ 0 & 0 & 0 \end{pmatrix}
\tag{6.40}
$$

and

$$
\langle 1/2 \| \tilde{\sigma}^{(1)} \| 1/2 \rangle = \sqrt{6}
\tag{6.41}
$$

Combining (6.36) through (6.41),

$$
\langle nkjm_j | a_{lL}(r) X_q^{(1l,L)} | p\kappa' j' m'_j \rangle = (-1)^{j - m_j + \lambda} \sqrt{2j + 1} \sqrt{2j' + 1} \sqrt{2L + 1} \sqrt{2\lambda + 1}
$$

$$
\times \sqrt{2\lambda' + 1} \sqrt{6} \begin{pmatrix} \lambda & l & \lambda' \\ 0 & 0 & 0 \end{pmatrix} \begin{pmatrix} j & L & j' \\ -m_j & q & m'_j \end{pmatrix}
$$

$$
\times \begin{Bmatrix} \lambda & \lambda' & l \\ \frac{1}{2} & \frac{1}{2} & 1 \\ j & j' & L \end{Bmatrix} \sum_\varsigma \langle nkj, \varsigma | a_{lL}(r) | p\kappa' j', -\varsigma \rangle \delta_{\lambda, j + \frac{1}{2} a \varsigma} \delta_{\lambda', j' - \frac{1}{2} a' \varsigma}
\tag{6.42}
$$

We can simplify (6.42) by invoking the following identity based on (C.73) (for $l = L + 1$) [10, 76],

$$
\begin{pmatrix} \lambda & L+1 & \lambda' \\ 0 & 0 & 0 \end{pmatrix} \begin{Bmatrix} \lambda & \lambda' & L+1 \\ \frac{1}{2} & \frac{1}{2} & 1 \\ j & j' & L \end{Bmatrix}
$$

$$
= (-1)^{\lambda + j + \frac{3}{2}} \frac{[(\lambda - j)(2j + 1) + (\lambda' - j')(2j' + 1) + L + 1]}{\sqrt{6(L+1)(2L+1)(2L+3)(2\lambda+1)(2\lambda'+1)}} \begin{pmatrix} j & L & j' \\ 1/2 & 0 & -1/2 \end{pmatrix}
\tag{6.43}
$$

Triangle rules inherent in the 9-j symbol in (6.43) tell us that $\lambda = j \pm 1/2$ and $\lambda' = j' \pm 1/2$. So, we can write,

$$[(\lambda - j)(2j + 1) + (\lambda' - j')(2j' + 1) + L + 1]$$

$$= \left[\overbrace{\varsigma \left(\frac{1}{2}\right)(2j + 1)}^{-\kappa} \overbrace{-\varsigma \left(\frac{1}{2}\right)(2j' + 1)}^{\kappa'} + L + 1 \right] \tag{6.44}$$

Giving us (for $l = L + 1$),

$$\begin{pmatrix} \lambda & L+1 & \lambda' \\ 0 & 0 & 0 \end{pmatrix} \begin{Bmatrix} \lambda & \lambda' & L+1 \\ \frac{1}{2} & \frac{1}{2} & 1 \\ j & j' & L \end{Bmatrix}$$

$$= (-1)^{\lambda + j + \frac{3}{2}} \frac{[\varsigma(\kappa' - \kappa) + L + 1]}{\sqrt{6(L+1)(2L+1)(2L+3)(2\lambda+1)(2\lambda'+1)}} \begin{pmatrix} j & L & j' \\ 1/2 & 0 & -1/2 \end{pmatrix} \tag{6.45}$$

We also have the identity (for $l = L - 1$) [10, 76],

$$\begin{pmatrix} \lambda & L-1 & \lambda' \\ 0 & 0 & 0 \end{pmatrix} \begin{Bmatrix} \lambda & \lambda' & L-1 \\ \frac{1}{2} & \frac{1}{2} & 1 \\ j & j' & L \end{Bmatrix}$$

$$= (-1)^{\lambda + j + \frac{3}{2}} \frac{[(\lambda - j)(2j + 1) + (\lambda' - j')(2j' + 1) - L]}{\sqrt{6L(2L+1)(2L-1)(2\lambda+1)(2\lambda'+1)}} \begin{pmatrix} j & L & j' \\ 1/2 & 0 & -1/2 \end{pmatrix}$$

$$= (-1)^{\lambda + j + \frac{3}{2}} \frac{[\varsigma(\kappa' - \kappa) - L]}{\sqrt{6L(2L+1)(2L-1)(2\lambda+1)(2\lambda'+1)}} \begin{pmatrix} j & L & j' \\ 1/2 & 0 & -1/2 \end{pmatrix} \tag{6.46}$$

And finally (for $l = L$) [10, 76],

$$\begin{pmatrix} \lambda & L & \lambda' \\ 0 & 0 & 0 \end{pmatrix} \begin{Bmatrix} \lambda & \lambda' & L \\ \frac{1}{2} & \frac{1}{2} & 1 \\ j & j' & L \end{Bmatrix} = (-1)^{\lambda + \lambda' + L} \frac{\overbrace{\left(-\frac{1}{2}\right)[(2j + 1) + (2j' + 1)]}^{\varsigma(\kappa' + \kappa)}}{\sqrt{6L(L+1)(2L+1)(2\lambda+1)(2\lambda'+1)}}$$

$$\times \begin{pmatrix} j & L & j' \\ 1/2 & 0 & -1/2 \end{pmatrix} \tag{6.47}$$

Combining (6.30) and (6.42, 6.43, 6.44, 6.45 and 6.46), we have, for the electric multipoles [66],

$$
\langle n\kappa j m_j | \left(\widetilde{\alpha} \cdot \vec{A} \right)^L_{q\,(e)} | p\kappa' j' m'_j \rangle = -q(-1)^{m_j+\frac{1}{2}}(i)^{L-1}\sqrt{2j+1}\sqrt{2j'+1}
$$

$$
\times \begin{pmatrix} j & L & j' \\ 1/2 & 0 & -1/2 \end{pmatrix} \begin{pmatrix} j & L & j' \\ -m_j & q & m'_j \end{pmatrix}
$$

$$
\times \left\{ \sum_\varsigma \left[\sqrt{\frac{(L+1)}{2L}} [\varsigma(\kappa'-\kappa)-L] \langle n\kappa j, \varsigma | j_{L-1}(\omega r) | p\kappa' j', -\varsigma \rangle \right. \right.
$$
$$
\left. \left. - \sqrt{\frac{L}{2(L+1)}} [\varsigma(\kappa'-\kappa)+L+1] \langle n\kappa j, \varsigma | j_{L+1}(\omega r) | p\kappa' j', -\varsigma \rangle \right] \delta_{\lambda,j+\frac{1}{2}a\varsigma}\delta_{\lambda' j'-\frac{1}{2}a'\varsigma} \right\}
$$

$$(6.48)$$

Combining (6.32), (6.42), and (6.47), we find, for the magnetic term [66],

$$
\langle n\kappa j m_j | \left(\widetilde{\alpha} \cdot \vec{A} \right)^L_{q\,(m)} | p\kappa' j' m'_j \rangle = (-1)^{m_j+\frac{1}{2}}(i)^L \sqrt{2j+1}\sqrt{2j'+1}
$$

$$
\times \begin{pmatrix} j & L & j' \\ 1/2 & 0 & -1/2 \end{pmatrix} \begin{pmatrix} j & L & j' \\ -m_j & q & m'_j \end{pmatrix}
$$

$$
\times \frac{(2L+1)}{\sqrt{2L(L+1)}} \sum_\varsigma [\varsigma(\kappa'+\kappa)]
$$

$$
\times \langle n\kappa j, \varsigma | j_L(\omega r) | p\kappa' j', -\varsigma \rangle \delta_{\lambda,j+\frac{1}{2}a\varsigma}\delta_{\lambda' j'-\frac{1}{2}a'\varsigma} \quad (6.49)
$$

where we have noted (realizing that $m_j + 1/2$ is an integer),

$$
(-1)^{j-m_j+\lambda}(-1)^{\lambda+j+\frac{3}{2}} = (-1)^{2(j+\lambda)-m_j+1+\frac{1}{2}} = (-1)^{2\left(j+j\pm\frac{1}{2}\right)-m_j+1+\frac{1}{2}}
$$
$$
= (-1)^{4j}(-1)^{\pm1+1}(-1)^{-m_j+\frac{1}{2}} = (-1)^{-m_j+\frac{1}{2}} = -(-1)^{-m_j-\frac{1}{2}} = -(-1)^{m_j+\frac{1}{2}}
$$

$$(6.50)$$

We now evaluate the term $\chi^{m'_j*}_{\kappa'}(p)\chi^{m''_j}_{\kappa''}(p)$ by first recalling,

$$
\chi^{m_s*}\chi^{m'_s} = \delta_{m_s,m_{s'}} \tag{6.51}
$$

which results in [cf. equation (6.1)],

$$
\chi^{m'_j*}_{\kappa'}(p)\chi^{m''_j}_{\kappa''}(p) = (-1)^{\nu'+\nu''-m'_j-m''_j}\sqrt{2j'+1}\sqrt{2j''+1}
$$

$$
\times \sum_{m_s} \begin{pmatrix} \nu' & \frac{1}{2} & j' \\ m'_j-m_s & m_s & -m'_j \end{pmatrix} \begin{pmatrix} \nu'' & \frac{1}{2} & j'' \\ m''_j-m_s & m_s & -m''_j \end{pmatrix}
$$

$$
\times Y^*_{\nu',m'_j-m_s}(\hat{r})Y_{\nu'',m''_j-m_s}(\hat{r}) \tag{6.52}
$$

We invoke the following identity based on (C.74) [3, 10],

$$
\begin{pmatrix} \nu' & \dfrac{1}{2} & j' \\ m'_j - m_s & m_s & -m'_j \end{pmatrix}
\begin{pmatrix} \nu'' & \dfrac{1}{2} & j'' \\ m''_j - m_s & m_s & -m''_j \end{pmatrix}
$$

$$
= \sum_{\wedge} (-1)^{\nu'+j'+j''+m'_j-m_s-m''_j} (2\wedge+1)
\begin{Bmatrix} j' & j'' & \wedge \\ \nu'' & \nu' & \dfrac{1}{2} \end{Bmatrix} \tag{6.53}
$$

$$
\times
\begin{pmatrix} \nu'' & \nu' & \wedge \\ m''_j - m_s & -(m'_j - m_s) & m_\wedge \end{pmatrix}
\begin{pmatrix} j' & j'' & \wedge \\ -m'_j & m''_j & m_\wedge \end{pmatrix}
$$

leaving us with,

$$
\chi^{m'_j\,*}_{\kappa'}(p)\chi^{m''_j}_{\kappa''}(p) = (-1)^{\nu'+\nu''-m'_j-m''_j}\sqrt{2j'+1}
$$

$$
\times \sqrt{2j''+1} \sum_{m_s,\wedge} Y^*_{\nu',m'_j-m_s}(\hat{r}) Y_{\nu'',m''_j-m_s}(\hat{r})(-1)^{\nu'+j'+j''+m'_j-m_s-m''_j}(2\wedge+1)
$$

$$
\times
\begin{Bmatrix} j' & j'' & \wedge \\ \nu'' & \nu' & \dfrac{1}{2} \end{Bmatrix}
\begin{pmatrix} \nu' & \nu'' & \wedge \\ m'_j - m_s & -(m'_j - m_s) & m_\wedge \end{pmatrix}
\begin{pmatrix} j' & j'' & \wedge \\ -m'_j & m''_j & m_\wedge \end{pmatrix}
$$

$$
\tag{6.54}
$$

The two 3-j symbols in (6.54) are incompatible with one another unless $m_\wedge = 0$ (this also means that $m''_j = m'_j$). Also, there are no other terms in m''_j anywhere else in the expression for the pd. differential cross-section (6.34) except for those in (6.54) above. So, we can "pull forward" that summation to write,

$$
\chi^{m'_j\,*}_{\kappa'}(p)\chi^{m''_j}_{\kappa''}(p) =
$$
$$
\sqrt{2j'+1}\sqrt{2j''+1}\sum_{\wedge}(2\wedge+1)\sum_{m_s,m''_j}
\underbrace{Y^*_{\nu',m'_j-m_s}}_{(-1)^{m'_j-m_s}Y_{\nu',-(m'_j-m_s)}}(\hat{r})\, Y_{\nu'',m''_j-m_s}(\hat{r})
$$

$$
\times (-1)^{2\nu'+\nu''-2m'_j+j'+j''-m_s}
\begin{Bmatrix} j' & j'' & \wedge \\ \nu'' & \nu' & \dfrac{1}{2} \end{Bmatrix}
\begin{pmatrix} \nu' & \nu'' & \wedge \\ \underset{=m'_j}{m''_j - m_s} & -(m'_j - m_s) & \underset{=0}{m_\wedge} \end{pmatrix}
$$

$$
\times
\begin{pmatrix} j' & j'' & \wedge \\ -m'_j & \underset{=m'_j}{m''_j} & \underset{=0}{m_\wedge} \end{pmatrix}
$$

$$
\tag{6.55}
$$

and then use the relation [16],

$$\sum_{m_s, m_j''} Y_{\nu', -(m_j'-m_s)}(\hat{r}) Y_{\nu'', m_j''-m_s}(\hat{r}) \begin{pmatrix} \nu' & \nu'' & \wedge \\ m_j''-m_s & -(m_j'-m_s) & m_\wedge \end{pmatrix}$$

$$= \sqrt{\frac{(2\nu''+1)}{4\pi}} \frac{(-1)^{\wedge-m_\wedge}}{(2\wedge+1)} \langle \hat{~}\|C^{(\nu'')}\|\nu'\rangle Y_{\wedge, m_\wedge} \tag{6.56}$$

to get,

$$\chi_{\kappa'}^{m_j'*}(p)\chi_{\kappa''}^{m_j''}(p) = \sqrt{2j'+1}\sqrt{2j''+1}\sum_{\wedge}(2\wedge+1)(-1)^{2\nu'+\nu''-m_j'+j'+j''}$$

$$\times \begin{Bmatrix} j' & j'' & \wedge \\ \nu'' & \nu' & \frac{1}{2} \end{Bmatrix} \begin{pmatrix} j' & j'' & \wedge \\ -m_j' & m_j'' & m_\wedge \\ \underbrace{}_{=m_j'} & \underbrace{}_{=0} & \end{pmatrix} \sqrt{\frac{(2\nu''+1)}{4\pi}} \frac{(-1)^{\wedge-m_\wedge}}{(2\wedge+1)} \langle \wedge\|C^{(\nu'')}\|\nu'\rangle Y_{\wedge, m_\wedge}$$

$$\tag{6.57}$$

and then use (2.198),

$$\langle \wedge\|C^{(\nu'')}\|\nu'\rangle = (-1)^{\wedge}\sqrt{2\wedge+1}\sqrt{2\nu'+1}\begin{pmatrix} \wedge & \nu'' & \nu' \\ 0 & 0 & 0 \end{pmatrix} \tag{6.58}$$

to get,

$$\chi_{\kappa'}^{m_j'*}(p)\chi_{\kappa''}^{m_j''}(p) = \sqrt{2j'+1}\sqrt{2j''+1}\sqrt{\frac{(2\nu''+1)}{4\pi}}(-1)^{2\nu'+\nu''-m_j'+j'+j''+2\wedge}$$

$$\times \sum_{\wedge} \begin{Bmatrix} j' & j'' & \wedge \\ \nu'' & \nu' & \frac{1}{2} \end{Bmatrix} \begin{pmatrix} j' & j'' & \wedge \\ -m_j' & m_j' & 0 \end{pmatrix} \sqrt{2\wedge+1}\sqrt{2\nu'+1}\begin{pmatrix} \wedge & \nu'' & \nu' \\ 0 & 0 & 0 \end{pmatrix} Y_{\wedge, 0}$$

$$\tag{6.59}$$

Example 6.5 Prove the relation in (6.56).

Recall the result of Problem 2.9,

$$\sum_{m_1, m_2} \begin{pmatrix} l_1 & l_2 & l \\ m_1 & m_2 & m \end{pmatrix} Y_{l_1 m_1}(\theta, \phi) Y_{l_2 m_2}(\theta, \phi)$$

$$= \sqrt{\frac{(2l_1+1)(2l_2+1)}{4\pi(2l+1)}}\begin{pmatrix} l_1 & l_2 & l \\ 0 & 0 & 0 \end{pmatrix} Y_{lm}(\theta, \phi)$$

We combine this result with (2.198) to get,

$$\sum_{m_1, m_2} \begin{pmatrix} l_1 & l_2 & l \\ m_1 & m_2 & m \end{pmatrix} Y_{l_1 m_1}(\theta, \phi) Y_{l_2 m_2}(\theta, \phi)$$

$$= \sqrt{\frac{(2l_2 + 1)}{4\pi}} \frac{(-1)^{l_1}}{(2l + 1)} \langle l_1 \| C^{(l_2)} \| l \rangle Y_{lm}(\theta, \phi)$$

which is equivalent to the relation of (6.56). ∎

Recalling (2.60),

$$Y_{\wedge,0} = \sqrt{\frac{(2 \wedge + 1)}{4\pi}} P_\wedge(\cos\theta) \tag{6.60}$$

which, when applied, gives [66],

$$\chi_{\kappa'}^{m_j'^*}(p)\chi_{\kappa''}^{m_j''}(p) = \sqrt{2j'+1}\sqrt{2j''+1}\sqrt{\frac{(2\nu''+1)}{4\pi}}(-1)^{2\nu'+\nu''-m_j'+j'+j''+2\wedge}$$

$$\times \sum_\wedge \begin{Bmatrix} j' & j'' & \wedge \\ \nu'' & \nu' & \frac{1}{2} \end{Bmatrix} \begin{pmatrix} j' & j'' & \wedge \\ -m_j' & m_j' & 0 \end{pmatrix}$$

$$\times \sqrt{2\wedge+1}\sqrt{2\nu'+1} \begin{pmatrix} \wedge & \nu'' & \nu' \\ 0 & 0 & 0 \end{pmatrix} \sqrt{\frac{(2\wedge+1)}{4\pi}} P_\wedge(\cos\theta) \tag{6.61}$$

We can take care of the 6-j symbol by the following [based on (C.73)] [10, 76],

$$\begin{Bmatrix} j' & j'' & \wedge \\ \nu'' & \nu' & \frac{1}{2} \end{Bmatrix} \begin{pmatrix} \wedge & \nu'' & \nu' \\ 0 & 0 & 0 \end{pmatrix} = \begin{Bmatrix} \nu'' & \nu' & \wedge \\ j' & j'' & \frac{1}{2} \end{Bmatrix} \begin{pmatrix} \nu'' & \nu' & \wedge \\ 0 & 0 & 0 \end{pmatrix}$$

$$= -\frac{1}{\sqrt{2\nu''+1}\sqrt{2\nu'+1}} \begin{pmatrix} j' & j'' & \wedge \\ 1/2 & -1/2 & 0 \end{pmatrix} \tag{6.62}$$

where $\nu'' + \nu' + \wedge$ =even integer (otherwise the 3-j symbol is zero), giving us our final result for $\chi_{\kappa'}^{m_j'^*}(p)\chi_{\kappa''}^{m_j''}(p)$,

$$\chi_{\kappa'}^{m_j'^*}(p)\chi_{\kappa''}^{m_j''}(p) = \frac{\sqrt{2j'+1}\sqrt{2j''+1}}{4\pi}(-1)^{\nu'-m_j'+j'+j''+\wedge+1}$$

$$\times \sum_\wedge (2\wedge+1) \begin{pmatrix} j' & j'' & \wedge \\ -m_j' & m_j' & 0 \end{pmatrix} \begin{pmatrix} j' & j'' & \wedge \\ 1/2 & -1/2 & 0 \end{pmatrix} P_\wedge(\cos\theta) \tag{6.63}$$

We are finally in the position of being able to evaluate the differential cross-section for photodetachment by substituting (6.48), (6.49), and (6.63) into (6.34),

$$
\begin{aligned}
\frac{d\sigma_{pd}}{d\Omega} = \frac{16\pi^2}{(2j+1)} \sum_{\substack{m_j, m_j' \\ L, L''j'j''}} & \frac{\sqrt{2j'+1}\sqrt{2j''+1}}{4\pi}(-1)^{\nu'-m_j'+j'+j''+\wedge+1}\sum_{\wedge}(2\wedge+1) \\
\times & \begin{pmatrix} j' & j'' & \wedge \\ -m_j' & m_j' & 0 \end{pmatrix}\begin{pmatrix} j' & j'' & \wedge \\ 1/2 & -1/2 & 0 \end{pmatrix}P_\wedge(\cos\theta)(-1)^{m_j+\frac{1}{2}}\sqrt{2j+1}\sqrt{2j'+1} \\
\times & \begin{pmatrix} j & L & j' \\ 1/2 & 0 & -1/2 \end{pmatrix}\begin{pmatrix} j & L & j' \\ -m_j & q & m_j' \end{pmatrix}\left[\sum_{\varsigma'}\langle n\kappa j, \varsigma'|a_{l;L}|p\kappa'j', -\varsigma'\rangle\right](-1)^{m_j'+\frac{1}{2}}\sqrt{2j+1} \\
\times & \sqrt{2j''+1}\begin{pmatrix} j & L'' & j'' \\ 1/2 & 0 & -1/2 \end{pmatrix}\begin{pmatrix} j & L'' & j'' \\ -m_j & q & m_j' \end{pmatrix}\left[\sum_{\varsigma''}\langle p\kappa''j'', \varsigma''|a_{f';L''}|n\kappa j, -\varsigma''\rangle\right]
\end{aligned}
\tag{6.64}
$$

where we have condensed the radial terms (for example) as,

$$
\langle n\kappa j, \varsigma'|a_{l;L}|p\kappa'j', -\varsigma'\rangle =
\left\{
\begin{aligned}
& (i)^L\frac{(2L+1)}{\sqrt{2L(L+1)}}[\varsigma'(\kappa'+\kappa)]\langle n\kappa j, \varsigma'|j_L(\omega r)|p\kappa'j', -\varsigma'\rangle \\
& -(i)^{L-1}\sqrt{\frac{(L+1)}{2L}}[\varsigma'(\kappa'-\kappa)-L]\langle n\kappa j, \varsigma'|j_{L-1}(\omega r)|p\kappa'j', -\varsigma'\rangle \\
& +(i)^{L-1}\sqrt{\frac{L}{2(L+1)}}[\varsigma'(\kappa'-\kappa)+L+1]\langle n\kappa j, \varsigma'|j_{L+1}(\omega r)|p\kappa'j', -\varsigma'\rangle
\end{aligned}
\right\}
\tag{6.65}
$$

We perform the sums over m_j and m_j' by first noting the following,

$$
\begin{aligned}
& \begin{pmatrix} j' & j'' & \wedge \\ -m_j' & m_j' & 0 \end{pmatrix}\begin{pmatrix} j & L & j' \\ -m_j & q & m_j' \end{pmatrix}\begin{pmatrix} j & L'' & j'' \\ -m_j & q & m_j' \end{pmatrix} \\
& = (-1)^{3(j+j'')}(-1)^{L''+\wedge}\begin{pmatrix} j' & j'' & \wedge \\ m_j' & -m_j' & 0 \end{pmatrix}\begin{pmatrix} L & j' & j \\ q & m_j' & -m_j \end{pmatrix}\begin{pmatrix} j & j'' & L'' \\ m_j & -m_j' & -q \end{pmatrix}
\end{aligned}
\tag{6.66}
$$

and by considering the identity [again based on (C.73)] [3, 10],

$$\sum_{m_j, m_j'} (-1)^{j'+j''+L''+\wedge} \begin{pmatrix} j' & j'' & \wedge \\ m_j' & -m_j' & 0 \end{pmatrix} \begin{pmatrix} L & j' & j \\ q & m_j' & -m_j \end{pmatrix} \begin{pmatrix} j & j'' & L'' \\ m_j & -m_j' & -q \end{pmatrix}$$

$$= (-1)^{-L+j+q} \begin{pmatrix} L'' & L & \wedge \\ -q & q & 0 \end{pmatrix} \begin{Bmatrix} L & j' & j \\ j'' & L'' & \wedge \end{Bmatrix}$$

$$\tag{6.67}$$

giving us [66],

$$\frac{d\sigma_{pd}}{d\Omega} = 4\pi \sum_{LL''j'j''\wedge} (-1)^{j+\frac{1}{2}+\wedge+q}(2j'+1)(2j''+1)(2\wedge+1)$$

$$\times \begin{pmatrix} L & L'' & \wedge \\ q & -q & 0 \end{pmatrix} \begin{pmatrix} j' & j'' & \wedge \\ 1/2 & -1/2 & 0 \end{pmatrix} \begin{pmatrix} j & L & j' \\ 1/2 & 0 & -1/2 \end{pmatrix}$$

$$\times \begin{pmatrix} j'' & L'' & j \\ 1/2 & 0 & -1/2 \end{pmatrix} \begin{Bmatrix} L & L'' & \wedge \\ j'' & j' & j \end{Bmatrix} P_\wedge(\cos\theta)$$

$$\times \left[\sum_{\varsigma', \varsigma''} |\langle n\kappa j, \varsigma' | a_{l=L,L\pm1; L}(r) | p\kappa' j', -\varsigma' \rangle \langle p\kappa'' j'', \varsigma'' | a_{l''=L'',L''\pm1; L''}(r) | n\kappa j, -\varsigma'' \rangle | \right]$$

$$\tag{6.68}$$

where we have also made liberal use of the symmetry properties of the 3-j symbols to arrive at our final phase factor. It is clear that this expression is more complicated than the non-relativistic expressions seen earlier, since it allows for additional interference between different multipole terms (except when $\wedge = 0$) [66].

We can realize the electric dipole approximation by setting $\exp\left(i\vec{k}\cdot\vec{r}\right) \approx 1$, which translates to setting $L = 1$ and $l = 0$ in (6.42) [c.f. (6.24) and (6.25)] [66]. Under these circumstances, the top term on the RHS of (6.65) is the magnetic dipole term, and the bottom term on the RHS of (6.65) is the electric quadrupole term. It is the middle term that represents the electric dipole. Setting $L, L'' = 1$; $l = 0$ in (6.68) and using the middle term on the RHS of (6.65),

$$\frac{d\sigma_{pd}}{d\Omega} = 4\pi \sum_{j'j''\wedge} (2j'+1)(2j''+1)(2\wedge+1) \begin{pmatrix} 1 & 1 & \wedge \\ q & -q & 0 \end{pmatrix} \begin{pmatrix} j' & j'' & \wedge \\ 1/2 & -1/2 & 0 \end{pmatrix}$$

$$\times \begin{pmatrix} j & 1 & j' \\ 1/2 & 0 & -1/2 \end{pmatrix} \begin{pmatrix} j'' & 1 & j \\ 1/2 & 0 & -1/2 \end{pmatrix} \begin{Bmatrix} 1 & 1 & \wedge \\ j'' & j' & j \end{Bmatrix} P_\wedge(\cos\theta)$$

$$\times \left[\sum_{\varsigma', \varsigma''} [\varsigma'(\kappa-\kappa') - 1][\varsigma''(\kappa-\kappa'') - 1] |\langle n\kappa j, \varsigma' | p\kappa' j', -\varsigma' \rangle \langle p\kappa'' j'', \varsigma'' | n\kappa j, -\varsigma'' \rangle | \right]$$

$$\tag{6.69}$$

where we have also noted that $j_0(kr) \to 1$ [c.f. (F.26)] in the long-wavelength limit (dipole approximation).

Triangle rules and symmetry properties of the 3-j and 6-j symbols in (6.69) limit the sum over $\wedge$ to the terms $\wedge = 0, 2$, giving us,

$$
\frac{d\bar{\sigma}_{pd}}{d\Omega} = 4\pi \sum_{j'j''} (-1)^{j+\frac{1}{2}} (2j'+1)(2j''+1)
$$

$$
\times \left\{
\begin{array}{l}
\begin{pmatrix} 1 & 1 & 0 \\ q & -q & 0 \end{pmatrix}
\begin{pmatrix} j' & j'' & 0 \\ 1/2 & -1/2 & 0 \end{pmatrix}
\begin{pmatrix} j & 1 & j' \\ 1/2 & 0 & -1/2 \end{pmatrix} \\[12pt]
\quad \times \begin{pmatrix} j'' & 1 & j \\ 1/2 & 0 & -1/2 \end{pmatrix}
\begin{Bmatrix} 1 & 1 & 0 \\ j'' & j' & j \end{Bmatrix} \\[12pt]
+5 \begin{pmatrix} 1 & 1 & 2 \\ q & -q & 0 \end{pmatrix}
\begin{pmatrix} j' & j'' & 2 \\ 1/2 & -1/2 & 0 \end{pmatrix}
\begin{pmatrix} j & 1 & j' \\ 1/2 & 0 & -1/2 \end{pmatrix} \\[12pt]
\quad \times \begin{pmatrix} j'' & 1 & j \\ 1/2 & 0 & -1/2 \end{pmatrix}
\begin{Bmatrix} 1 & 1 & 2 \\ j'' & j' & j \end{Bmatrix} P_2(\cos\theta)
\end{array}
\right\}
$$

$$
\times \left[\sum_{\varsigma',\varsigma''} [\varsigma'(\kappa-\kappa')-1][\varsigma''(\kappa-\kappa'')-1] |\langle n\kappa j, \varsigma'|p\kappa'j', -\varsigma'\rangle \langle p\kappa''j'', \varsigma''|n\kappa j, -\varsigma''\rangle| \right]
$$

$$(6.70)$$

For linearly-polarized photons, we set $q = 0$ in the above expression,

$$
\frac{d\bar{\sigma}_{pd}}{d\Omega} = 4\pi \sum_{j'j''} (-1)^{j+\frac{1}{2}} (2j'+1)(2j''+1)
$$

$$
\times \left\{
\begin{array}{l}
\underbrace{\begin{pmatrix} 1 & 1 & 0 \\ 0 & 0 & 0 \end{pmatrix}}_{-\sqrt{\frac{1}{3}}}
\underbrace{\begin{pmatrix} j' & j'' & 0 \\ 1/2 & -1/2 & 0 \end{pmatrix}}_{(-1)^{j'+1/2}\sqrt{\frac{1}{(2j'+1)}}\delta_{j'j''}}
\begin{pmatrix} j & 1 & j' \\ 1/2 & 0 & -1/2 \end{pmatrix} \\[20pt]
\quad \times \begin{pmatrix} j'' & 1 & j \\ 1/2 & 0 & -1/2 \end{pmatrix}
\underbrace{\begin{Bmatrix} 1 & 1 & 0 \\ j'' & j' & j \end{Bmatrix}}_{(-1)^{j+j'+1}\sqrt{\frac{1}{3(2j'+1)}}\delta_{j'j''}} \\[20pt]
+5 \underbrace{\begin{pmatrix} 1 & 1 & 2 \\ 0 & 0 & 0 \end{pmatrix}}_{\sqrt{\frac{2}{15}}}
\begin{pmatrix} j' & j'' & 2 \\ 1/2 & -1/2 & 0 \end{pmatrix}
\begin{pmatrix} j & 1 & j' \\ 1/2 & 0 & -1/2 \end{pmatrix} \\[20pt]
\quad \times \begin{pmatrix} j'' & 1 & j \\ 1/2 & 0 & -1/2 \end{pmatrix}
\begin{Bmatrix} 1 & 1 & 2 \\ j'' & j' & j \end{Bmatrix} P_2(\cos\theta)
\end{array}
\right\}
$$

$$
\times \left[\sum_{\varsigma',\varsigma''} [\varsigma'(\kappa-\kappa')-1][\varsigma''(\kappa-\kappa'')-1] |\langle n\kappa j, \varsigma'|p\kappa'j', -\varsigma'\rangle \langle p\kappa''j'', \varsigma''|n\kappa j, -\varsigma''\rangle| \right]
$$

$$\Rightarrow \frac{d\sigma_{pd}}{d\Omega} = 4\pi \sum_{j'j''} (-1)^{2(j+j')+1} \frac{(2j'+1)(2j''+1)}{4\pi}$$

$$\times \left\{ \begin{array}{l} \dfrac{1}{3(2j'+1)} \begin{pmatrix} j & 1 & j' \\ 1/2 & 0 & -1/2 \end{pmatrix} \begin{pmatrix} \times j'' & 1 & j \\ 1/2 & 0 & -1/2 \end{pmatrix} \\[2em] +5\sqrt{\dfrac{2}{15}} \begin{pmatrix} j' & j'' & 2 \\ 1/2 & -1/2 & 0 \end{pmatrix} \begin{pmatrix} j & 1 & j' \\ 1/2 & 0 & -1/2 \end{pmatrix} \begin{pmatrix} j'' & 1 & j \\ 1/2 & 0 & -1/2 \end{pmatrix} \\[2em] \times \begin{Bmatrix} 1 & 1 & 2 \\ j'' & j' & j \end{Bmatrix} P_2(\cos\theta) \end{array} \right\}$$

$$\times \left[\sum_{\varsigma',\varsigma''} [\varsigma'(\kappa-\kappa')-1][\varsigma''(\kappa-\kappa'')-1] |\langle n\kappa j, \varsigma'|p\kappa'j', -\varsigma'\rangle \langle p\kappa''j'', \varsigma''|n\kappa j, -\varsigma''\rangle| \right]$$

$$(6.71)$$

Again the 3-j symbols come to our rescue, limiting the sums over j' and j'' to the values $j', j'' = j, j \pm 1$. We can now evaluate all nine possible combinations of terms using tabulated values of 3-j and 6-j symbols, which allow us to calculate the following factors [3, 7, 10],

For $j' = j + 1, j'' = j - 1$,

$$(-1)^{2(2j+1)+1} (2j+3)(2j-1)$$

$$\times \left\{ \begin{array}{l} \dfrac{1}{3(2j+3)} \begin{pmatrix} j & 1 & j+1 \\ 1/2 & 0 & -1/2 \end{pmatrix} \begin{pmatrix} j-1 & 1 & j \\ 1/2 & 0 & -1/2 \end{pmatrix} \\[2em] -5\sqrt{\dfrac{2}{15}} \begin{pmatrix} j+1 & j-1 & 2 \\ 1/2 & -1/2 & 0 \end{pmatrix} \begin{pmatrix} j & 1 & j+1 \\ 1/2 & 0 & -1/2 \end{pmatrix} \begin{pmatrix} j-1 & 1 & j \\ 1/2 & 0 & -1/2 \end{pmatrix} \\[2em] \times \begin{Bmatrix} 1 & 1 & 2 \\ j-1 & j+1 & j \end{Bmatrix} P_2(\cos\theta) \end{array} \right\}$$

$$\times \left[\sum_{\varsigma',\varsigma''} [\varsigma'(\kappa-\kappa')-1][\varsigma''(\kappa-\kappa'')-1] |\langle n\kappa j, \varsigma'|p\kappa'j', -\varsigma'\rangle \langle p\kappa''j-1, \varsigma''|n\kappa j, -\varsigma''\rangle| \right]$$

$$= \left[\frac{-(2j-1)}{12\sqrt{j(j+1)}} + \frac{(2j+3)(2j-1)}{16j(j+1)} P_2(\cos\theta) \right] |\mathfrak{R}_{j+1}\mathfrak{R}_{j-1}^*|$$

$$(6.72a)$$

For $j' = j - 1, j'' = j + 1$,

$$(-1)^{2(2j-1)+1}(2j-1)(2j+3)$$

$$\times \left\{ \begin{array}{l} \dfrac{1}{3(2j+3)} \begin{pmatrix} j & 1 & j-1 \\ 1/2 & 0 & -1/2 \end{pmatrix} \begin{pmatrix} j+1 & 1 & j \\ 1/2 & 0 & -1/2 \end{pmatrix} \\[3mm] -5\sqrt{\dfrac{2}{15}} \begin{pmatrix} j-1 & j+1 & 2 \\ 1/2 & -1/2 & 0 \end{pmatrix} \begin{pmatrix} j & 1 & j-1 \\ 1/2 & 0 & -1/2 \end{pmatrix} \begin{pmatrix} j+1 & 1 & j \\ 1/2 & 0 & -1/2 \end{pmatrix} \\[3mm] \times \begin{Bmatrix} 1 & 1 & 2 \\ j+1 & j-1 & j \end{Bmatrix} P_2(\cos\theta) \end{array} \right\}$$

$$\times \left[\sum_{\varsigma',\varsigma''} [\varsigma'(\kappa-\kappa')-1][\varsigma''(\kappa-\kappa'')-1] |\langle n\kappa j, \varsigma'|p\kappa'j', -\varsigma'\rangle \langle p\kappa''j+1, \varsigma''|n\kappa j, -\varsigma''\rangle| \right]$$

$$= \left[\frac{(2j-1)}{12\sqrt{j(j+1)}} + \frac{(2j+3)(2j-1)}{16j(j+1)} P_2(\cos\theta) \right] |\mathfrak{R}_{j-1}\mathfrak{R}_{j+1}^*|$$

$$(6.72b)$$

For $j' = j + 1, j'' = j$,

$$(-1)^{2(2j+3)+1}(2j+3)(2j+1)$$

$$\times \left\{ \begin{array}{l} \dfrac{1}{3(2j+3)} \begin{pmatrix} j+1 & 1 & j \\ 1/2 & 0 & -1/2 \end{pmatrix} \begin{pmatrix} j & 1 & j \\ 1/2 & 0 & -1/2 \end{pmatrix} \\[3mm] -5\sqrt{\dfrac{2}{15}} \begin{pmatrix} j+1 & j & 2 \\ 1/2 & -1/2 & 0 \end{pmatrix} \begin{pmatrix} j & 1 & j+1 \\ 1/2 & 0 & -1/2 \end{pmatrix} \begin{pmatrix} j & 1 & j \\ 1/2 & 0 & -1/2 \end{pmatrix} \\[3mm] \times \begin{Bmatrix} 1 & 1 & 2 \\ j & j+1 & j \end{Bmatrix} P_2(\cos\theta) \end{array} \right\}$$

$$\times \left[\sum_{\varsigma',\varsigma''} [\varsigma'(\kappa-\kappa')-1][\varsigma''(\kappa-\kappa'')-1] |\langle n\kappa j, \varsigma'|p\kappa'j', -\varsigma'\rangle \langle p\kappa''j, \varsigma''|n\kappa j, -\varsigma''\rangle| \right]$$

$$= \left[\frac{-(2j+1)}{12\sqrt{2j(j+1)}} + \frac{(2j+3)}{16j(j+1)^2} P_2(\cos\theta) \right] |\mathfrak{R}_{j+1}\mathfrak{R}_j^*|$$

$$(6.72c)$$

For $j' = j$, $j'' = j + 1$,

$$(-1)^{2(2j)+1}(2j+1)(2j+3)$$

$$\times \left\{ \begin{array}{c} \dfrac{1}{3(2j+3)} \begin{pmatrix} j & 1 & j \\ 1/2 & 0 & -1/2 \end{pmatrix} \begin{pmatrix} j+1 & 1 & j \\ 1/2 & 0 & -1/2 \end{pmatrix} \\[2em] -5\sqrt{\dfrac{2}{15}} \begin{pmatrix} j & j+1 & 2 \\ 1/2 & -1/2 & 0 \end{pmatrix} \begin{pmatrix} j & 1 & j \\ 1/2 & 0 & -1/2 \end{pmatrix} \begin{pmatrix} j+1 & 1 & j \\ 1/2 & 0 & -1/2 \end{pmatrix} \\[2em] \times \begin{Bmatrix} 1 & 1 & 2 \\ j+1 & j & j \end{Bmatrix} P_2(\cos\theta) \end{array} \right\}$$

$$\times \left[\sum_{\varsigma',\varsigma''} [\varsigma'(\kappa - \kappa') - 1][\varsigma''(\kappa - \kappa'') - 1] |\langle n\kappa j, \varsigma'|p\kappa'j', -\varsigma'\rangle\langle p\kappa''j+1, \varsigma''|n\kappa j, -\varsigma''\rangle| \right]$$

$$= \left[\frac{(2j+1)}{12\sqrt{2j(j+1)}} + \frac{(2j+3)}{16j(j+1)^2} P_2(\cos\theta) \right] \left| \mathfrak{R}_j \mathfrak{R}_{j+1}^* \right|$$

$$(6.72d)$$

For $j', j'' = j$,

$$(-1)^{2(2j)+1}(2j+1)^2$$

$$\times \left\{ \begin{array}{c} \dfrac{1}{3(2j+1)} \begin{pmatrix} j & 1 & j \\ 1/2 & 0 & -1/2 \end{pmatrix}^2 \\[2em] -5\sqrt{\dfrac{2}{15}} \begin{pmatrix} j & j & 2 \\ 1/2 & -1/2 & 0 \end{pmatrix} \begin{pmatrix} j & 1 & j \\ 1/2 & 0 & -1/2 \end{pmatrix}^2 \begin{Bmatrix} 1 & 1 & 2 \\ j & j & j \end{Bmatrix} P_2(\cos\theta) \end{array} \right\}$$

$$\times \left[\sum_{\varsigma',\varsigma''} [\varsigma'(\kappa - \kappa') - 1][\varsigma''(\kappa - \kappa'') - 1] |\langle n\kappa j, \varsigma'|p\kappa'j', -\varsigma'\rangle\langle p\kappa''j, \varsigma''|n\kappa j, -\varsigma''\rangle| \right]$$

$$= \left[\frac{1}{12j(j+1)} - \frac{(2j-1)(2j+3)}{48j^2(j+1)^2} P_2(\cos\theta) \right] \mathfrak{R}_j^2$$

$$(6.72e)$$

For $j', j'' = j + 1$,

$$(-1)^{2(2j+3)+1}(2j+3)^2$$

$$\times \left\{ \begin{array}{l} \dfrac{1}{3(2j+3)} \begin{pmatrix} j & 1 & j+1 \\ 1/2 & 0 & -1/2 \end{pmatrix} \begin{pmatrix} j+1 & 1 & j \\ 1/2 & 0 & -1/2 \end{pmatrix} \\[3mm] -5\sqrt{\dfrac{2}{15}} \begin{pmatrix} j+1 & j+1 & 2 \\ 1/2 & -1/2 & 0 \end{pmatrix} \begin{pmatrix} j & 1 & j+1 \\ 1/2 & 0 & -1/2 \end{pmatrix} \begin{pmatrix} j+1 & 1 & j \\ 1/2 & 0 & -1/2 \end{pmatrix} \\[3mm] \times \begin{Bmatrix} 1 & 1 & 2 \\ j+1 & j+1 & j \end{Bmatrix} P_2(\cos\theta) \end{array} \right\}$$

$$\times \left[\sum_{\varsigma',\varsigma''} [\varsigma'(\kappa-\kappa')-1][\varsigma''(\kappa-\kappa'')-1] |\langle n\kappa j, \varsigma'|p\kappa'j', -\varsigma'\rangle \langle p\kappa''j+1, \varsigma''|n\kappa j, -\varsigma''\rangle| \right]$$

$$= \left[\frac{(2j+3)}{12(j+1)} + \frac{(2j+5)(2j+3)}{48(j+1)^2} P_2(\cos\theta) \right] \mathfrak{R}_{j+1}^2$$

$$(6.72\text{f})$$

For $j' = j, j'' = j - 1$,

$$(-1)^{2(2j)+1}(2j+1)(2j-1)$$

$$\times \left\{ \begin{array}{l} \dfrac{1}{3(2j+1)} \begin{pmatrix} j & 1 & j \\ 1/2 & 0 & -1/2 \end{pmatrix} \begin{pmatrix} j-1 & 1 & j \\ 1/2 & 0 & -1/2 \end{pmatrix} \\[3mm] -5\sqrt{\dfrac{2}{15}} \begin{pmatrix} j & j-1 & 2 \\ 1/2 & -1/2 & 0 \end{pmatrix} \begin{pmatrix} j & 1 & j \\ 1/2 & 0 & -1/2 \end{pmatrix} \begin{pmatrix} j-1 & 1 & j \\ 1/2 & 0 & -1/2 \end{pmatrix} \\[3mm] \times \begin{Bmatrix} 1 & 1 & 2 \\ j-1 & j & j \end{Bmatrix} P_2(\cos\theta) \end{array} \right\}$$

$$\times \left[\sum_{\varsigma',\varsigma''} [\varsigma'(\kappa-\kappa')-1][\varsigma''(\kappa-\kappa'')-1] |\langle n\kappa j, \varsigma'|p\kappa'j', -\varsigma'\rangle \langle p\kappa''j-1, \varsigma''|n\kappa j, -\varsigma''\rangle| \right]$$

$$= \left[\frac{(2j+1)}{12j\sqrt{(2j+1)(j+1)}} + \frac{(2j-1)}{16j^2(j+1)} P_2(\cos\theta) \right] |\mathfrak{R}_j \mathfrak{R}_{j-1}^*|$$

$$(6.72\text{g})$$

For $j' = j - 1, j'' = j,$

$$(-1)^{2(2j-1)+1}(2j-1)(2j+1)$$

$$\times \left\{ \begin{array}{l} \dfrac{1}{3(2j+1)} \begin{pmatrix} j & 1 & j-1 \\ 1/2 & 0 & -1/2 \end{pmatrix} \begin{pmatrix} j & 1 & j \\ 1/2 & 0 & -1/2 \end{pmatrix} \\[2em] -5\sqrt{\dfrac{2}{15}} \begin{pmatrix} j-1 & j & 2 \\ 1/2 & -1/2 & 0 \end{pmatrix} \begin{pmatrix} j & 1 & j-1 \\ 1/2 & 0 & -1/2 \end{pmatrix} \begin{pmatrix} j & 1 & j \\ 1/2 & 0 & -1/2 \end{pmatrix} \\[2em] \times \begin{Bmatrix} 1 & 1 & 2 \\ j & j-1 & j \end{Bmatrix} P_2(\cos\theta) \end{array} \right\}$$

$$\times \left[\sum_{\varsigma',\varsigma''} [\varsigma'(\kappa-\kappa')-1][\varsigma''(\kappa-\kappa'')-1]|\langle n\kappa j, \varsigma'|p\kappa'j', -\varsigma'\rangle\langle p\kappa''j, \varsigma''|n\kappa j, -\varsigma''\rangle| \right]$$

$$= \left[\frac{-(2j+1)}{12j\sqrt{(2j+1)(j+1)}} + \frac{(2j-1)}{16j^2(j+1)} P_2(\cos\theta) \right] |\mathfrak{R}_{j-1}\mathfrak{R}_j^*|$$

$$(6.72\text{h})$$

For $j', j'' = j - 1,$

$$(-1)^{2(2j-1)+1}(2j-1)^2$$

$$\times \left\{ \begin{array}{l} \dfrac{1}{3(2j-1)} \begin{pmatrix} j & 1 & j-1 \\ 1/2 & 0 & -1/2 \end{pmatrix} \begin{pmatrix} j-1 & 1 & j \\ 1/2 & 0 & -1/2 \end{pmatrix} \\[2em] -5\sqrt{\dfrac{2}{15}} \begin{pmatrix} j-1 & j-1 & 2 \\ 1/2 & -1/2 & 0 \end{pmatrix} \begin{pmatrix} j & 1 & j-1 \\ 1/2 & 0 & -1/2 \end{pmatrix} \begin{pmatrix} j-1 & 1 & j \\ 1/2 & 0 & -1/2 \end{pmatrix} \\[2em] \times \begin{Bmatrix} 1 & 1 & 2 \\ j-1 & j-1 & j \end{Bmatrix} P_2(\cos\theta) \end{array} \right\}$$

$$\times \left[\sum_{\varsigma',\varsigma''} [\varsigma'(\kappa-\kappa')-1][\varsigma''(\kappa-\kappa'')-1]|\langle n\kappa j, \varsigma'|p\kappa'j', -\varsigma'\rangle\langle p\kappa''j-1, \varsigma''|n\kappa j, -\varsigma''\rangle| \right]$$

$$= \left[\frac{(2j-1)}{12j} + \frac{(2j-1)(2j-3)}{48j^2} P_2(\cos\theta) \right] \mathfrak{R}_{j-1}^2$$

$$(6.72\text{i})$$

We use the symbol $\mathfrak{R}_j$ for the radial portion of the matrix element, written below in terms of the large and small radial components of the radial Dirac equation [66, 73],

$$\mathfrak{R}_{j'} = \sum_{\varsigma'}[\varsigma'(\kappa - \kappa') - 1]|\langle n\kappa j, \varsigma'|p\kappa'j', -\varsigma'\rangle|\exp(i\delta_{j'})$$

$$= -i\{(\kappa - \kappa' - 1)\langle Q_{j'}|P_j\rangle + (\kappa - \kappa' + 1)\langle P_{j'} \mid Q_j\rangle\}\exp(i\delta_{j'}) \tag{6.73}$$

Substituting (6.72) and (6.73) into (6.71) results in [67],

$$\frac{d\sigma_{pd}}{d\Omega} = \frac{\sigma_{pd}}{4\pi}[1 + \beta P_2(\cos\theta)] \tag{6.74}$$

where [66, 67, 77],

$$\sigma_{pd} = 16\pi^2\left[\frac{2j-1}{12j}\mathfrak{R}_{j-1}^2 + \frac{1}{12j(j+1)}\mathfrak{R}_j^2 + \frac{(2j+3)}{12(j+1)}\mathfrak{R}_{j+1}^2\right] \tag{6.75a}$$

and the relativistic asymmetry parameter is (in the dipole approximation) [66, 67, 77],

$$\beta =$$
$$\left\{\begin{array}{l}\dfrac{(2j-3)(2j-1)}{48j^2}\mathfrak{R}_{j-1}^2 - \dfrac{(2j-1)(2j+3)}{48j^2(j+1)^2}\mathfrak{R}_j^2 + \dfrac{(2j+5)(2j+3)}{48(j+1)^2}\mathfrak{R}_{j+1}^2 \\[2mm] + \dfrac{(2j-1)}{8j^2(j+1)}\left|\mathfrak{R}_j\mathfrak{R}_{j-1}^*\right| + \dfrac{(2j+3)}{8j(j+1)^2}\left|\mathfrak{R}_j\mathfrak{R}_{j+1}^*\right| + \dfrac{(2j+3)(2j-1)}{8j(j+1)}\left|\mathfrak{R}_{j-1}\mathfrak{R}_{j+1}^*\right|\end{array}\right\}$$
$$\times\left(\frac{(2j-1)}{12j}\mathfrak{R}_{j-1}^2 + \frac{1}{12j(j+1)}\mathfrak{R}_j^2 + \frac{(2j+3)}{12(j+1)}\mathfrak{R}_{j+1}^2\right)^{-1}$$

$$\tag{6.75b}$$

Other expressions for the relativistic version of β (using different normalization schemes) have also been proposed [69, 77].

It now remains only to show that the above expression for the relativistic asymmetry parameter reduces to the familiar non-relativistic expression in the appropriate limit. That is the subject of the following example.

Example 6.6 Show that (6.75b) reduces to the appropriate expression in the non-relativistic limit.

First, we shift the energy by a constant amount: $E = W + mc^2$. Equations (6.11a) and (6.11b) now appear as,

$$\frac{\partial P_{n\kappa}(r)}{\partial r} = -\frac{\kappa}{r}P_{n\kappa}(r) + \frac{1}{c\hbar}\left(W - V(r) + 2mc^2\right)Q_{n\kappa}(r) \qquad \text{(6.11a-modified)}$$

$$\frac{\partial Q_{n\kappa}(r)}{\partial r} = \frac{\kappa}{r}Q_{n\kappa}(r) - \frac{1}{c\hbar}(W - V(r))P_{n\kappa}(r) \qquad \text{(6.11b-modified)}$$

In the limit $c \to \infty$, we have from (6.11a-modified),

$$Q \sim \frac{1}{2c}\left(P' + \frac{\kappa}{r}P\right) \tag{6.76}$$

Taking the derivative of (6.11a-modified) after realizing in the non-relativistic limit that $(W - V)/c << 2mc$,

$$P'' = -\frac{\kappa}{r}P' + \frac{\kappa}{r^2}P + \frac{2mc}{\hbar}Q' \tag{6.77}$$

Substituting from (6.11b-modified) for Q' [73],

$$P'' = -\frac{\kappa}{r}P' + \frac{\kappa}{r^2}P + \frac{2mc}{\hbar}\left[\frac{\kappa}{r}Q - \frac{1}{c\hbar}(W - V(r))P\right] \tag{6.78}$$

We use (6.76) for Q in (6.78),

$$P'' = -\frac{\kappa}{r}P' + \frac{\kappa}{r^2}P + \frac{2mc}{\hbar}\left\{\frac{\kappa}{r}\left[\frac{\hbar}{2mc}\left(P' + \frac{\kappa}{r}P\right)\right] - \frac{1}{c\hbar}(W - V(r))P\right\} \tag{6.79}$$

And so, to within $O(1/c^2)$ and to within an additive constant we find [66, 74],

$$P'' - \frac{l(l+1)}{r^2}P + \frac{2m}{\hbar^2}\left(V(r) - E + mc^2\right)P = 0 \tag{6.80}$$

where we have used (6.4) and (6.5) to note that $\kappa(\kappa + 1) = l(l + 1)$. We see that the large component solves the non-relativistic radial Schrödinger's equation, and so reduces to the non-relativistic radial wave function in the non-relativistic limit. If we substitute the results of (6.76) and (6.80) into (6.73), we find radial integrals of the type [66],

$$\mathcal{R}_{l_0\pm1} = \langle P_{l_0\pm1}| \mp \frac{d}{dr} + \frac{2l + 1 \pm 1}{2r}|P_{l_0}\rangle \exp\left(i\delta_{l\pm1}\right) \tag{6.81}$$

This expression is the same as the integral form seen in (4.39d) if we note that the operator appearing here is the familiar dipole operator in its "velocity" form, while the form of the dipole operator appearing in (4.39d) is in its "length" form. Specifically, we have the following relationships between the relativistic and the non-relativistic radial matrix elements [66],

$$\begin{aligned} a = -1 &\Rightarrow \mathcal{R}_{j-1} = \mathcal{R}_j \to -\mathcal{R}_{l_0-1}; \; \mathcal{R}_{j+1} \to \mathcal{R}_{l_0+1} \\ a = +1 &\Rightarrow \mathcal{R}_{j-1} \to -\mathcal{R}_{l_0-1}; \; \mathcal{R}_{j+1} = \mathcal{R}_j \to \mathcal{R}_{l_0+1} \end{aligned} \tag{6.82}$$

So, for example, if we let $j = l + 1/2$; $a = +1$, and substitute the appropriate parts from (6.82) into (6.75b), we get,

$$\beta = \left\{ \begin{aligned} &\frac{(2l_0-2)(2l_0)}{48(l_0+1/2)^2}\mathfrak{R}^2_{l_0-1} - \frac{(2l_0)(2l_0+4)}{48(l_0+1/2)^2(l_0+3/2)^2}\mathfrak{R}^2_{l_0+1} \\ &+ \frac{(2l_0+4)(2l_0+6)}{48(l_0+3/2)^2}\mathfrak{R}^2_{l_0+1} - \frac{(2l_0)}{8(l_0+1/2)^2(l_0+3/2)}\left|\mathfrak{R}_{l_0+1}\mathfrak{R}^*_{l_0-1}\right| \\ &+ \frac{(2l_0+4)}{8(l_0+1/2)(l_0+3/2)^2}\mathfrak{R}^2_{l_0+1} - \frac{(2l_0+4)(2l_0)}{8(l_0+1/2)(l_0+3/2)}\left|\mathfrak{R}_{l_0-1}\mathfrak{R}^*_{l_0+1}\right| \end{aligned} \right\}$$

$$\times \left(\frac{2l_0}{12(l_0+1/2)}\mathfrak{R}^2_{l_0-1} + \frac{1}{12(l_0+1/2)(l_0+3/2)}\mathfrak{R}^2_{l_0+1} + \frac{(2l_0+4)}{12(l_0+3/2)}\mathfrak{R}^2_{l_0+1} \right)^{-1}$$

$$= \frac{1}{3(2l_0+1)^2}\left\{ \begin{aligned} &l_0(l_0-1)\mathfrak{R}^2_{l_0-1} + \frac{(l_0+2)}{(2l_0+3)^2}\begin{bmatrix} -4l_0 + (l_0+3)(2l_0+1)^2 \\ +6(2l_0+1)\end{bmatrix}\mathfrak{R}^2_{l_0+1} \\ &- \frac{6l_0}{(2l_0+3)}[1 + (l_0+2)(2l_0+1)]\left|\mathfrak{R}_{l_0+1}\mathfrak{R}^*_{l_0-1}\right| \end{aligned} \right\}$$

$$\times \left\{ \frac{1}{3(2l_0+1)}\left[l_0\mathfrak{R}^2_{l_0-1} + (l+1)\mathfrak{R}^2_{l_0+1} \right] \right\}^{-1}$$

$$= \frac{l_0(l_0-1)\mathfrak{R}^2_{l_0-1} + (l_0+1)(l_0+2)\mathfrak{R}^2_{l_0+1} - 6l_0(l_0+1)\left|\mathfrak{R}_{l_0+1}\mathfrak{R}^*_{l_0-1}\right|}{(2l_0+1)\left[l_0\mathfrak{R}^2_{l_0-1} + (l_0+1)\mathfrak{R}^2_{l_0+1} \right]}$$

$$(6.83)$$

This result is identical to the form in (4.39b). ∎

Similarly, we find that (6.75a) reduces to (4.39a) if we account for the appropriate definition (normalization) for the non-relativistic cross-section (see Problem 6.9) [66, 67]. So, our relativistic analysis agrees with our non-relativistic one in the appropriate limit.

One thing example 6.6 shows us is that a change in coupling schemes cannot, in and of itself, lead to a change in the asymmetry parameter. But it does show that the value of β will deviate from the non-relativistic value when the matrix elements (and relative phase shifts) are different in the relativistic theory [66]. For example, in the case of s-electron detachment, we find by substituting $j = 1/2$ into (6.75b),

$$\beta = \frac{2\mathfrak{R}^2_{p_{3/2}} + 4\left|\mathfrak{R}_{p_{3/2}}\mathfrak{R}^*_{p_{1/2}}\right|}{\mathfrak{R}^2_{p_{1/2}} + 2\mathfrak{R}^2_{p_{3/2}}} \tag{6.84}$$

It is only when the two radial integrals in (6.84) are equal (and the relative phase shifts are identical) that we recover the familiar (non-relativistic) result $\beta(l_0 = 0) = 2$. Furthermore, if the two radial matrix elements $\mathfrak{R}_{p_{3/2}}$ and $\mathfrak{R}_{p_{1/2}}$ were to go to zero at the same photon energy, then β would have no meaning (since there would be a zero cross-section). Fortunately, it is well-known that the two matrix elements go to zero at slightly different energies, which is why a non-zero minimum is observed at the "Cooper minimum" in certain photodetachment and photoionization spectra [66]. If $\mathfrak{R}_{p_{1/2}}$ alone is zero, then (6.84) tells us that $\beta = 1$. If $\mathfrak{R}_{p_{3/2}}$ alone is zero, then (6.84) tells us that $\beta = 0$. So, if rapid oscillations in β are observed near the (non-zero) minimum of certain photoelectron spectra, it may be an indication of (possibly large) deviations from the predictions of the non-relativistic theory for that species. These effects may also be present in systems having angular momentum of some value other than $j = 1/2$; certainly (6.75b) seems to predict that, in many cases, β will be different for the two components of any l value. For example, if two photoelectrons from two states split by spin-orbit coupling have two different kinetic energies when detached by photons of the same energy (because their pd. potentials are slightly different, for example), then it may be necessary to account for spin-orbit effects even in a non-relativistic approximation, especially near pd. cross-section minima [67, 78–86]. This effect can be further magnified because the two electrons have different radial wave functions in the relativistic theory, and hence, different dipole matrix elements with the continuum orbitals. So, while it may be obvious that relativistic descriptions are necessary for high-Z elements and for high photon energies, it may also be necessary to consider relativistic effects for lighter species at low photon energies, even within the dipole approximation [51, 66–69].

Problems

6.1. Spherical Bessel functions.

In App. F, we developed a series representation of the spherical Bessel functions by first identifying an integral form for these functions in the complex plane. In this problem, we will use standard series solution methods to find an expression for the spherical Bessel functions.

The standard form of Bessel's equation is,

$$\frac{d^2y}{dx^2} + \frac{1}{x}\frac{dy}{dx} + \left(1 - \frac{\nu^2}{x^2}\right)y = 0$$

The parameter v is a given number, so we can consider $v \geq 0$ without loss of generality. When $v = p$, p=integer, we refer to the above equation as Bessel's equation of order p.

(a) Use the Method of Frobenius (MOF) to construct a series solution for the above equation about the regular singular point $x = 0$.
(b) Denote the solutions to Bessel's equation as $J_\nu(x)$. For $\nu\neq$integer, $J_\nu(x)$ and $J_{-\nu}(x)$ are linearly independent. The general solution to Bessel's equation is then,

$$y(x) = C_1 J_\nu(x) + C_2 J_{-\nu}(x)$$

For $v = p$, p=integer, $J_p(x)$ and $J_{-p}(x)$ are no longer linearly independent. To prove this, show that $J_{-p}(x) = (-1)^p J_p(x)$.

Note: To get the second linearly independent solution of Bessel's equation of integer order, we define,

$$N_\nu(x) = \frac{J_\nu(x)\cos\nu\pi - J_{-\nu}(x)}{\sin\pi\nu}$$

$N_\nu(x)$ is called the Neumann function or the Bessel function of the second kind. Thus, for the case p=integer, the general solution of Bessel's equation of integer order is,

$$y(x) = C_1 J_p(x) + C_2 N_p(x)$$

Note that $J_\nu(x)$ and $N_\nu(x)$ are linearly independent even if $\nu\neq$integer. Also note that $N_\nu(x)$ is not finite at $x = 0$, but is still needed for solutions in regions "away" from origin.

The differential equation for the spherical Bessel functions arises from expressing the Helmholtz equation in spherical coordinates. The differential equation for the spherical Bessel functions is,

$$r^2R'' + 2rR' + \left[k^2r^2 - l(l+1)\right]R = 0; l = \text{integer}; \text{prime denotes } d/dr.$$

(c) Let $R(r) = r^{-1/2}S(r)$ and show that the above differential equation can be written as Bessel's equation of order $l+\frac{1}{2}$, and that the solutions are therefore,

$$R(r) = r^{-1/2}\left[c_1 J_{l+1/2}(kr) + c_2 N_{l+1/2}(kr)\right]$$

(d) The standard representations of the spherical Bessel functions of the first and second kind are,

$$j_l(x) \equiv \sqrt{\frac{\pi}{2x}} J_{l+1/2}(x) \text{ and } n_l(x) \equiv \sqrt{\frac{\pi}{2x}} N_{l+1/2}(x).$$

Using your results from parts (a) and (c), develop a series representation for $j_l(x)$. Show that it matches (F.26).

6.2. The Pauli spin matrices.

There are several ways to deduce the forms of the spin matrices of (6.8). One way is based on physical arguments (see, for example, Feynman, Leighton, and Sands, *The Feynman Lectures of Physics: The New Millennium Edition, Vol III: Quantum Mechanics*, Basic Books, New York, 2010). Another way to construct the matrices is by assuming they obey the commutation rules of (2.1).

(a) Consider the set of 2x2 spin matrices which satisfy the commutation relations of (2.1),

$$\left[S_l, S_j\right] = i\hbar\sum_k \varepsilon_{ljk}S_k; \quad l,j,k = x\rightleftarrows1, y\rightleftarrows2, z\rightleftarrows3$$

Since the eigenvalues of these matrices are $\pm\hbar/2$, it is conventional to introduce the Pauli matrices σ_k such that $S_k \equiv \frac{\hbar}{2}\sigma_k$, giving us,

$$\left[\sigma_l, \sigma_j\right] = 2i\sum_k \varepsilon_{ljk}\sigma_k; \quad l,j,k = x\rightleftarrows1, y\rightleftarrows2, z\rightleftarrows3$$

Given $\sigma_3 = \begin{pmatrix} 1 & 0 \\ 0 & -1 \end{pmatrix}$, find the general Hermitian matrices σ_1 and σ_2 that (along with σ_3) satisfy the commutation rules above.

(b) Using your results from part (a), establish the following relations:

(i) $\sigma_i\sigma_j = i\sum_k \varepsilon_{ijk}\sigma_k + \delta_{ij}I_2$ where I_2 is the unit matrix for the 2x2 space.

(ii) $\sigma_i\sigma_j = -\sigma_j\sigma_i$

(iii) $\sigma_1\sigma_1\sigma_3 = iI_2$

(iv) $\text{Tr}\sigma_i = 0$

(v) $\text{Det}\sigma_i = 1$

(vi) $\sigma_i^2 = I_2$

(vii) $\text{Tr}(\sigma_i\sigma_j) = 2\delta_{ij}$

(viii) $\{\sigma_i, \sigma_j\} = 2\delta_{ij}I_2$ where $\{\sigma_i, \sigma_j\} \equiv \sigma_i\sigma_j + \sigma_j\sigma_i$

(ix) $\tilde{\sigma} \times \tilde{\sigma} = 2i\tilde{\sigma}$

(x) $\left(\tilde{\sigma} \cdot \vec{A}\right)\left(\tilde{\sigma} \cdot \vec{B}\right) = i\left(\tilde{\sigma} \times \vec{A}\right) \cdot \vec{B} + \left(\vec{A} \cdot \vec{B}\right)I_2 = i\tilde{\sigma} \cdot \left(\vec{A} \times \vec{B}\right) + \left(\vec{A} \cdot \vec{B}\right)I_2$ where $\vec{A}$ and $\vec{B}$ are not spin operators.

(xi) For an arbitrary vector $\vec{\alpha}$, show that,

$$e^{i\vec{\alpha}\cdot\tilde{\sigma}} = \cos|\alpha| + i\frac{\vec{\alpha} \cdot \tilde{\sigma}}{|\alpha|} \sin|\alpha|$$

6.3. The Dirac matrices are defined in (6.21). Prove the following properties of the Dirac matrices,

(a) $\{\alpha_i, \alpha_j\} = 2\delta_{ij}I_4$

(b) $\{\alpha_i, \tilde{\beta}\} = 0; \tilde{\beta} \equiv \begin{pmatrix} I_2 & 0 \\ 0 & -I_2 \end{pmatrix}$

(c) $(\alpha_i)^2 = \tilde{\beta}^2 = I_4$

(d) $\tilde{\alpha} \times \tilde{\alpha} = 2i\tilde{\sigma}_4$ where $\tilde{\sigma}_4 \equiv \hat{x}\sigma_{4x} + \hat{y}\sigma_{4y} + \hat{z}\sigma_{4z}; \sigma_{4i} = \begin{pmatrix} \sigma_i & 0 \\ 0 & \sigma_i \end{pmatrix}$

(e) $\left(\tilde{\alpha} \cdot \vec{A}\right)\left(\tilde{\alpha} \cdot \vec{B}\right) = i\left(\tilde{\sigma}_4 \times \vec{A}\right) \cdot \vec{B} + \left(\vec{A} \cdot \vec{B}\right)I_4 = i\tilde{\sigma}_4 \cdot \left(\vec{A} \times \vec{B}\right) + \left(\vec{A} \cdot \vec{B}\right)I_4$

(f) $\tilde{\gamma}_5\tilde{\sigma}_4 = -\tilde{\alpha}$ and $\tilde{\gamma}_5\tilde{\alpha} = -\tilde{\sigma}_4$ where $\tilde{\gamma}_5 \equiv \begin{pmatrix} 0 & -I_2 \\ -I_2 & 0 \end{pmatrix}$.

6.4. Prove the following,

(a) $\vec{\nabla} = \hat{r}\left(\hat{r} \cdot \vec{\nabla}\right) - \hat{r} \times \left(\hat{r} \times \vec{\nabla}\right) = \hat{r}\frac{\partial}{\partial r} - \frac{i}{\hbar}\frac{\hat{r}}{|r|} \times \vec{L}$

(b) From part (a), we can say,

$$\tilde{\alpha} \cdot \vec{p} = -i\hbar\tilde{\alpha} \cdot \vec{\nabla} = -i\hbar\tilde{\alpha} \cdot \hat{r}\frac{\partial}{\partial r} - \frac{1}{|r|}\tilde{\alpha} \cdot \left(\vec{r} \times \vec{L}\right)$$

Thus, show that,

$$\tilde{\alpha} \cdot \vec{p} = -i\hbar\alpha_r\frac{\partial}{\partial r} + \frac{i}{|r|}\alpha_r\left(\tilde{\beta}K - \hbar\right)$$

where $\alpha_r = \tilde{\alpha} \cdot \hat{r}$ and the operator $\tilde{\beta}K = \vec{L} \cdot \tilde{\sigma}_4 + \hbar$.

6.5. The Dirac Hamiltonian for a free particle is,

$$H_0 = c\tilde{\alpha} \cdot \vec{p} + \tilde{\beta}mc^2$$

The eigenvalue equation $H_0\psi = E\psi$ can therefore be written as,

$$\left(E - c\tilde{\alpha} \cdot \vec{p} - \tilde{\beta}mc^2\right)\Psi = 0$$

If we identify $E = i\hbar\frac{\partial}{\partial t}$ and $p = -i\hbar\vec{\nabla}$, then the time-dependent Dirac equation is,

$$\left(i\hbar\frac{\partial}{\partial t} + ci\hbar\tilde{\alpha} \cdot \vec{\nabla} - \tilde{\beta}mc^2\right)\Psi\left(\vec{r},t\right) = 0$$

Solve this equation to find $\Psi\left(\vec{r},t\right)$. Assume the form $\Psi\left(\vec{r},t\right) = \tilde{u}e^{i\left(\vec{k}\cdot\vec{r}-\omega t\right)}$ where the elements of $\tilde{u}$ are just numbers. Remember that, since $\tilde{\alpha}$ and $\tilde{\beta}$ are 4×4 matrices, $\tilde{u}$ must be in the form of a 4×1 column matrix.

6.6. The Dirac equation for a spherically symmetric potential is,

$$\left[c\tilde{\alpha} \cdot \vec{p} + \tilde{\beta}mc^2 + V(r)\right]\psi_{n\kappa m_j}(r) = E\psi_{n\kappa m_j}(r)$$

Substitute the following form into the Dirac equation,

$$\psi_{n\kappa m_j}\left(\vec{r}\right) = \begin{bmatrix} F_{n\kappa}(r)\chi_\kappa^{m_j}(\hat{r}) \\ iG_{n\kappa}(r)\chi_{-\kappa}^{m_j}(\hat{r}) \end{bmatrix}$$

and using results from the previous problem, show that the functions of (6.10) solve the resulting radial eq. (6.11).

6.7. Interaction of an electron with the electromagnetic field in the Dirac formalism.

(a) An expression for the operator of the time-rate-of-change of an observable O can be found from the Heisenberg equation of motion [6],

$$\left[\frac{\partial O}{\partial t}\right]_{op} = \frac{\partial O}{\partial t} + \frac{i}{\hbar}[H, O]$$

where H is the Hamiltonian. The Dirac Hamiltonian for a free particle is,

$$H_0 = c\tilde{\alpha} \cdot \vec{p} + \tilde{\beta}mc^2$$

Use the Heisenberg equation of motion to find the form of the "velocity" operator for the Dirac Hamiltonian of a free particle.

(b) The Hamiltonian for the interaction of an electron with an electromagnetic field H_{int} is realized by replacing the canonical momentum operator $\vec{p}$ with the kinetic momentum operator $\vec{p} - \frac{e}{c}\vec{A}$ in the Dirac Hamiltonian for a free particle, so that,

$$H = c\tilde{\alpha} \cdot \left(\vec{p} - \frac{e}{c}\vec{A}\right) + \tilde{\beta}mc^2 = H_0 + H_{\text{int}}; \; H_{\text{int}} = c\tilde{\alpha} \cdot \frac{e}{c}\vec{A}$$

Show that this interaction Hamiltonian reduces to the correct form given in chapter four [in (4.4a), for example] in the non-relativistic limit.

(c) Use the Heisenberg equation of motion to find $\left[\frac{\partial \vec{p}}{\partial t}\right]_{op}$ and $\left[\frac{\partial \vec{A}}{\partial t}\right]_{op}$ for the Dirac Hamiltonian,

$$H = c\tilde{\alpha} \cdot \left(\vec{p} - \frac{e}{c}\vec{A}(\vec{r})\right) + V(\vec{r}) + \tilde{\beta}mc^2$$

(d) Use the above results along with Newton's second law to find the force on an electron.

6.8. The Dirac equation in the presence of an electromagnetic field.

From Problem 6.5 above, the Dirac equation for a free particle can be written as,

$$\left(E - c\tilde{\alpha} \cdot \vec{p} - \tilde{\beta}mc^2\right)\Psi = 0$$

In the presence of an electromagnetic field, we make the replacements,

$$E \rightarrow E - e\Phi \text{ and } \vec{p} \rightarrow \vec{p} - \frac{e}{c}\vec{A} \Rightarrow c\vec{p} \rightarrow c\vec{p} - e\vec{A}.$$

giving us,

$$\left[(E - e\Phi) - \tilde{\alpha} \cdot \left(c\vec{p} - e\vec{A}\right) - \tilde{\beta}mc^2\right]\Psi = 0$$

Multiply from the left by $\left[(E - e\Phi) + \tilde{\alpha} \cdot \left(c\vec{p} - e\vec{A}\right) + \tilde{\beta}mc^2\right]$. Manipulate the result to show that the Dirac equation in the presence of an electromagnetic field can be written as,

$$\left[(E - e\Phi)^2 - \left(c\vec{p} - e\vec{A} \right)^2 - m^2c^4 + \hbar ec\left(\tilde{\sigma}_4 \cdot \vec{B} \right) - ei\hbar c\left(\tilde{\alpha} \cdot \vec{E} \right) \right] \Psi = 0$$

Give a physical interpretation of the factors on the LHS of this equation.

Hint: To interpret the last two terms on the LHS, take the non-relativistic limit.

6.9. Show that (6.75a) reduces to the appropriate form in the non-relativistic limit.

6.10. Relativistic photon-electron scattering.

The energy of a photon can be expressed as $E = h\nu$, where h is Planck's constant and ν is the frequency of the photon. The momentum of a photon is expressed as $h\nu/c$. Show that, if the photon scatters from a free electron (of mass m_e), the scattered photon has an energy,

$$E' = E\left[1 + \frac{E}{m_e c^2}(1 - \cos\theta) \right]^{-1}$$

Show also that the electron acquires a (relativistic) kinetic energy,

$$K = \frac{E^2}{m_e c^2}\left[\frac{1 - \cos\theta}{1 + \frac{E}{m_e c^2}(1 - \cos\theta)} \right]$$

Potentially useful formula:

The total relativistic energy of a particle of mass m, speed v, and relativistic 3-momentum $\vec{p}^{rel}$ is,

$$E_{TOT} = \sqrt{(mc^2)^2 + \left(c\vec{p}^{rel} \right)^2} = mc^2\gamma_v;\, \gamma_v = \left(1 - v^2/c^2 \right)^{-1/2}$$

Use your results to show that a free electron can neither absorb nor emit a photon.

6.11. Matrix elements of tensor products [3, 7, 16].

   Equations (6.38) and (D.15) give the matrix elements of a product of tensor operators from two different angular momentum spaces. In this problem, you will find the matrix elements of the products of different types of tensor operators.

(a) Find the matrix elements of an operator that is a scalar product of two tensor operators, each of which act on separate parts of a coupled system.

   Hint: The scalar product $T_1^{(k)} \cdot T_2^{(k)}$ is related to the tensor contraction $\left[T_1^{(k)} \odot T_2^{(k)} \right]_0^{(0)}$ by,

$$T_1^{(k)} \cdot T_2^{(k)} = (-1)^k (2k+1)^{1/2} \left[T_1^{(k)} \odot T_2^{(k)} \right]_0^{(0)}$$

(b) Find the reduced matrix elements of a tensor operator that acts only on space one of a two-space coupled system.
(c) Find the reduced matrix elements of a tensor operator that acts only on space two of a two-space coupled system.
(d) Find the reduced matrix elements of an operator that is the product of two tensor operators that act in a space that is not decomposable into two subspaces.

6.12. Tensor products and the hyperfine interaction in hydrogen [8, 11, 15].

   The hyperfine interaction in hydrogen results from the interaction of the magnetic moment of the proton $\vec{m}_I$ (due to its intrinsic spin $\vec{I}$) with the magnetic field created by the electron. We can use the results of Problem 3.8 to model the magnetic field produced by the electron's magnetic dipole moment $\vec{m}_s$ (due to its intrinsic spin $\vec{S}$). In other words, because the dipole term in the magnetic multipole expansion is the dominant term, the magnetic interaction due to the spin of the two particles can be approximated as a magnetic dipole-dipole interaction. To this interaction we must add the effect of the magnetic field due to the electron's (nonrelativistic) orbital motion $\vec{L}$. The Hamiltonian for the hyperfine interaction can then be modeled after the classical formula for the energy of interaction of the proton's magnetic moment with the magnetic field $\vec{B}$ produced by the electron evaluated at the location of the proton,

$$U = -\vec{m}_I \cdot \vec{B}$$

(a) Show that the Hamiltonian for the hyperfine interaction can be written as,

$$H_{hf} = -\frac{\mu_0}{4\pi}\left\{\frac{e}{m_e r^3}\vec{L}\cdot\vec{m}_I + \frac{1}{r^3}\left[3\left(\vec{m}_s\cdot\hat{r}\right)\left(\vec{m}_I\cdot\hat{r}\right) - \vec{m}_s\cdot\vec{m}_I\right] + \frac{8\pi}{3}\vec{m}_s\cdot\vec{m}_I\delta\left(\vec{r}\right)\right\}$$

where $\hat{r}$ is a unit vector that points from the proton to the electron. Note that the charge e is negative and that $\vec{m}_s$ points in the opposite direction of the electron's spin. The results of Problem 3.8(d) might be helpful.

Rewrite this Hamiltonian in terms of the angular momentum operators involved. Take the gyromagnetic ratio of the electron to be $g_e = 2$, and let the gyromagnetic ratio of the proton be g_p. Otherwise, treat the proton as a point particle (i.e., ignore the magnetic field that may exist inside the proton due to its internal structure). Also, ignore any relativistic effects.

(b) By first noting that $\hat{r} = C_q^{(1)}$ (see note in Problem 4.3), show that,

$$\left[\vec{S} - 3\left(\vec{S}\cdot\hat{r}\right)\hat{r}\right]_q^{(1)} = \sqrt{10}\left[S_{q_1}^{(1)} \otimes C_{q_2}^{(2)}\right]_q^{(1)}$$

where $C_q^{(k)}$ is defined in (4.11).

(c) Use your results from parts (a) and (b), and the results from Problem 6.11 to find the hyperfine splitting for hydrogen states of $l \neq 0$. That is, find the matrix elements,

$$\left\langle n, \left([ls]_j, I\right)F, m_F \middle| H_{hf} \middle| n, \left([l's]_j, I\right)F', m_F'\right\rangle$$

where F is the eigenvalue of the total angular momentum operator,

$$\vec{F} = \vec{L} + \vec{S} + \vec{I} = \vec{J} + \vec{I}$$

There is an elementary method for evaluating these matrix elements, but in this problem you will show how tensor methods deliver the same results. You will need tabulated values of some 3-j, 6-j, and 9-j symbols found in [3, 10, 89]. You will also need the quantity $\langle 1/r^3 \rangle$ found in Problem 5.7. Show that the matrix elements are diagonal in j, l, F, m_F.

6.13. Tensor products and the Zeeman effect in hydrogen [6, 8, 15].

The Hamiltonian H_Z for the Zeeman effect in hydrogen describes the interaction of the atom with an externally-imposed constant magnetic field $\vec{B}$,

$$H_Z = -\left(\vec{m}_L + \vec{m}_S + \vec{m}_I\right)\cdot\vec{B}$$

where $\vec{m}_L, \vec{m}_S, \vec{m}_I$ are the magnetic dipole moments due to the electron orbital motion $\vec{L}$, the electron spin $\vec{S}$, and the proton spin $\vec{I}$, respectively. Note that we are neglecting higher-order terms in the magnetic multipole expansion and other small effects.

(a) Make the case that we can neglect the interaction of the external magnetic field with the spin of the proton and write the remaining Zeeman Hamiltonian in terms of the resultant angular momentum operators.

(b) Use your results from Problem 6.11 to find the matrix elements,

$$\left\langle \left([ls]_j, I\right) F, m_F \middle| H_Z \middle| \left([l's]_{j'}, I\right) F', m'_F \right\rangle$$

where F is the eigenvalue of the total angular momentum operator,

$$\vec{F} = \vec{L} + \vec{S} + \vec{I} = \vec{J} + \vec{I}$$

Chapter 7
Angular Momentum Transfer Theory

A more general way of examining photoelectron angular distributions is via angular momentum transfer theory. Angular momentum transfer theory analyzes the pd process by focusing on the net angular momentum deposited by the photon into the target. The angular momentum transfer is equal to the difference between the angular momentum input to the target and the angular momentum output from the target, which (in this theory) is equal to the photoelectron's final-state orbital angular momentum. Angular momentum transfer theory can account for non-isotropic effects (among other phenomena) and so should be considered a more complete theory (in that respect) than some of the treatments presented earlier.

Consider a standard, one-photon photodetachment reaction in which a photon collides with an unpolarized target and detaches an electron (we use a photodetachment reaction as an example, but we could just as easily have used a photoionization process):

$$A^-\left(L_{A^-}, S_{A^-}, j_{A^-}, \pi_{A^-}\right) + \gamma\left(j_\gamma, m_\gamma, \pi_\gamma\right) \rightarrow A(L_A, S_A, j_A, \pi_A)$$
$$+ e^-\left(l, s, j_e, \pi_e = (-1)^l\right) \qquad (7.1)$$

where spin S, orbital angular momentum L, total angular momentum j, and parity π are labeled with respect to the initial anionic state A^-, and residual neutral core A, and l, s, j_e label the photoelectron's post-reaction orbital, spin, and total angular momentum, respectively. The total angular momentum, projection quantum number, and parity of the photon γ are similarly labeled. We are working here only with states which are well-characterized by these quantum numbers, that is, by states that are defined in the LS-coupling approximation. Note that the parity of the photoelectron

---

The original version of this chapter was revised. The correction to this chapter is available at https://doi.org/10.1007/978-3-031-08027-2_11

© The Author(s), under exclusive license to Springer Nature Switzerland AG 2022, Corrected Publication 2023
V. T. Davis, *Introduction to Photoelectron Angular Distributions*, Springer Tracts in Modern Physics 286, https://doi.org/10.1007/978-3-031-08027-2_7

π_e is a function of its post-pd-reaction orbital angular momentum l. The photodetachment interaction can be characterized by the total angular momentum $\vec{J}$ which is conserved in the reaction

$$\vec{J} = \vec{j}_{A^-} + \vec{j}_\gamma = \vec{j}_A + \vec{j}_e \qquad (7.2)$$

or by the angular momentum transferred during the pd reaction to the residual neutral core

$$\vec{j}_t \equiv \vec{j}_\gamma - \vec{l} = \left(\vec{j}_A + \vec{s} \right) - \vec{j}_{A^-} \qquad (7.3)$$

Here we assume that the target A^- is unpolarized (i.e., its orientation has not been preselected), and the angular momentum/orientation of the residual neutral core (and the photoelectron spin-which can be regarded as still coupled to the residual core) remain unobserved. One advantage of angular momentum transfer theory is that it allows for averaging over magnetic quantum numbers of reactants whose orientation is not observed. The transferred angular momentum $\vec{j}_t$ can either be used to eliminate $\vec{J}$ entirely or used in tandem with it. In any case, one can always introduce one or the other via the appropriate recoupling transformation, as will be shown below.

Given the conditions stated above, the differential cross-section for photodetachment $d\sigma_{pd}/d\Omega$ is proportional to the scattering matrix for the reaction, averaged over all possible initial target orientations, and summed over all possible (undetected) final states of the residual neutral core [87, 88]:

$$\frac{d\sigma_{pd}}{d\Omega} \propto \frac{1}{(2j_{A^-}+1)} \sum_{m_{A^-},m_A} \left| \sum_{lm} Y_{lm}(\theta,\phi) \left\langle lm, j_A m_A \left| S_q^{(k)} \right| j_{A^-} m_{A^-} \right\rangle \right|^2 \qquad (7.4)$$

Equation (7.4) is structured so that $S(k,q)$ represents the $2q+1$ spherical tensor operators responsible for the pd process. The initial state is characterized by the symbol $|i\rangle = |j_{A^-} m_{A^-}\rangle$, keeping in mind that the properties of the photon are contained in $S(k,q)$. The final state is given by $|f\rangle = |lm, j_A m_A\rangle$, with the caveat that the spin of the photoelectron, since it remains unobserved, is considered to be coupled to the residual neutral core. The quantity in the bracket represents the probability amplitude for photoelectron emission into the direction (θ,ϕ) while in the state l, m and we notice that it has therefore been expanded in a series of spherical harmonics $Y_{lm}(\theta,\phi)$. As stated above, the differential cross-section for photodetachment is then proportional to the square of the probability amplitude, averaged over all initial possible target orientations, and summed over all possible undetected final states. In short, the essence of angular momentum transfer theory lies in performing the indicated sums over the magnetic quantum numbers and expanding the results into a series of spherical harmonics, which serve as convenient functions with which to represent the final angular distribution [87, 88].

We begin our operation by recoupling to the basis of the total angular momentum $\vec{J}$ [c.f. (2.119)] and executing the indicated square modulus:

$$\frac{d\sigma_{pd}}{d\Omega} \propto \frac{1}{(2j_{A^-}+1)} \sum_{m_{A^-},m_A} \left| \sum_{lm} Y_{lm}(\theta,\phi) \sum_{JM} \langle l,j_A;m,m_A|J,M\rangle \langle JM|S_q^{(k)}|j_A\text{-}m_{A^-}\rangle \right|^2$$

$$= \frac{1}{(2j_{A^-}+1)} \sum_{m_{A^-},m_A} \sum_{lm} \sum_{l'm'} \sum_{JM} \sum_{J'M'} Y_{lm} Y_{l'm'}^* \langle l,j_A;m,m_A|J,M\rangle \langle l',j_A;m',m_A|J',M'\rangle$$

$$\times \langle JM|S_q^{(k)}|j_A\text{-}m_{A^-}\rangle \langle J'M'|S_q^{(k)}|j_A\text{-}m_{A^-}\rangle^*$$

$$(7.5)$$

Operating on the matrix elements via the Wigner-Eckart theorem [3],

$$\langle JM|S_q^{(k)}|j_A\text{-}m_{A^-}\rangle \langle J'M'|S_q^{(k)}|j_A\text{-}m_{A^-}\rangle^*$$

$$= (-1)^{J-M} \begin{pmatrix} J & k & j_{A^-} \\ -M & q & m_{A^-} \end{pmatrix} (-1)^{J'-M'} \begin{pmatrix} J' & k & j_{A^-} \\ -M' & q & m_{A^-} \end{pmatrix} \langle J\|S^{(k)}\|j_{A^-}\rangle \langle J'\|S^{(k)}\|j_{A^-}\rangle^*$$

$$(7.6)$$

and then recasting all C-G coefficients as 3-j symbols [c.f. (2.144a)], gives us [3, 89]

$$\frac{d\sigma_{pd}}{d\Omega} \propto \frac{1}{(2j_{A^-}+1)} \sum_{JJ'} \sqrt{(2J+1)} \sqrt{(2J'+1)} \langle J\|S^{(k)}\|j_{A^-}\rangle \langle J'\|S^{(k)}\|j_{A^-}\rangle^*$$

$$\times \sum_{lm}\sum_{l'm'}\sum_{m_{A^-},m_A MM'} \times \begin{bmatrix} Y_{lm} Y_{l'm'}^* (-1)^{l+l'-2j_A+J+J'} \\[4pt] \times \begin{pmatrix} l & j_A & J \\ m & m_A & -M \end{pmatrix} \begin{pmatrix} l' & j_A & J' \\ m' & m_A & -M' \end{pmatrix} \\[6pt] \times \underbrace{\begin{pmatrix} J & k & j_{A^-} \\ -M & q & m_{A^-} \end{pmatrix}}_{} \qquad \underbrace{\begin{pmatrix} J' & k & j_{A^-} \\ -M' & q & m_{A^-} \end{pmatrix}}_{} \\[6pt] (-1)^{j_{A^-}+k+J}\begin{pmatrix} j_{A^-} & k & J \\ -m_{A^-} & -q & M \end{pmatrix} (-1)^{j_{A^-}+k+J'}\begin{pmatrix} j_{A^-} & k & J' \\ -m_{A^-} & -q & M' \end{pmatrix} \end{bmatrix}$$

$$(7.7)$$

where we also used (2.145c) as indicated. We recouple from the $\vec{J}$ basis to the $\vec{j}_t$ basis as follows, using (C.74) [3, 87–89]

$$
\begin{pmatrix} l & j_A & J \\ m & m_A & -M \end{pmatrix} \begin{pmatrix} j_{A^-} & k & J \\ -m_{A^-} & -q & M \end{pmatrix}
$$

$$
= \sum_{j_t} \left[\begin{array}{l} (2j_t + 1)(-1)^{l+j_A-J+j_{A^-}+k+j_t-m+m_{A^-}} \\ \times \begin{Bmatrix} l & j_A & J \\ j_{A^-} & k & j_t \end{Bmatrix} \begin{pmatrix} k & j & j_t \\ -q & m & m_t \end{pmatrix} \begin{pmatrix} j_A & j_{A^-} & j_t \\ m_A & -m_{A^-} & -m_t \end{pmatrix} \end{array} \right] \quad (7.8\text{a})
$$

$$
\begin{pmatrix} l' & j_A & J' \\ m' & m_A & -M' \end{pmatrix} \begin{pmatrix} j_{A^-} & k & J' \\ -m_{A^-} & -q & M' \end{pmatrix}
$$

$$
= \sum_{j'_t} \left[\begin{array}{l} (2j'_t + 1)(-1)^{l'+j_A-J'+j_{A^-}+k+j'_t-m'+m_{A^-}} \\ \times \begin{Bmatrix} l' & j_A & J' \\ j_{A^-} & k & j'_t \end{Bmatrix} \begin{pmatrix} k & l' & j'_t \\ -q & m' & m'_t \end{pmatrix} \begin{pmatrix} j_A & j_{A^-} & j'_t \\ m_A & -m_{A^-} & -m'_t \end{pmatrix} \end{array} \right] \quad (7.8\text{b})
$$

Combining (7.7) and (7.8a and 7.8b),

$$
\frac{d\sigma_{pd}}{d\Omega} \propto \frac{1}{(2j_{A^-}+1)} \sum_{JJ'} \sqrt{(2J+1)} \sqrt{(2J'+1)} \langle J \| S^{(k)} \| j_{A^-} \rangle \langle J' \| S^{(k)} \| j_{A^-} \rangle^*
$$

$$
\times \sum_{lm} \sum_{l'm'} \sum_{m_{A^-}, m_A} \sum_{MM'} Y_{lm} Y^*_{l'm'} (-1)^{l+l'-2j_A+J+J'}
$$

$$
\times \sum_{j_t j'_t} \left[\begin{array}{l} (-1)^{j_{A^-}+k+J}(-1)^{j_{A^-}+k+J'}(2j_t+1)(2j'_t+1) \\[2mm] \times(-1)^{l+l'-J-J'+2j_A+2k+j_t+j'_t-m-m'} \overbrace{(-1)^{2(j_{A^-}+m_{A^-})}}^{=1 \text{ for all cases}} \\[2mm] \times \begin{Bmatrix} l & j_A & J \\ j_{A^-} & k & j_t \end{Bmatrix} \begin{Bmatrix} l' & j_A & J' \\ j_{A^-} & k & j'_t \end{Bmatrix} \begin{pmatrix} k & l & j_t \\ -q & m & m_t \end{pmatrix} \begin{pmatrix} k & l' & j'_t \\ -q & m' & m'_t \end{pmatrix} \\[4mm] \times \underbrace{\begin{pmatrix} j_A & j_{A^-} & j_t \\ m_A & -m_{A^-} & -m_t \end{pmatrix}}_{(-1)^{j_A+j_{A^-}+j_t}\begin{pmatrix} j_A & j_{A^-} & j_t \\ -m_A & m_{A^-} & m_t \end{pmatrix}} \underbrace{\begin{pmatrix} j_A & j_{A^-} & j'_t \\ m_A & -m_{A^-} & m'_t \end{pmatrix}}_{(-1)^{j_A+j_{A^-}+j'_t}\begin{pmatrix} j_A & j_{A^-} & j'_t \\ -m_A & m_{A^-} & m'_t \end{pmatrix}} \end{array} \right]
$$

$$
(7.9)
$$

Using the orthogonality properties of the 3-j symbols (see Problem 2.8) [3, 10, 89]:

$$\sum_{m_{A^-},m_A} \begin{pmatrix} j_A & j_{A^-} & j_t \\ -m_A & m_{A^-} & m_t \end{pmatrix} \begin{pmatrix} j_A & j_{A^-} & j_t' \\ -m_A & m_{A^-} & m_t' \end{pmatrix} = (2j_t + 1)^{-1}\delta_{j_t,j_t'}\delta_{m_t,m_t'}$$

(7.10)

The recoupling in (7.8) results in the fact that there are no longer any terms in M and M' in (7.9). The sums over those terms will therefore serve only to bring in the multiplicative factors $(2J + 1)(2J' + 1)$, leaving us with

$$\frac{d\sigma_{pd}}{d\Omega} \propto \frac{1}{(2j_{A^-} + 1)} \sum_{JJ'} (2J + 1)^{\frac{3}{2}}(2J' + 1)^{\frac{3}{2}} \langle J\|S^{(k)}\|j_{A^-}\rangle\langle J'\|S^{(k)}\|j_{A^-}\rangle^*$$

$$\times \sum_{lm}\sum_{l'm'} Y_{lm}Y_{l'm'}^*(-1)^{2l+2l'}$$

$$\times \sum_{j_t} \left[\begin{array}{c} (-1)^{j_{A^-}+k+J}(-1)^{j_{A^-}+k+J'}(2j_t + 1)(-1)^{2j_{A^-}+2j_A+2k+4j_t-m-m'} \\ \times \begin{Bmatrix} l & j_A & J \\ j_A & k & j_t \end{Bmatrix} \begin{Bmatrix} l' & j_A & J' \\ j_{A^-} & k & j_t \end{Bmatrix} \begin{pmatrix} k & l & j_t \\ -q & m & m_t \end{pmatrix} \begin{pmatrix} k & l' & j_t \\ -q & m' & m_t \end{pmatrix} \end{array} \right]$$

(7.11)

From the two 3-j symbols, we observe that $\{m_t = q - m; m_t = q - m'\} \Rightarrow m = m'$ which allows us to eliminate the sum over m':

$$\frac{d\sigma_{pd}}{d\Omega} \propto \frac{1}{(2j_{A^-} + 1)} \sum_{JJ'} (2J + 1)^{\frac{3}{2}}(2J' + 1)^{\frac{3}{2}} \langle J\|S\|j_{A^-}\rangle\langle J'\|S^{(k)}\|j_{A^-}\rangle^*$$

$$\times \sum_{l'm} Y_{lm}Y_{l'm}^* \underbrace{(-1)^{2(l+l')}}_{=1;\, l,l' = \text{integer}}$$

$$\times \sum_{j_t} \left[\begin{array}{c} (-1)^{j_{A^-}+k+J}(-1)^{j_{A^-}+k+J'}(2j_t + 1) \overbrace{(-1)^{2(j_{A^-}+j_A+j_t)}}^{=1} \overbrace{(-1)^{2(k+j_t-m)}}^{=1} \\ \times \begin{Bmatrix} l & j_A & J \\ j_{A^-} & k & j_t \end{Bmatrix} \begin{Bmatrix} l' & j_A & J' \\ j_{A^-} & k & j_t \end{Bmatrix} \begin{pmatrix} k & l & j_t \\ -q & m & q-m \end{pmatrix} \begin{pmatrix} k & l' & j_t \\ -q & m & q-m \end{pmatrix} \end{array} \right]$$

(7.12)

We make use of the following identity, which is easily derived using (2.155) and (2.60) [see also Problem 4.6(a)] [3, 87–89]:

$$Y_{lm}Y_{l'm}^* = Y_{lm}(-1)^m Y_{l'-m}$$

$$= \sum_L (-1)^m \frac{\sqrt{(2l + 1)(2l' + 1)}}{4\pi} \langle l, l'; 0, 0|L, 0\rangle\langle l, l'; m, -m|L, 0\rangle P_L(\cos\theta)$$

(7.13)

to get

$$\frac{d\sigma_{pd}}{d\Omega} \propto \frac{1}{(2j_{A^-}+1)} \sum_{JJ'} (2J+1)^{\frac{3}{2}}(2J'+1)^{\frac{3}{2}} \langle J \| S^{(k)} \| j_{A^-} \rangle \langle J' \| S^{(k)} \| j_{A^-} \rangle^*$$

$$\times \sum_{l'm}\sum_{L} (-1)^m \frac{\sqrt{(2l+1)(2l'+1)}}{4\pi} \langle l,l';0,0|L,0\rangle \quad \underbrace{\langle l,l';m,-m|L,0\rangle}_{=(-1)^{l-l'}\begin{pmatrix} l & l' & L \\ m & -m & 0 \end{pmatrix}\sqrt{2L+1}} \quad P_L(\cos\theta)$$

$$\times \sum_{j_t} \left[\underbrace{\overbrace{(-1)^{j_{A^-}-k-J}\overbrace{(-1)^{j_{A^-}-k-J'}}}^{(-1)^{j_{A^-}-+k+J}(-1)^{j_{A^-}-+k+J'}}}_{} \times (2j_t+1) \begin{Bmatrix} l & j_A & J \\ j_A & k & j_t \end{Bmatrix} \begin{Bmatrix} l' & j_A & J' \\ j_{A^-} & k & j_t \end{Bmatrix} \right.$$

$$\left. \times \begin{pmatrix} k & l & j_t \\ -q & m & q-m \end{pmatrix} \underbrace{\begin{pmatrix} k & l' & j_t \\ -q & m & q-m \end{pmatrix}}_{\begin{pmatrix} l' & k & j_t \\ -m & q & m-q \end{pmatrix}} \right]$$

$$\tag{7.14}$$

Now to eliminate the sum over the last remaining magnetic quantum number by noting that what should be a triple sum in (7.15) below [c.f. (C.73)] in our case reduces to the sum over one term, m (see problem 7.1) [3, 87–89],

$$\sum_m (-1)^m \begin{pmatrix} k & l & j_t \\ -q & m & q-m \end{pmatrix} \begin{pmatrix} l' & k & j_t \\ -m & q & m-q \end{pmatrix} \begin{pmatrix} l & l' & L \\ m & -m & 0 \end{pmatrix}$$

$$= (-1)^{2k+l-j_t+l'+L+q} \underbrace{\begin{Bmatrix} k & l & j_t \\ l' & k & L \end{Bmatrix}}_{\begin{Bmatrix} k & k & L \\ l & j' & j_t \end{Bmatrix}} \underbrace{\begin{pmatrix} k & k & L \\ -q & q & 0 \end{pmatrix}}_{(-1)^{2k+L}\begin{Bmatrix} k & k & L \\ q & -q & 0 \end{Bmatrix}} \tag{7.15}$$

Putting it all together,

$$\frac{d\sigma_{pd}}{d\Omega} \propto \frac{1}{(2j_{A^-}+1)}\sum_{ll'}\sum_{j_t}\sum_{J}(-1)^{j_{A^-}-k-J}(2J+1)^{\frac{3}{2}}\langle J\|S^{(k)}\|j_{A^-}\rangle \begin{Bmatrix} l & j_A & J \\ j_{A^-} & k & j_t \end{Bmatrix}$$

$$\times \sum_{J'}(-1)^{j_{A^-}-k-J'}(2J'+1)^{\frac{3}{2}}\langle J'\|S^{(k)}\|j_{A^-}\rangle^* \begin{Bmatrix} l' & j_A & J' \\ j_{A^-} & k & j_t \end{Bmatrix}$$

$$\times \sum_{L}(-1)^{-j_t+q}\frac{\sqrt{(2l+1)(2l'+1)}}{4\pi}\langle l,0;l',0|L,0\rangle P_L(\cos\theta)(2j_t+1)$$

$$\times \underbrace{(-1)^{2(2k+L)}}_{=1}\underbrace{(-1)^{2l}}_{=1:\,l\text{ integer}} \begin{Bmatrix} k & k & L \\ l & l' & j_t \end{Bmatrix}\underbrace{\begin{Bmatrix} k & k & L \\ q & -q & 0 \end{Bmatrix}\sqrt{(2L+1)}}_{(-1)^{k-k}\langle k,k;q,-q|L,0\rangle}$$

(7.16)

The differential cross-section is usually presented in the following form [87, 88]:

$$\frac{d\sigma_{pd}}{d\Omega} = \frac{K}{(2j_{A^-}+1)}\sum_{j_t}\sum_{ll'}\langle J\|\bar{S}(j_t)\|j_{A^-}\rangle\langle j_{A^-}\|\bar{S}^\dagger(j_t)\|J'\rangle\Theta(j_t;kq;ll';\theta) \quad (7.17)$$

where the scattering amplitudes/matrices are

$$\langle J\|\bar{S}(j_t)\|j_{A^-}\rangle \equiv \sum_{J}(-1)^{j_{A^-}-k-J}(2J+1)^{\frac{3}{2}}\langle J\|S^{(k)}\|j_{A^-}\rangle \begin{Bmatrix} l & j_A & J \\ j_{A^-} & k & j_t \end{Bmatrix} \quad (7.18a)$$

$$\langle j_{A^-}\|\bar{S}^\dagger(j_t)\|J'\rangle \equiv \sum_{J'}(-1)^{j_{A^-}-k-J'}(2J'+1)^{\frac{3}{2}}\langle J'\|S^{(k)}\|j_{A^-}\rangle^* \begin{Bmatrix} l' & j_A & J' \\ j_{A^-} & k & j_t \end{Bmatrix}$$

(7.18b)

and $\Theta(j_t;kq;ll';\theta)$ is called the geometrical function

$$\Theta(j_t;kq;ll';\theta) \equiv \sum_{L}(-1)^{-j_t+q}\frac{\sqrt{(2l+1)(2l'+1)}}{4\pi}$$

$$\times (2j_t+1)\langle l,l';0,0|L,0\rangle\langle k,k;q,-q|L,0\rangle P_L(\cos\theta)\begin{Bmatrix} k & k & L \\ l & l' & j_t \end{Bmatrix}$$

(7.18c)

The constant of proportionality K is chosen in terms of the wavelength of the photon λ (divided by 2π) to match boundary conditions between incoming and outgoing states [87, 88]

$$K = 3\pi\lambdabar^2; \lambdabar \equiv \lambda/2\pi \tag{7.19}$$

Notice from (7.17) that the differential cross-section has taken the form of an incoherent sum over the angular momentum transfer quantum number [3, 87, 88]:

$$\frac{d\sigma_{pd}}{d\Omega} = \sum_{j_t} \frac{d\sigma(j_t, \theta)}{d\Omega} \tag{7.20}$$

To recap, we have now a general expression for the photodetachment process of (7.1) in terms of separate components characterized by alternate magnitudes of $\vec{j}_t$. The allowed values of $\vec{j}_t$ must be consistent with the conservation of the total angular momentum $\vec{J}$ and the conservation of parity [87, 88, 90, 91][1]:

$$\pi_{A-}\pi_\gamma = \pi_A\pi_e \tag{7.21}$$

For most purposes, the most interesting and applicable event is the low-energy pd event ($E_\gamma \leq 100\,\text{eV}$), for which the electric dipole approximation is valid and for which the incident photon carries an angular momentum $j_\gamma = 1$ and parity $\pi_\gamma = -1$ to the target atom. For this case (7.2) and (7.3) read [90]

$$\vec{J} = \vec{j}_{A-} + \vec{1} = \vec{j}_A + \vec{j}_e \tag{7.22}$$

and

$$\vec{j}_t \equiv \vec{1} - \vec{l} = \left(\vec{j}_A + \vec{s}\right) - \vec{j}_{A-} \tag{7.23}$$

In addition, because the electric dipole interaction is spin-independent (under the LS-coupling approximation), the angular momentum imparted by the photon affects only the orbital angular momentum of the system. As a result, the total spin $\vec{S} = \vec{S}_{A-}$ and orbital angular momentum $\vec{L}$ are conserved separately [90]:

$$\vec{S}_{A-} = \vec{S}_A + \vec{s}; \quad \vec{L} = \vec{L}_{A-} + \vec{1} = \vec{L}_A + \vec{l} \tag{7.24}$$

and (7.3) reduces to [90]

$$\vec{j}_t \equiv \vec{1} - \vec{l} = \vec{L}_A - \vec{L}_{A-} \tag{7.25}$$

---

[1]The fact that photoelectron angular distributions we have seen up to this point have no linear term in $\cos\theta$ is due to parity conservation [1].

Equation (7.21) now takes on a particularly meaningful form for our discussion [90]:

$$\pi_{A-}\underbrace{\pi_{\gamma}}_{-1} = \pi_A\underbrace{\pi_e}_{(-1)^l} \Rightarrow \pi_{A-}(-1)(-1) = \pi_A(-1)^l(-1) \Rightarrow \pi_{A-} = \pi_A(-1)^{l+1}$$

$$\Rightarrow \pi_{A-}\pi_A = \underbrace{(\pi_A)^2}_{=1}(-1)^{l+1}$$

$$\Rightarrow \pi_{A-}\pi_A = (-1)^{l+1} \tag{7.26}$$

In the dipole approximation, and under LS-coupling, allowed values of j_t are determined by (7.25) and (7.26) [90].

For a photodetachment caused by a linearly polarized photon, we have $k = 1$; $q = 0$. The geometrical function now reads

$$\Theta(j_t; 10; ll'; \theta) = (-1)^{j_t}\frac{\sqrt{(2l+1)(2l'+1)}}{4\pi}(2j_t+1)$$
$$\times \sum_L \langle l, l'; 0, 0|L, 0\rangle\langle 1, 1; 0, 0|L, 0\rangle P_L(\cos\theta)\begin{Bmatrix} 1 & 1 & L \\ l & l' & j_t \end{Bmatrix} \tag{7.27}$$

Triangular conditions and symmetry properties of the 6-j symbol limit the sum to just two terms, $L = 0, 2$,

$$\Theta(j_t; 10; ll'; \theta) = (-1)^{j_t}\frac{\sqrt{(2l+1)(2l'+1)}}{4\pi}(2j_t+1)$$

$$\times \left[\underbrace{\langle 1,0;1,0|0,0\rangle}_{=-\frac{1}{\sqrt{3}}}\underbrace{\langle l,0;l',0|0,0\rangle}_{=(-1)^l(2l+1)^{-\frac{1}{2}}}\underbrace{P_0(\cos\theta)}_{=1}\overbrace{\begin{Bmatrix} 1 & 1 & 0 \\ l & l' & j_t \end{Bmatrix}}^{=(-1)^{j_t+l+1}\sqrt{\frac{1}{3(2l+1)}}\delta_{ll'}} \right.$$

$$\left. + \underbrace{\langle 1,0;1,0|2,0\rangle}_{=\sqrt{\frac{2}{3}}}\langle l,0;l',0|2,0\rangle P_2(\cos\theta)\begin{Bmatrix} 1 & 1 & 2 \\ l & l' & j_t \end{Bmatrix} \right] \tag{7.28}$$

Parity-unfavored transitions are defined as transitions for which $j_t + l + (j_\gamma = 1)=$ odd, or $j_t = l, l'$. For this case, (7.26) implies [87, 88, 90–93]

$$\pi_{A-}\pi_A = (-1)^{j_t+1} \tag{7.29}$$

and the geometric function becomes

$$\Theta(j_t;10;j_tj_t;\theta) = (-1)^{j_t}\frac{(2j_t+1)^2}{4\pi}\left[\begin{array}{c}\left(-\frac{1}{\sqrt{3}}\right)\langle j_t,j_t;0,0|0,0\rangle\begin{Bmatrix}1 & 1 & 0 \\ j_t & j_t & j_t\end{Bmatrix} \\ +\sqrt{\frac{2}{3}}\langle j_t,j_t;0,0|2,0\rangle P_2(\cos\theta)\begin{Bmatrix}1 & 1 & 2 \\ j_t & j_t & j_t\end{Bmatrix}\end{array}\right]$$

$$(7.30)$$

Tabulated values of C-G coefficients and 6-j symbols [see also (C.75) and (C.76)] allow us to compute the following [3, 7, 10]:

$$\langle j_t,j_t;0,0|0,0\rangle = (-1)^{j_t}(2j_t+1)^{-\frac{1}{2}} \tag{7.31a}$$

$$\langle j_t,j_t;0,0|2,0\rangle = (-1)^{j_t}\sqrt{\frac{5}{(2j_t+1)}}\frac{(-j_t)(j_t+1)}{\sqrt{(2j_t+3)(j_t+1)(j_t)(2j_t-1)}} \tag{7.31b}$$

$$\begin{Bmatrix}1 & 1 & 0 \\ j_t & j_t & j_t\end{Bmatrix} = (-1)\frac{1}{\sqrt{3}}\frac{1}{\sqrt{(2j_t+1)}} \tag{7.31c}$$

$$\begin{Bmatrix}1 & 1 & 2 \\ j_t & j_t & j_t\end{Bmatrix} = (-1)^{2j_t+1}\frac{2[6-8j_t(j_t+1)]}{\sqrt{5!(2j_t+3)(2j_t+2)(2j_t+1)(2j_t)(2j_t-1)}} \tag{7.31d}$$

Example 7.1
Verify (7.31a).

Equation (7.31a) can be verified by combining (2.146) and (2.144a):

$$\begin{pmatrix}j & j & 0 \\ m & -m & 0\end{pmatrix} = \langle j,j;m,-m|j_3,0\rangle = (2j+1)^{-\frac{1}{2}}(-1)^{j-m}$$

$$\Rightarrow \langle j_t,j_t;0,0|0,0\rangle = (-1)^{j_t}(2j_t+1)^{-\frac{1}{2}}$$

∎

Combining (7.30) and (7.31a, 7.31b, 7.31c and 7.31d),

$$
\Theta(j_t; 10; j_t j_t; \theta) = (-1)^{j_t}
$$

$$
\times \frac{(2j_t + 1)^2}{4\pi}
\begin{bmatrix}
\left(-\frac{1}{\sqrt{3}}\right)(-1)^{j_t+1} \frac{1}{\sqrt{(2j_t + 1)}} \frac{1}{\sqrt{3}} \frac{1}{\sqrt{(2j_t + 1)}} \\[2ex]
+ \sqrt{\frac{2}{3}}(-1)^{3j_t+1} \sqrt{\frac{5}{(2j_t + 1)}} \frac{(-j_t)(j_t + 1)}{\sqrt{(2j_t + 3)(j_t + 1)(j_t)(2j_t - 1)}} \\[2ex]
\times \frac{2[6 - 8j_t(j_t + 1)]}{\sqrt{5!(2j_t + 3)(2j_t + 2)(2j_t + 1)(2j_t)(2j_t - 1)}} P_2(\cos\theta)
\end{bmatrix}
$$

$$(7.32)$$

which, after some algebra, reduces to [88]

$$
\Theta(j_t; 10; j_t j_t; \theta) = \frac{(2j_t + 1)}{12\pi}[1 - P_2(\cos\theta)]
\tag{7.33}
$$

Parity-favored transitions are defined as transitions for which $j_t + l + (j_\gamma = 1) =$ even, and which consist of two functions diagonal in $l, l'; l = l' = j_t \pm 1$ and one interference term $l = j_t + 1;\ l' = j_t - 1$. For this case, (7.26) implies [87, 88, 90–93]

$$
\pi_{A-}\pi_A = (-1)^{j_t}
\tag{7.34}
$$

Tabulated values of C-G coefficients and 6-j symbols allow us to compute the following factors [3, 7, 10]:

$$
\langle j_t + 1, j_t + 1; 0, 0 | 2, 0 \rangle = (-1)^{j_t} \sqrt{\frac{5(j_t + 1)(j_t + 2)}{(2j_t + 1)(2j_t + 3)(2j_t + 5)}}
\tag{7.35a}
$$

$$
\langle j_t - 1, j_t - 1; 0, 0 | 2, 0 \rangle = (-1)^{j_t} \sqrt{\frac{5(j_t - 1)(j_t)}{(2j_t - 3)(2j_t - 1)(2j_t + 1)}}
\tag{7.35b}
$$

$$
\langle j_t + 1, j_t - 1; 0, 0 | 2, 0 \rangle = (-1)^{j_t-1} \sqrt{\frac{15(j_t + 1)(j_t)}{2(2j_t - 1)(2j_t + 1)(2j_t + 3)}}
\tag{7.35c}
$$

$$
\begin{Bmatrix} 1 & 1 & 2 \\ j_t + 1 & j_t + 1 & j_t \end{Bmatrix} = (-1)^{2j_t+2} \frac{2(j_t + 2)(2j_t + 5)}{\sqrt{5!(2j_t + 5)(2j_t + 4)(2j_t + 3)(2j_t + 2)(2j_t + 1)}}
\tag{7.35d}
$$

$$\left\{\begin{matrix} 1 & 1 & 2 \\ j_t - 1 & j_t - 1 & j_t \end{matrix}\right\} = (-1)^{2j_t} \frac{2(j_t - 1)(2j_t - 3)}{\sqrt{5!(2j_t + 1)(2j_t)(2j_t - 1)(2j_t - 2)(2j_t - 3)}} \tag{7.35e}$$

$$\left\{\begin{matrix} 1 & 1 & 2 \\ j_t + 1 & j_t - 1 & j_t \end{matrix}\right\} = [5(2j_t + 1)]^{-1/2} \tag{7.35f}$$

Combining (7.28) and the appropriate iterations of (7.35) (and after grinding through some algebra), we get [87, 88]

$$\Theta(j_t; 10; j_t + 1 j_t + 1; \theta) = \frac{(2j_t + 1)}{12\pi} \left[1 + \frac{(j_t + 2)}{(2j_t + 1)} P_2(\cos\theta)\right] \tag{7.36a}$$

$$\Theta(j_t; 10; j_t - 1 j_t - 1; \theta) = \frac{(2j_t + 1)}{12\pi} \left[1 + \frac{(j_t - 1)}{(2j_t + 1)} P_2(\cos\theta)\right] \tag{7.36b}$$

$$\Theta(j_t; 10; j_t + 1 j_t - 1; \theta) = \frac{-1}{4\pi} \left[\sqrt{(j_t)(j_t + 1)} P_2(\cos\theta)\right] \tag{7.36c}$$

the last term being the interference term.

Combining (7.17), (7.18a, 7.18b and 7.18c), (7.19), (7.33), and (7.36a, 7.36b and 7.36c), and accounting for all the possible values of l, l' in the summation $\sum\limits_{ll'}^{j_t, j_t \pm 1}$, we get, for a photodetachment caused by a linearly polarized photon:

$$\frac{d\sigma_{pd}}{d\Omega} = \frac{3\pi\lambda^2}{(2j_{A^-} + 1)} \sum_{j_t} \left\{ \begin{array}{l} |\bar{S}_+(j_t)|^2 \frac{(2j_t + 1)}{12\pi}\left[1 + \frac{(j_t + 2)}{(2j_t + 1)}P_2(\cos\theta)\right] \\ + |\bar{S}_-(j_t)|^2 \frac{(2j_t + 1)}{12\pi}\left[1 + \frac{(j_t - 1)}{(2j_t + 1)}P_2(\cos\theta)\right] \\ - [\bar{S}_+(j_t)\bar{S}_-^*(j_t) + \bar{S}_-(j_t)\bar{S}_+^*(j_t)]\frac{3}{12\pi}\sqrt{(j_t)(j_t + 1)}P_2(\cos\theta) \\ + |\bar{S}_0(j_t)|^2 \frac{(2j_t + 1)}{12\pi}[1 + (-1)P_2(\cos\theta)] \end{array} \right\}$$

$$= \frac{3\pi\lambda^2}{12\pi(2j_{A^-} + 1)} \sum_{j_t} \left\{ \begin{array}{l} \left(|\bar{S}_+(j_t)|^2 + |\bar{S}_-(j_t)|^2\right)(2j_t + 1) + P_2(\cos\theta) \\ \times \left[\begin{array}{l} |\bar{S}_+(j_t)|^2(j_t + 2) + |\bar{S}_-(j_t)|^2(j_t - 1) \\ -3\sqrt{(j_t)(j_t + 1)}(\bar{S}_+(j_t)\bar{S}_-^*(j_t) + \bar{S}_-(j_t)\bar{S}_+^*(j_t)) \end{array}\right] \\ + |\bar{S}_0(j_t)|^2(2j_t + 1)[1 + (-1)P_2(\cos\theta)] \end{array} \right\}$$

$$\Rightarrow \frac{d\sigma_{pd}}{d\Omega} = \frac{3\pi\lambda^2}{12\pi(2j_{A^-}+1)}$$

$$\times \sum_{j_t} \left\{ \begin{array}{l} \left[\left(|\bar{S}_+(j_t)|^2 + |\bar{S}_-(j_t)|^2 \right)(2j_t+1) \right] \\ \times \left[1 + \frac{\left[\begin{array}{l} |\bar{S}_+(j_t)|^2(j_t+2) + |\bar{S}_-(j_t)|^2(j_t-1) - \\ 3\sqrt{(j_t)(j_t+1)}\left(\bar{S}_+(j_t)\bar{S}_-^*(j_t) + \bar{S}_-(j_t)\bar{S}_+^*(j_t) \right) \end{array} \right]}{\left(|\bar{S}_+(j_t)|^2 + |\bar{S}_-(j_t)|^2 \right)(2j_t+1)} P_2(\cos\theta) \right] \\ + |\bar{S}_0(j_t)|^2(2j_t+1)[1+(-1)P_2(\cos\theta)] \end{array} \right\}$$

(7.37)

Now we do a little rearranging

$$\frac{d\sigma_{pd}}{d\Omega} = \frac{1}{4\pi}\sum_{j_t} \left\{ \begin{array}{l} \frac{\pi\lambda^2(2j_t+1)}{(2j_{A^-}+1)}\left[|\bar{S}_+(j_t)|^2 + |\bar{S}_-(j_t)|^2 \right] \\ \times \left[1 + \frac{\left[\begin{array}{l} |\bar{S}_+(j_t)|^2(j_t+2) + |\bar{S}_-(j_t)|^2(j_t-1) - \\ 3\sqrt{(j_t)(j_t+1)}\left(\bar{S}_+(j_t)\bar{S}_-^*(j_t) + \bar{S}_-(j_t)\bar{S}_+^*(j_t) \right) \end{array} \right]}{\left(|\bar{S}_+(j_t)|^2 + |\bar{S}_-(j_t)|^2 \right)(2j_t+1)} P_2(\cos\theta) \right] \\ + \frac{\pi\lambda^2(2j_t+1)}{(2j_{A^-}+1)}|\bar{S}_0(j_t)|^2[1+(-1)P_2(\cos\theta)] \end{array} \right\}$$

$$= \frac{1}{4\pi}\sum_{j_t} \left\{ \sigma(j_t)_{fav}\left[1+\beta(j_t)_{fav}P_2(\cos\theta)\right] + \sigma(j_t)_{unf}\left[1+\beta(j_t)_{unf}P_2(\cos\theta)\right] \right\}$$

(7.38)

where [91–93].

$$\sigma(j_t)_{fav} = \frac{\pi\lambda^2(2j_t+1)}{(2j_{A^-}+1)}\left[|\bar{S}_+(j_t)|^2 + |\bar{S}_-(j_t)|^2 \right]$$

(7.39a)

$$\sigma(j_t)_{unf} = \frac{\pi\lambda^2(2j_t+1)}{(2j_{A^-}+1)}|\bar{S}_0(j_t)|^2$$

(7.39b)

$$\beta(j_t)_{fav} = \frac{\left[|\overline{S}_+(j_t)|^2(j_t+2) + |\overline{S}_-(j_t)|^2(j_t-1) - 3\sqrt{(j_t)(j_t+1)}\left(\overline{S}_+(j_t)\overline{S}_-^*(j_t) + \overline{S}_-(j_t)\overline{S}_+^*(j_t)\right)\right]}{\left(|\overline{S}_+(j_t)|^2 + |\overline{S}_-(j_t)|^2\right)(2j_t+1)}$$

$$\tag{7.39c}$$

$$\beta(j_t)_{unf} = -1 \tag{7.39d}$$

and $|\overline{S}_\pm(j_t)|$ and $|\overline{S}_0(j_t)|$ represent photodetachment scattering amplitudes for photoelectron escape angular momenta $l = j_t \pm 1$ and $l = j_t$, respectively.

Let us extend our manipulation of (7.38) a little further:

$$
\begin{aligned}
\frac{d\sigma_{pd}}{d\Omega} &= \frac{1}{4\pi}\sum_{j_t}\left\{\sigma(j_t)_{fav}\left[1 + \beta(j_t)_{fav}P_2(\cos\theta)\right] + \sigma(j_t)_{unf}\left[1 + \beta(j_t)_{unf}P_2(\cos\theta)\right]\right\} \\
&= \frac{1}{4\pi}\sum_{j_t}\left\{\left[\sigma(j_t)_{fav} + \sigma(j_t)_{fav}\beta(j_t)_{fav}P_2(\cos\theta)\right]\right. \\
&\qquad\qquad \left. + \left[\sigma(j_t)_{unf} + \sigma(j_t)_{unf}\beta(j_t)_{unf}P_2(\cos\theta)\right]\right\} \\
&= \frac{1}{4\pi}\left\{\overbrace{\sum_{j_t}^{fav}\sigma(j_t)_{fav} + \sum_{j_t}^{unf}\sigma(j_t)_{unf}}^{\sigma = \sum\sigma(j_t)} + \left[\sum_{j_t}^{fav}\sigma(j_t)_{fav}\beta(j_t)_{fav}\right.\right. \\
&\qquad\qquad \left.\left. + \sum_{j_t}^{unf}\sigma(j_t)_{unf}\beta(j_t)_{unf}\right]P_2(\cos\theta)\right\} \\
&= \frac{\sigma}{4\pi}[1 + \beta P_2\cos(\theta)]
\end{aligned}
$$

$$\tag{7.40}$$

where [90, 92, 93]

$$\sigma = \sum_{j_t}\sigma(j_t) \tag{7.41}$$

and

$$\beta = \frac{\sum\limits_{j_t}\sigma(j_t)\beta(j_t)}{\sum\limits_{j_t}\sigma(j_t)} \tag{7.42}$$

Equation (7.42) shows the asymmetry parameter β is actually an average over $\beta(j_t)$, weighted by the corresponding partially integrated pd cross-sections $\sigma(j_t)$ [91–94]. Note also that the integrated cross-section of equation (7.41) is a function only of the squared moduli of the amplitudes $\left|\overline{S}_\pm(j_t)\right|$ and $\left|\overline{S}_0(j_t)\right|$, whereas the asymmetry parameter depends not only on the moduli but also on the phases of the scattering amplitudes via the interference terms. For any particular pd process, a range of values of j_t might be allowed, the dynamics of the pd process being determined by the relative contributions of the different j_t components. But it is important to remember that each angular momentum transfer component has a characteristic angular distribution *independent* of the dynamics of the pd reaction [91].

Parity-unfavored transitions result in an asymmetry parameter of $\beta = -1$ and a $\sin^2\theta$ distribution of photoelectrons in the dipole approximation [see (3.37) and (3.38) and Fig. 3.1], regardless of reaction dynamics and reactant constituent makeup. Parity-unfavored transitions can arise from secondary angular momentum transfers between the photoelectron and the residual core resulting from effects not accounted for in simpler models. These effects include electrostatic repulsions/interactions between different terms within configurations (e.g., orbit-orbit), and relativistic effects such as spin-orbit coupling (which serve to break down the *LS*-coupling approximation). At this point, it is important to realize that the photoelectron must interact with a *nonzero* core angular momentum to experience a torque which will allow for an exchange of angular momentum [87, 88, 91].

Although the angular distribution of photoelectrons is energy-independent for parity-unfavored transitions, parity-favored transitions present a different story. The expressions for β in (7.39c) and (7.42) show that, in general, the asymmetry parameter for parity-favored transitions is energy-dependent, the energy dependence being hidden in the scattering amplitudes $\overline{S}(j_t)$. The energy dependence arises because the photoelectron orbital angular momentum can take on two different values, $l = j_t \pm 1$, and the interference between the scattering amplitudes for these two values leads to the energy dependence [90]. However, since β is expressed as a ratio of these energy-dependent scattering amplitudes, there always exists the chance that the amplitudes may cancel out, leaving β as an energy-independent constant. Recall that there are, in general, two possible values of j_t. But (7.25) implies that j_t will be restricted to a single value when either L_{A-} or L_A is zero. It is in these cases, in which only one value of j_t is allowed, that the cross-sections $\sigma(j_t)$ may cancel, leaving $\beta = \beta(j_t)$. For example, in the special case of $j_t = 0$, l takes on the single value of $+1$ [due to angular momentum conservation—see equation (7.25)], in which case we have (noting that $\left|\overline{S}_-(j_t)\right|^2 = 0$) $\beta(j_t = 0) = 2$ from (7.39c) [see also the discussion after (4.39) and the discussion after (7.69b)]. Various types of approximations lead to other predictions of energy-independent asymmetry parameters (e.g., from s-subshell photodissociations); deviations of $\beta(j_t)$ from constancy in these cases provide a sensitive measure of electron-correlation forces, which exist beyond independent-particle approximations [87, 88].

Equations (7.39a, 7.39b, 7.39c and 7.39d), (7.40), (7.41), and (7.42) are the main results of angular momentum transfer theory. At this point, we now attempt to obtain the *LS*-coupling form of the scattering amplitude from the scattering matrix of (7.18).

Then the general results of (7.39–7.42) can be compared to the more restricted results of (4.39) via a series of simplifying assumptions to check if the two theories are compatible in the appropriate limit(s). We start first by expressing the scattering amplitude in (7.18a) as a sum of reduced dipole matrix elements for linearly polarized photons [90–93]:

$$
\langle J \| \bar{S}(j_t) \| j_{A-} \rangle \equiv \sum_J (-1)^{j_{A-}-k-J}(2J+1)^{\frac{3}{2}} \langle J \| S^{(k)} \| j_{A-} \rangle \begin{Bmatrix} l & j_A & J \\ j_{A-} & k & j_t \end{Bmatrix}
$$

$$
= n(\lambda) \sum_J (-1)^{j_{A-}-1-J}(2J+1)^{\frac{3}{2}} \left[\frac{1}{(2J+1)} \right] \langle ([j_A s]_{j_{As}}, l)J \| D^{(1)} \| \alpha_- j_{A-} \rangle \begin{Bmatrix} l & j_A & J \\ j_{A-} & 1 & j_t \end{Bmatrix}
$$

$$
\tag{7.43}
$$

where the factor $1/(2J + 1)$ accounts for the multiplicity in the $\vec{J}$ states and $n(\lambda) = 4\pi\alpha\hbar\omega/3\lambda^2$ is the normalization factor needed to match incoming and outgoing waves. The α_{A-} are the quantum numbers needed to uniquely specify the initial state. This form is still completely general (in the LS-coupling approximation). Nevertheless, this form of the reduced dipole matrix elements is not convenient for our purposes. For one thing, the final state is defined in terms of quantum numbers appropriate for the core-photoelectron system at infinite separations. To account for the photon-absorption part of the pd process, it is convenient to express the reduced dipole matrix elements in terms of the real transition amplitudes reflecting short-range electron-core interactions [91–93]:

$$
\langle ([j_A s]_{j_{As}}, l)J \| D^{(1)} \| \alpha_A j_{A-} \rangle = (-i)^l e^{i\sigma(j_A l)} \sum_\alpha e^{i\delta(\alpha)} \langle ([j_A s]_{j_{As}}, l)J | \alpha J \rangle \langle \alpha J \| D^{(1)} \| \alpha_- j_{A-} \rangle
$$

$$
\tag{7.44}
$$

The phase factor $(-i)^l e^{i\sigma(j_A l)} e^{i\delta(\alpha)}$ reflects the change from incoming-wave to standing-wave normalization and also reflects photoelectron phase shifts associated with short-range electron-core interactions in the Coulomb field. The $\langle ([j_A s]_{j_{As}}, l)J | \alpha J \rangle$ are the recoupling coefficients needed to connect the eigenchannel coupling scheme (represented by the electron-target eigenchannel states $\langle \alpha J |$) with the core-photoelectron dissociation-channel coupling scheme and are evaluated next.

In the LS-coupling approximation, $\alpha_{A-} = L_A\text{-}S_A\text{-}$ and $\alpha = (L_A l)_L(S_A s)_S$. Implicit in this scheme are the following relations;

$$
\vec{J} = \vec{j}_A + \vec{j}_e = \vec{j}_{As} + \vec{l} = \vec{L} + \vec{S} \tag{7.45a}
$$

$$\vec{J}_{As} = \vec{j}_A + \vec{s} = \vec{L}_A + \vec{S} \tag{7.45b}$$

$$S = S_{A^-} \tag{7.45c}$$

in which case the recoupling coefficient becomes

$$\langle([j_As]_{j_{As}},l)J|\alpha J\rangle \rightarrow \langle([j_As]_{j_{As}},l)J|([L_Al]_L,[S_As]_S)J\rangle \tag{7.46}$$

Note from equations (7.45a,b) that we must first account for the various ways to couple angular momenta together to get $\vec{j}_{As}$ and $\vec{J}$:

$$\langle([j_As]_{j_{As}},l)J|([L_Al]_L,[S_As]_S)J\rangle \rightarrow \langle([L_AS_A]_{j_A},s)j_{As}|(L_A,[S_As]_S)j_{As}\rangle$$
$$\times \langle([SL_A]_{j_{As}},l)J|(S,[L_Al]_L)J\rangle \tag{7.47}$$

We evaluate (7.47) using the relationship between recoupling coefficients and 6-j symbols from (2.171) and the symmetry properties of the 6-j symbols [3, 16];

$$\langle([L_AS_A]_{j_A},s)j_{As}|(L_A,[S_As]_S)j_{As}\rangle\langle([SL_A]_{j_{As}},l)J|(S,[L_Al]_L)J\rangle$$
$$= (-1)^{L_A+S_A+s+j_{As}}\sqrt{(2j_A+1)(2S+1)}\begin{Bmatrix} L_A & S_A & j_A \\ s & j_{As} & S \end{Bmatrix}$$
$$\times(-1)^{\overbrace{S+l+L_A+J}^{=L+J_{As}}}\sqrt{(2j_{As}+1)(2L+1)}\begin{Bmatrix} L & S & J \\ j_{As} & l & L_A \end{Bmatrix} \tag{7.48}$$

Let us pause now and recap where we are so far by combining (7.43), (7.44), and (7.48):

$$\langle J\|\vec{S}(j_t)\|j_{A^-}\rangle = n(\lambda)\sum_J (-1)^{\overbrace{j_{A^-}-1-J}^{=J-j_{A^-}-1}}\sqrt{(2J+1)}(-i)^l e^{i\sigma(j_Al)}$$
$$\times\sum_{\alpha\to L,S} e^{i\delta(\alpha)}(-1)^{(L_A+l+L)+(S_A+s+S)+2j_{As}}\sqrt{(2j_A+1)(2S+1)(2j_{As}+1)(2L+1)}$$
$$\times\begin{Bmatrix} L_A & S_A & j_A \\ s & j_{As} & S \end{Bmatrix}\begin{Bmatrix} L & S & J \\ j_{As} & l & L_A \end{Bmatrix}\begin{Bmatrix} l & j_A & J \\ j_{A^-} & 1 & j_t \end{Bmatrix}\langle([L_Al]_L,[S_As]_S)J\|D^{(1)}\|(S_{A^-},L_{A^-})j_{A^-}\rangle \tag{7.49}$$

We now further identify a consequence of the *LS*-coupling-driven definition of α_{A^-} and α. The initial atomic configuration consists of a single open shell l_0^N where l_0 is the orbital angular momentum we have defined earlier, and N is the occupation number. The configuration of the final state is of the form $l_0^{N-1}l$. We can clarify this further by rewriting (7.1) as [92, 93]:

$$A^- \left([l_0^N L_{A^-} S_{A^-}] J_{A^-}, \pi_{A^-} \right) + \gamma \left(j_\gamma = 1, \pi_\gamma = -1 \right)$$
$$\rightarrow A \left([l_0^{N-1} L_A S_A] J_A, \pi_A \right) + e^- \left(l, s, j_e, \pi_e = (-1)^l \right) \tag{7.50a}$$

Or, putting it another way,

$$L_{A^-} = \overline{L}_{A^-} + l_0 \tag{7.50b}$$

$$S_{A^-} = \overline{S}_{A^-} + s \tag{7.50c}$$

The reduced dipole matrix now reads [92, 93]

$$\left\langle ([L_A l]_L, [S_A s]_S) J \| D^{(1)} \| (S_{A^-}, L_{A^-}) j_{A^-} \right\rangle$$
$$\rightarrow \left\langle ([L_A l]_L, [S_A s]_S) J \| D^{(1)} \| \left([\overline{L}_{A^-} l_0]_{L_{A^-}}, [\overline{S}_{A^-} s]_{S_{A^-}} \right) j_{A^-} \right\rangle \tag{7.51}$$

The operator is a tensor product of two separate tensor operators that act on both the spin and the orbital angular momenta, which are separate tensor spaces. We can thus separate out the spin-dependence by using (D.15) as follows [3, 92, 93]:

$$\left\langle ([L_A l]_L, [S_A s]_S) J \| \left(D^{(1)} = \left[D_L^{(k_1)} \otimes D_S^{(k_2)} \right]^{(1)} \right) \| \left([\overline{L}_{A^-} l_0]_{L_{A^-}}, [\overline{S}_{A^-} s]_{S_{A^-}} \right) j_{A^-} \right\rangle$$
$$= \sqrt{(2J+1)(2j_{A^-}+1)(3)} \begin{Bmatrix} L & L_{A^-} & k_1 \\ S & S_{A^-} & k_2 \\ J & j_{A^-} & 1 \end{Bmatrix} \left\langle L \| D_L^{(k_1)} \| L_{A^-} \right\rangle \left\langle S \| D_S^{(k_2)} \| S_{A^-} \right\rangle$$

$$\tag{7.52}$$

From the symmetry properties of the 9-j symbols, we have

$$\begin{Bmatrix} L & L_{A^-} & k_1 \\ S & S_{A^-} & k_2 \\ J & j_{A^-} & 1 \end{Bmatrix} = \begin{Bmatrix} J & L & S \\ j_{A^-} & L_{A^-} & S_{A^-} \\ 1 & k_1 & k_2 \end{Bmatrix} \tag{7.53}$$

Therefore,

$$\left\langle ([L_A l]_L, [S_A s]_S) J \| D^{(1)} \| \left([\overline{L}_{A^-} l_0]_{L_{A^-}}, [\overline{S}_{A^-} s]_{S_{A^-}} \right) j_{A^-} \right\rangle$$
$$= \sqrt{(2J+1)(2j_{A^-}+1)(3)} \begin{Bmatrix} J & L & S \\ j_{A^-} & L_{A^-} & S_{A^-} \\ 1 & k_1 & k_2 \end{Bmatrix} \left\langle L \| D_L^{(k_1)} \| L_{A^-} \right\rangle \left\langle S \| D_S^{(k_2)} \| S_{A^-} \right\rangle$$

$$\tag{7.53}$$

But the pd process is spin-independent under the dipole approximation, which means that $D_S^{(k_2)}$ is the same as the identity operator. It also means that $k_2 = 0$ and the 9-j symbol in (7.53) reduces to a 6-j symbol via (D.17), leaving us with

$$
\left\langle \left([L_A l]_L, [S_A s]_S\right) J \left\| D^{(1)} \right\| \left(\left[\overline{L}_{A^-} l_0\right]_{L_{A^-}}, \left[\overline{S}_{A^-} s\right]_{S^-} \right) j_{A^-} \right\rangle
$$

$$
= (-1)^{L+S+j_{A^-}+1} \sqrt{\frac{(2J+1)(2j_{A^-}+1)}{(2S+1)}}
$$

$$
\times \begin{Bmatrix} J & j_{A^-} & 1 \\ L_{A^-} & L & S \end{Bmatrix} \left\langle (L_A, l)L \left\| D^{(k_1)} \right\| (\overline{L}_{A^-}, l_0)L_{A^-} \right\rangle \underbrace{\left\langle S \| 1 \| S_{A^-} \right\rangle}_{\sqrt{(2S+1)}\delta_{S,S_{A^-}} \delta_{\overline{S}_{A^-}, S_A}} \delta_{1k_1}
$$

(7.54)

$$
\Rightarrow \left\langle \left([L_A l]_L, [S_A s]_S\right) J \left\| D^{(1)} \right\| \left(\left[\overline{L}_{A^-} l_0\right]_{L_{A^-}}, \left[\overline{S}_{A^-} s\right]_{S^-} \right) j_{A^-} \right\rangle
$$

$$
= (-1)^{L+S+j_{A^-}+1} \sqrt{(2J+1)(2j_{A^-}+1)} \begin{Bmatrix} J & j_{A^-} & 1 \\ L_{A^-} & L & S \end{Bmatrix}
$$

(7.55)

$$
\left\langle (L_A, l)L \left\| D^{(1)} \right\| (\overline{L}_{A^-}, l_0)L_{A^-} \right\rangle \delta_{S,S_{A^-}} \delta_{\overline{S}_{A^-}, S_A}
$$

where we also used (2.203). Note that this action will also serve to collapse the sum over S in (7.49).

The form of the remaining operator tells us that it purports to act on the orbital angular momenta of the core and the orbital angular momenta of the electron, again two separate tensor spaces. However, the pd process does not affect the core (although the escaping photoelectron does interact with it, as we shall see), so that portion of the operator is again equivalent to the identity operator. We can therefore follow the same logic as we did in (7.52, 7.53, 7.54 and 7.55) above in decomposing the operator into its two separate tensor spaces [3, 16, 92, 93]:

$$
\left\langle (L_A, l)L \| D^{(1)} \| (\overline{L}_{A^-}, l_0)L_{A^-} \right\rangle = (-1)^{L_A+l_0+1+L} \sqrt{(2L+1)(2L_{A^-}+1)}
$$

$$
\times \begin{Bmatrix} L & L_{A^-} & 1 \\ l_0 & l & L_A \end{Bmatrix} \left\langle l \| D^{(1)} \| l_0 \right\rangle \delta_{L_A, \overline{L}_{A^-}}
$$

(7.56)

where we also used the symmetry properties of the 6-j symbols.

We can now easily separate the radial and angular parts of the reduced dipole matrix elements, since they are now a sum of N one-electron operators acting on the initial states [c.f. (4.10)] [3, 90, 92, 93]:

$$\langle l\|D^{(1)}\|l_0\rangle \to \left\langle l \left\| \sum_j^N \vec{r}_j^{(1)} \right\| l_0 \right\rangle = \sqrt{N}\langle l\|C^{(1)}\|l_0\rangle \mathfrak{R}_{E_c l}^{L_A S_A L}$$

$$= \sqrt{N}(-1)^l \sqrt{(2l+1)(2l_0+1)} \begin{pmatrix} l & 1 & l_0 \\ 0 & 0 & 0 \end{pmatrix} \mathfrak{R}_{E_c l}^{L_A S_A L} \qquad (7.57)$$

where we also used (2.198). $\sqrt{N}$ is a weight factor due to the presence of the N electrons in the initial state, E_c is the photoelectron kinetic energy, and $\mathfrak{R}_{E_c l}^{L_A S_A L}$ is the radial dipole integral:

$$\mathfrak{R}_{E_c l}^{L_A S_A L} = \int_0^\infty P_{n l_0} r P_{E_c l}^{L_A S_A l}(r) dr \qquad (7.58)$$

$P_{n l_0}$ and $P_{E_c l}^{L_A S_A l}(r)$ are the radial wave functions (multiplied by r) of the initial and final orbitals of the photoelectron, respectively. Note that these wave functions depend dynamically on the orbital angular momentum and spin quantum numbers of the initial state, final state, and the residual core (via the dynamical coupling of the photoelectron orbital motion to the orbital motion of the residual core), which means that, at this point, the interactions can still be anisotropic (noncentral) [90, 92, 93].

Putting together the results of (7.49), (7.55), (7.56), and (7.57),

$$\langle J\|\overline{S}(j_t)\|j_{A-}\rangle = n(\lambda)\sqrt{N}(-i)^l e^{i\sigma(j_A l)}$$
$$\times \sqrt{(2j_A+1)(2S+1)(2j_{As}+1)(2L_{A-}+1)(2J_{A-}+1)(2l+1)(2l_0+1)}$$
$$\times (-1)^{S_A+s+S} \begin{Bmatrix} L_A & S_A & j_A \\ s & j_{As} & S \end{Bmatrix} \begin{pmatrix} l & 1 & l_0 \\ 0 & 0 & 0 \end{pmatrix}$$
$$\times \sum_{\alpha \to L} e^{i\delta_{E_c l}^{L_A S_A l}} (-1)^{2j_{As}+S+L+l} \underbrace{(-1)^{\overbrace{2L_A+l_0+1}^{=l}+l+2L}}_{=1} (2L+1) \begin{Bmatrix} L & L_{A-} & 1 \\ l_0 & l & L_A \end{Bmatrix} \mathfrak{R}_{E_c l}^{L_A S_A L}$$
$$\times \sum_J (-1)^J (2J+1) \begin{Bmatrix} l & j_A & J \\ j_{A-} & 1 & j_t \end{Bmatrix} \begin{Bmatrix} L & S & J \\ j_{As} & l & L_A \end{Bmatrix} \begin{Bmatrix} J & j_{A-} & 1 \\ L_{A-} & L & S \end{Bmatrix} \qquad (7.59)$$

It is interesting to see the changed character of the phase factor $e^{i\delta(\alpha)} \to e^{i\delta_{E_c l}^{L_A S_A L}}$. It now reflects the proper dependence on all the appropriate quantum numbers under the LS-coupling approximation. But none of the phase factors (nor the radial integral) depend on the total angular momentum. That means we can evaluate the sum over J analytically via (C.77) [3, 10]

$$\sum_J (-1)^J (2J+1) \begin{Bmatrix} l & j_A & J \\ j_{A^-} & 1 & j_t \end{Bmatrix} \begin{Bmatrix} L & S & J \\ j_{As} & l & L_A \end{Bmatrix} \begin{Bmatrix} J & L_{A^-} & 1 \\ L_{A^-} & L & S \end{Bmatrix}$$

$$= (-1)^{-L_A - j_t - L_{A^-} - 1 - L - l - j_{A^-} - S - J_{As}} \begin{Bmatrix} L_A & j_t & L_{A^-} \\ 1 & L & l \end{Bmatrix} \begin{Bmatrix} L_A & j_t & L_{A^-} \\ j_{A^-} & S & j_{As} \end{Bmatrix}$$

(7.60)

Substituting the result of (7.60) into (7.59),

$$\langle J \| \overline{S}(j_t) \| j_{A^-} \rangle = n(\hbar)\sqrt{N}(-i)^l e^{i\sigma(j_A l)} Q(j_t, j_A, j_{As}) \sqrt{(2J_{A^-}+1)(2l+1)(2l_0+1)}$$

$$\times (-1)^{(S_A + s + S) + (j_{As} - j_{A^-} - j_t) + \left(\overbrace{-L_A - L_{A^-} - 1}^{=L_A + L_{A^-} + 1} \right)} \begin{Bmatrix} l & 1 & l_0 \\ 0 & 0 & 0 \end{Bmatrix}$$

$$\times \sum_L e^{i\delta_{Ecl}^{L_A S_A L}} (2L+1) \begin{Bmatrix} L & L_{A^-} & 1 \\ l_0 & l & L_A \end{Bmatrix} \begin{Bmatrix} L_A & j_t & L_{A^-} \\ 1 & L & 1 \end{Bmatrix} \Re_{Ecl}^{L_A S_A L}$$

(7.61)

where [92, 93]

$$Q(j_t, j_A, j_{As}) \equiv \sqrt{(2L_{A^-}+1)(2S+1)(2j_A+1)(2j_{As}+1)}$$

$$\times \begin{Bmatrix} L_A & S_A & j_A \\ s & j_{As} & S \end{Bmatrix} \begin{Bmatrix} L_A & j_t & L_{A^-} \\ j_{A^-} & S & j_{As} \end{Bmatrix}$$

(7.62)

All the dependence on j_{As} is contained in the factor $(-1)^{j_{As} - j_{A^-}} Q(j_t, j_A, j_{As})$. The square of this term (with phase +1) enters the expression of the total cross-section and asymmetry parameter of (7.41) and (7.42), each of which contains (an implicit) summation over j_{As} [91].[2] The quantity

$$\overline{Q}^2(j_t, j_A) \equiv \sum_{j_{As}} Q^2(j_t, j_A, j_{As})$$

(7.63)

gives the statistical weight with which photodetachment probability for a given j_t is distributed among the possible fine-structure levels j_A since [92, 93]

---

[2] Coupling the photoelectron spin to the core introduces an additional summation over j_{As}, with the result that the expressions in equations (7.39a–7.39d), (7.41), and (7.42) must, in general, also be summed over j_{As}. In many cases, however, only one value of j_{As} occurs in a particular reaction.

$$\sum_{j_A} \overline{Q}^2(j_t, j_A) = 1 \tag{7.64}$$

If we now make our first approximation that the fine-structure of the residual core is not resolved, then we can take $\sigma(j_A l)$ to be independent of j_A and thus remove the dependence of the total cross-section and asymmetry parameter of (7.41) and (7.42) on j_A by an application of (7.64) [92, 93].

We can reduce (7.61) further by introducing more approximations. If we ignore all electron-electron interchannel scattering (which serves to populate final states from the ground state via all possible open and closed channels), then all factors which depend on the initial state (such as $\sqrt{(2J_{A-} + 1)}$) will be fixed for a particular pd process and will thus cancel in the numerator and denominator of (7.41) and (7.42). We may then write [90]

$$\langle J \| \overline{S}(j_t) \| j_{A-} \rangle \propto (-i)^l e^{i\sigma_l} \sqrt{(2l+1)(2l_0+1)} \begin{Bmatrix} l & 1 & l_0 \\ 0 & 0 & 0 \end{Bmatrix}$$

$$\times \sum_L e^{i\delta_{Ecl}^{L_A S_A L}} (2L+1) \begin{Bmatrix} L & L_{A-} & 1 \\ l_0 & l & L_A \end{Bmatrix} \begin{Bmatrix} L_A & j_t & L_{A-} \\ l & L & l \end{Bmatrix} \mathfrak{R}_{Ecl}^{L_A S_A L} \tag{7.65}$$

Another simplifying assumption is to neglect the multiplet structure of the pd reaction. In this approximation, the phase shifts and the radial wave functions are assumed to be a function only of the configuration (n_0, l_0, l), and not the term quantum numbers (L_A, S_A, L). Recall that the coupling of the photoelectron to the residual core is reflected in the dependence of the radial integral $\mathfrak{R}_{Ecl}^{L_A S_A L}$ and the phase shifts $e^{i\delta_{Ecl}^{L_A S_A L}}$ on the total orbital angular momentum of the core-electron complex, L [a coupling which determines the dynamical weights with which transition amplitudes for alternative values of the total orbital angular momentum $\vec{L} = \vec{L}_A + \vec{l}$ superimpose in (7.61)]. This simplifying approximation removes this dependence (the weights become independent of L, and hence L_A) and thus renders the electron-core interaction isotropic [as in the standard central-potential model of equations (4.39)], which means [90, 92, 93]

$$e^{i\delta_{Ecl}^{L_A S_A L}} \mathfrak{R}_{Ecl}^{L_A S_A L} \rightarrow e^{i\delta_l} \mathfrak{R}_l \tag{7.66}$$

The sum over L can now be performed analytically via (C.78) [3, 10]

$$\sum_L (2L+1) \begin{Bmatrix} L & L_{A-} & 1 \\ l_0 & l & L_A \end{Bmatrix} \begin{Bmatrix} L_A & j_t & L_{A-} \\ 1 & L & l \end{Bmatrix}$$

$$= \sum_L (2L+1) \begin{Bmatrix} L_{A-} & 1 & L \\ l & L_A & l_0 \end{Bmatrix} \begin{Bmatrix} L_{A-} & 1 & L \\ l & L_A & j_t \end{Bmatrix} = (2l_0+1)^{-1} \delta_{j_t, l_0} \tag{7.67}$$

This approximation also means that j_t, whose values had heretofore been determined by the 6-j symbol in (7.61), is now restricted to the value $j_t = l_0$, giving us [92, 93]

$$\langle J\|\overline{S}(j_t = l_0)\|j_{A^-}\rangle \propto (-i)^l e^{i\sigma_l} e^{i\delta_l} \sqrt{\frac{(2l+1)}{(2l_0+1)}} \begin{pmatrix} l & 1 & l_0 \\ 0 & 0 & 0 \end{pmatrix} \mathfrak{R}_l \qquad (7.68)$$

Or, using (4.24),

$$\overline{S}_+(j_t = l_0) \propto (-i)^{l_0+1}(-1)^{l_0+1} e^{i\sigma_{l_0+1}} e^{i\delta_{l_0+1}} \sqrt{\frac{(l_0+1)}{(2l_0+1)}} \mathfrak{R}_{l_0+1} \qquad (7.69a)$$

$$\overline{S}_-(j_t = l_0) \propto (-i)^{l_0}(-1)^{l_0-1} e^{i\sigma_{l_0-1}} e^{i\delta_{l_0-1}} \sqrt{\frac{l_0}{(2l_0+1)}} \mathfrak{R}_{l_0-1} \qquad (7.69b)$$

Substitution of these two expressions for the scattering amplitude into (7.39c) results in (4.39b)—as it should.

Example 7.2 We prove the last assertion by substituting (7.69a) and (7.69b) into (7.39c):

$$\beta(j_t) = \frac{\left[\left|\sqrt{\frac{(l_0+1)}{(2l_0+1)}}\mathfrak{R}_{l_0+1}\right|^2(j_t+2) + \left|\sqrt{\frac{l_0}{(2l_0+1)}}\mathfrak{R}_{l_0-1}\right|^2(j_t-1)\right.}{\left.-3\sqrt{(j_t)(j_t+1)}\left\{\begin{array}{l}\left[(-i)^{l_0+1}(-1)^{l_0+1}e^{i\sigma_{l_0+1}}e^{i\delta_{l_0+1}}\sqrt{\frac{(l_0+1)}{(2l_0+1)}}\mathfrak{R}_{l_0+1}\right]\\ \times\left[(i)^{l_0}(-1)^{l_0-1}e^{-i\sigma_{l_0-1}}e^{-i\delta_{l_0-1}}\sqrt{\frac{l_0}{(2l_0+1)}}\mathfrak{R}_{l_0-1}\right]\\ +\left[(-i)^{l_0}(-1)^{l_0-1}e^{i\sigma_{l_0-1}}e^{i\delta_{l_0-1}}\sqrt{\frac{l_0}{(2l_0+1)}}\mathfrak{R}_{l_0-1}\right]\\ \times\left[(i)^{l_0+1}(-1)^{l_0+1}e^{-i\sigma_{l_0+1}}e^{-i\delta_{l_0+1}}\sqrt{\frac{(l_0+1)}{(2l_0+1)}}\mathfrak{R}_{l_0+1}\right]\end{array}\right\}\right]}{\left(\left|\sqrt{\frac{(l_0+1)}{(2l_0+1)}}\mathfrak{R}_{l_0+1}\right|^2 + \left|\sqrt{\frac{l_0}{(2l_0+1)}}\mathfrak{R}_{l_0-1}\right|^2\right)(2j_t+1)}$$

$$\Rightarrow \beta(j_t = l_0) = \frac{\left[\begin{array}{l} (l_0+1)(l_0+2)\mathfrak{R}^2_{l_0+1} + l_0(l_0-1)\mathfrak{R}^2_{l_0-1} \\ -3\sqrt{l_0(l_0+1)}\left\{ \begin{array}{l} \left[(-i)^{l_0+1}e^{i\sigma_{l_0+1}}e^{i\delta_{l_0+1}}\sqrt{(l_0+1)}\mathfrak{R}_{l_0+1}\right] \\ \times\left[(i)^{l_0}e^{-i\sigma_{l_0-1}}e^{-i\delta_{l_0-1}}\sqrt{l_0}\mathfrak{R}_{l_0-1}\right] \\ +\left[(-i)^{l_0}e^{i\sigma_{l_0-1}}e^{i\delta_{l_0-1}}\sqrt{l_0}\mathfrak{R}_{l_0-1}\right] \\ \times\left[(i)^{l_0+1}e^{-i\sigma_{l_0+1}}e^{-i\delta_{l_0+1}}\sqrt{(l_0+1)}\mathfrak{R}_{l_0+1}\right] \end{array} \right\} \end{array} \right]}{(2l_0+1)\left[(l_0+1)\mathfrak{R}^2_{l_0+1} + l_0\mathfrak{R}^2_{l_0-1}\right]}$$

$$\Rightarrow \beta(l_0) = \frac{\left[\begin{array}{l} (l_0+1)(l_0+2)\mathfrak{R}^2_{l_0+1} + l_0(l_0-1)\mathfrak{R}^2_{l_0-1} - 3l_0(l_0+1)\mathfrak{R}_{l_0+1}\mathfrak{R}_{l_0-1} \\ \times\left\{ -i\left[e^{i(\delta_{l_0+1}+\sigma_{l_0+1}-\delta_{l_0-1}-\sigma_{l_0-1})} - e^{-i(\delta_{l_0+1}+\sigma_{l_0+1}-\delta_{l_0-1}-\sigma_{l_0-1})} \right] \right\} \end{array} \right]}{(2l_0+1)\left[(l_0+1)\mathfrak{R}^2_{l_0+1} + l_0\mathfrak{R}^2_{l_0-1}\right]}$$

$$\Rightarrow \beta = \frac{\left[\begin{array}{l} (l_0+1)(l_0+2)\mathfrak{R}^2_{l_0+1} + l_0(l_0-1)\mathfrak{R}^2_{l_0-1} - 3l_0(l_0+1)\mathfrak{R}_{l_0+1}\mathfrak{R}_{l_0-1} \\ \times\left\{ 2\sin\left[(\delta_{l_0+1}-\delta_{l_0-1}) + (\sigma_{l_0+1}-\sigma_{l_0-1})\right] \right\} \end{array} \right]}{(2l_0+1)\left[(l_0+1)\mathfrak{R}^2_{l_0+1} + l_0\mathfrak{R}^2_{l_0-1}\right]}$$

We are free to set the angle $\sigma_{l_0+1} - \sigma_{l_0-1} = \pi/2$, giving us our result [which matches (4.39b) to within a multiplicative factor]:

$$\beta = \frac{l_0(l_0-1)\mathfrak{R}^2_{l_0-1} + (l_0+1)(l_0+2)\mathfrak{R}^2_{l_0+1} - 6l_0(l_0+1)\mathfrak{R}_{l_0-1}\mathfrak{R}_{l_0+1}\cos(\delta_{l_0+1}-\delta_{l_0-1})}{(2l_0+1)\left[l_0\mathfrak{R}^2_{l_0-1} + (l_0+1)\mathfrak{R}^2_{l_0+1}\right]}$$

■

The phase factors (which match boundary conditions between incoming and outgoing waves in the Coulomb field, and which—prior to these approximations—reflected angular momentum coupling in the photoelectron-neutral complex [92, 93]) and proportionality constants do not matter since they will appear in equal powers in the numerator and denominator and, hence, will cancel. An interesting result of (7.67) bears repeating. This equation tells us that, due to geometrical effects, only the single value of $j_t = l_0$ is allowed. This is the value one would expect from the central-potential model, since, in a one-electron picture, $\vec{l}_0 + \vec{1} = \vec{l}$. Combine this with (7.25), and the result is $\vec{j}_t = -\vec{l}_0$. As a consequence, we see that these approximations end up treating only the first portion of the pd process—the one-electron photoabsorption—and ignore the subsequent scattering of the photo-electron in the field of the residual core during its escape. It is during this latter stage that the only good quantum number is the total orbital angular momentum L, and both the photoelectron and the core may change their orbital angular momenta orientation and magnitude during the (inelastic) scattering process, provided that $\vec{L} = \vec{L}_{A-} + \vec{1}$. It is by this mechanism only that angular momentum transfers other than $j_t = l_0$ are possible [90]. In other words, the full theory [reflected in (7.38), (7.39a, 7.39b, 7.39c and 7.39d), and (7.61)], which accounts for anisotropic photoelectron-core coupling, differs from the more restricted theory of (4.39b) in that the resulting angular distributions depend not only on alternative values of $\vec{l}$ but also on alternative values of $\vec{L}$. While only a third of the terms in (7.39c) will depend on the Coulomb phase shift difference, all three terms, $\left| \overline{S}_{\pm}(j_t) \right|$ and $\left| \overline{S}_{+}(j_t) \right| \times \left| \overline{S}_{-}^{*}(j_t) \right| + \text{c.c.}$, depend on phase-shift differences of alternative pairs of electron-core LS-coupled channels $(L_A l)L$. By contrast, the more restricted theory of equation (4.39b) has only a single interference term $\delta_{l_0+1} - \delta_{l_0-1}$ between two independent-particle channels $l = l_0 \pm 1$. The difference in phase shifts $\delta_{E_c l}^{L_A S_A L}$ for alternative channels $(L_A l)L$ can serve to measure the deviation from the independent-particle model, and hence, the impact of non-isotropic and electron correlation forces.[3]

In summary, we can imagine the pd reaction as a two-part process. During the first part, the photon is absorbed and imparts $j_\gamma = 1$ unit of orbital angular momentum to the photoelectron, yielding an "interim" orbital angular momentum $\vec{l}' = \vec{l}_0 + \vec{j}_\gamma$. At this point, the angular momentum transferred to the target is $\vec{j}_t' = \vec{j}_\gamma - \vec{l}' = -\vec{l}_0$, giving a single value of the magnitude, $j_t' = l_0$. Furthermore, due to parity conservation, $l' = l_0 \pm 1$, and hence, $j_t' = l_0$ is a parity-favored angular momentum transfer. During the next stage of the pd reaction (the "escape" of the photoelectron), additional angular momentum transfers can occur via interactions between the orbital angular momentum of the photoelectron and the net orbital motion of the core electrons (we are considering only spin-independent interactions here). These interactions can produce a dynamical coupling between $\vec{l}'$ and $\vec{L}_A$. The orbital angular momentum of the photoelectron can thereby change from $\vec{l}'$ to $\vec{l}$, in which case the angular momentum transferred is no longer $j_t' = l_0$ but is now $\vec{j}_t = \vec{j}_\gamma - 1 = \vec{j}_t' - \vec{k}$ where [92, 93]

$$\vec{k} = \vec{l} - \vec{l}' = \vec{L}_A - \vec{L}_A' = \vec{j}_t - \vec{j}_t' \tag{7.70}$$

---

[3]It is interesting to note that, for photodetachment from a closed-shell anion, $L_{A-} = 0$, the sum over L in equation (7.61) collapses to a single term $L = j_\gamma = 1$ and j_t is restricted to the single value $j_t = l_0$. It is for this reason that the model reflected in (4.39b) has been so successful when applied to these systems [92, 93].

represents the aggregate of the angular momentum exchanges between the photo-electron and the core. Even if the magnitudes of $\vec{l}'$ and $\vec{L}_A'$ remain unchanged, they will precess about $\vec{L}$ in such a way as to generate a change in the magnitude of $\vec{j}_t$, while simultaneously conserving $\vec{L}$ (it is because of the dependence of these interactions on $\vec{L}$ that they are anisotropic). The more restricted theory of (4.39b) treats the core as a mere spectator to the whole pd reaction, thereby ignoring the second part of the process altogether. That is why only the single, parity-favored reaction $j_t = l_0$ arises in the restricted theory, and why, for other than some closed-shell systems and very light open-shell systems, actual measured photoelectron angular distributions will differ significantly from the predictions of the restricted theory [92, 93].

Example 7.3 Let us apply angular momentum transfer analysis to the simplest of systems, the photoionization of hydrogen in its ground state [95].

The reaction is

$$H1s\left(^2S_{1/2}\right) + \gamma \rightarrow \left[H^+\left(^1S_0\right) + e_p^-\right]_{\left(^{2S+1}L\right)} \tag{7.71}$$

Angular momentum conservation (7.24) tells us

$$\vec{L} = \vec{L}_c + l = \vec{0} + \vec{1} \Rightarrow L = 1 \tag{7.72}$$

Equation (7.23) gives us the allowed angular momentum transfer values:

$$\vec{j}_t = \left(\vec{j}_{H^+} + \vec{s}\right) - \vec{j}_H \Rightarrow j_t = 0, 1 \tag{7.73}$$

Parity conservation in the dipole approximation (7.26) implies

$$\pi_H \pi_{H^+} = (-1)^{l+1} \tag{7.74}$$

where

$$\pi_H \pi_{H^+} = (-1)^0 (-1)^0 = (+)(+) = (+)$$

$$\Rightarrow l = odd = 1 \quad \text{(as expected)} \tag{7.75}$$

From (7.29), we find which of our values of j_t are parity-unfavored

$$\pi_H \pi_{H^+} = (+) = (-1)^{j_t+1} \Rightarrow (j_t)_{unf} = 1 \tag{7.76}$$

And from (7.34), we find which of our values of j_t are parity-favored

$$\pi_H \pi_{H^+} = (+) = (-1)^{j_t} \Rightarrow (j_t)_{fav} = 0 \tag{7.77}$$

Recalling (7.39a, 7.39b) and remembering that $\left|\overline{S}_\pm(j_t)\right|$ and $\left|\overline{S}_0(j_t)\right|$ represent (for this example) photoionization amplitudes for photoelectron escape angular momenta $l = j_t \pm 1$ and $l = j_t$ respectively, we see that for $\sigma(j_t = 0)_{fav}$, $\left|\overline{S}_-(0)\right| = 0$ because $l = -1$ is not allowed.

So, from (7.39a, 7.39b), we have

$$\sigma(0)_{fav} = C\left|\overline{S}_+(0)\right|^2 \qquad \sigma(1)_{unf} = 3C\left|\overline{S}_0(1)\right|^2 \tag{7.78}$$

where C is a common factor that will cancel when we compute β. Speaking of β, we now proceed to find it. From (7.39c, 7.39d),

$$\beta(j_t = 0)_{fav} = 2 \qquad \beta(j_t = 1)_{unf} = -1 \tag{7.79}$$

(7.42) gives us the asymmetry parameter [95]

$$\beta = \frac{2\sigma(0) - \sigma(1)}{\sigma(0) + \sigma(1)} \tag{7.80}$$

If we wish to account for spin-orbit interactions, we can compute $\left|\overline{S}_\pm(j_t)\right|$ and $\left|\overline{S}_0(j_t)\right|$ either by comparing (7.78) and (7.80) to (6.84) or by applying (7.59). Either way, we find [95, 96]

$$\sigma(0) \propto \left|\overline{S}_+(0)\right|^2 = \frac{2}{3}\left[\mathfrak{R}_{1/2}^2 + 4\mathfrak{R}_{3/2}^2 + 4\mathfrak{R}_{1/2}\mathfrak{R}_{3/2}\cos\left(\delta_{3/2} - \delta_{1/2}\right)\right] \tag{7.81a}$$

$$\sigma(1)_{uf} \propto 3\left|\overline{S}_0(1)\right|^2 = \frac{4}{3}\left[\mathfrak{R}_{1/2}^2 + \mathfrak{R}_{3/2}^2 - 2\mathfrak{R}_{1/2}\mathfrak{R}_{3/2}\cos\left(\delta_{3/2} - \delta_{1/2}\right)\right] \tag{7.81b}$$

where $\mathfrak{R}_j$ are the radial dipole matrix elements for the two possible final states with phases δ_j. To a very good approximation, $\mathfrak{R}_j$ and δ_j for the hydrogen atom are independent of j. As a result, from equation (7.81b), we see that $\sigma(1)_{uf} = 0$, giving us [from (7.80)] an asymmetry parameter $\beta = 2$. ∎

Example 7.4 As another example, we consider an s-subshell photoionization process of the type

$$As^2p^q\left(^{2S_0+1}L_0\right) + \gamma \rightarrow \left[A^+sp^q\left(^{2S_c+1}L_c\right) + e_{E_{cl}}^-\right]_{\left(^{2S+1}L\right)} \tag{7.82}$$

Specifically, we examine the inner-shell photoionization from the ground state of Cl [90, 94, 95]

$$\mathrm{Cl}3s^2 3p^5 \left(^2P\right) + \gamma \rightarrow \left[\mathrm{Cl}^+ 3s3p^5 \left(^{1,3}P\right) + e_p^-\right]_{\left(^2S,^2P,^2D\right)} \tag{7.83}$$

Angular momentum conservation (7.24) tells us

$$\vec{L} = \vec{L}_c + l = \vec{1} + \vec{1} \Rightarrow L = 0, 1, 2 \tag{7.84}$$

hence the term symbols $(^2S, ^2P, ^2D)$.

Equation (7.25) gives us the allowed angular momentum transfer values:

$$\vec{j}_t = \vec{1} - \vec{l} = \vec{1} - \vec{1} \Rightarrow j_t = 0, 1, 2 \tag{7.85}$$

Parity conservation (7.26) implies

$$\pi_A \pi_c = (-1)^{l+1} \tag{7.86}$$

where (remembering that s-electrons do not contribute to the parity)

$$\pi_A \pi_c = (-1)^{\left(\sum l_i\right)_A} (-1)^{\left(\sum l_i\right)_c} = (-1)^5 (-1)^5 = (-)(-) = (+)$$

$$\Rightarrow l = odd = 1 \tag{7.87}$$

which checks with what we already knew. From (7.29), we find which of our values of j_t are parity-unfavored:

$$\pi_A \pi_c = (-)(-) = (+) = (-1)^{j_t + 1} \Rightarrow (j_t)_{unf} = 1 \tag{7.88}$$

And from (7.34), we find which of our values of j_t are parity-favored:

$$\pi_A \pi_c = (-)(-) = (+) = (-1)^{j_t} \Rightarrow (j_t)_{fav} = 0, 2 \tag{7.89}$$

Recalling (7.39a, 7.39b) and remembering that $\left|\bar{S}_\pm(j_t)\right|$ and $\left|\bar{S}_0(j_t)\right|$ represent (for this example) photoionization amplitudes for photoelectron escape angular momenta $l = j_t \pm 1$ and $l = j_t$ respectively, we see that for $\sigma(j_t = 0)_{fav}$, $\left|\bar{S}_-(0)\right| = 0$ because $l = -1$ is not allowed. We also see that for $\sigma(j_t = 2)_{fav}$, $\left|\bar{S}_+(2)\right| = 0$ because $l = 3$ is also forbidden by angular momentum conservation in this approximation.

So, from (7.39a, 7.39b), we have

$$\sigma(0)_{fav} = C\left|\bar{S}_+(0)\right|^2 \quad \sigma(2)_{fav} = 5C\left|\bar{S}_-(2)\right|^2 \quad \sigma(1)_{uf} = 3C\left|\bar{S}_0(1)\right|^2 \tag{7.90}$$

where C is a common factor that will cancel when we compute β. Speaking of β, we now proceed to find it.

From (7.39c, 7.39d),

$$\beta(j_t = 0)_{fav} = 2 \qquad \beta(j_t = 2)_{fav} = 1/5 \qquad \beta(j_t = 1)_{uf} = -1 \qquad (7.91)$$

Equation (7.42) gives us the asymmetry parameter [90]:

$$\beta = \frac{2\sigma(0) + (1/5)\sigma(2) - \sigma(1)}{\sigma(0) + \sigma(2) + \sigma(1)} \qquad (7.92)$$

We use (7.65) to find the scattering amplitudes [90]:

$$\bar{S}_+(0) \propto \sqrt{2}\underbrace{\begin{pmatrix} 1 & 1 & 0 \\ 0 & 1 & 1 \end{pmatrix}}_{-1/3} \times \left[\begin{array}{l} \underbrace{\begin{Bmatrix} 0 & 1 & 1 \\ 0 & 1 & 1 \end{Bmatrix}}_{1/3}\underbrace{\begin{Bmatrix} 1 & 0 & 1 \\ 1 & 0 & 1 \end{Bmatrix}}_{1/3}\exp\left(-i\delta_{e_p^-}^{1S_c0}\right)\mathfrak{R}_{e_p^-}^{1S_c0} \\[4mm] +3\underbrace{\begin{Bmatrix} 1 & 1 & 1 \\ 0 & 1 & 1 \end{Bmatrix}}_{-1/3}\underbrace{\begin{Bmatrix} 1 & 0 & 1 \\ 1 & 1 & 1 \end{Bmatrix}}_{-1/3}\exp\left(-i\delta_{e_p^-}^{1S_c1}\right)\mathfrak{R}_{e_p^-}^{1S_c1} \\[4mm] +5\underbrace{\begin{Bmatrix} 2 & 1 & 1 \\ 0 & 1 & 1 \end{Bmatrix}}_{1/3}\underbrace{\begin{Bmatrix} 1 & 0 & 1 \\ 1 & 2 & 1 \end{Bmatrix}}_{1/3}\exp\left(-i\delta_{e_p^-}^{1S_c2}\right)\mathfrak{R}_{e_p^-}^{1S_c2} \end{array} \right]$$

$$\Rightarrow \bar{S}_+(0) = \frac{C}{3}\left[\exp\left(-i\delta_{e_p^-}^{1S_c0}\right)\mathfrak{R}_{e_p^-}^{1S_c0} + 3\exp\left(-i\delta_{e_p^-}^{1S_c1}\right)\mathfrak{R}_{e_p^-}^{L1S_c1} + 5\exp\left(-i\delta_{e_p^-}^{1S_c2}\right)\mathfrak{R}_{e_p^-}^{1S_c2} \right]$$

$$(7.93a)$$

$$\bar{S}_0(1) \propto \sqrt{2}\underbrace{\begin{pmatrix} 1 & 1 & 0 \\ 0 & 1 & 1 \end{pmatrix}}_{-1/3} \times \left[\begin{array}{l} \underbrace{\begin{Bmatrix} 0 & 1 & 1 \\ 0 & 1 & 1 \end{Bmatrix}}_{1/3}\underbrace{\begin{Bmatrix} 1 & 1 & 1 \\ 1 & 0 & 1 \end{Bmatrix}}_{-1/3}\exp\left(-i\delta_{e_p^-}^{1S_c0}\right)\mathfrak{R}_{e_p^-}^{1S_c0} \\[4mm] +3\underbrace{\begin{Bmatrix} 1 & 1 & 1 \\ 0 & 1 & 1 \end{Bmatrix}}_{-1/3}\underbrace{\begin{Bmatrix} 1 & 1 & 1 \\ 1 & 1 & 1 \end{Bmatrix}}_{1/6}\exp\left(-i\delta_{e_p^-}^{1S_c1}\right)\mathfrak{R}_{e_p^-}^{1S_c1} \\[4mm] +5\underbrace{\begin{Bmatrix} 2 & 1 & 1 \\ 0 & 1 & 1 \end{Bmatrix}}_{1/3}\underbrace{\begin{Bmatrix} 1 & 1 & 1 \\ 1 & 2 & 1 \end{Bmatrix}}_{1/6}\exp\left(-i\delta_{e_p^-}^{1S_c2}\right)\mathfrak{R}_{e_p^-}^{1S_c2} \end{array} \right]$$

$$\Rightarrow \overline{S}_0(1) = \frac{C}{6} \left[2 \exp\left(-i\delta_{e_p^-}^{1S_c0}\right) \mathfrak{R}_{e_p^-}^{1S_c0} + 3 \exp\left(-i\delta_{e_p^-}^{1S_c1}\right) \mathfrak{R}_{e_p^-}^{1S_c1} - 5 \exp\left(-i\delta_{e_p^-}^{1S_c2}\right) \mathfrak{R}_{e_p^-}^{1S_c2} \right]$$

$$(7.93b)$$

and similarly,

$$\overline{S}_-(2) = \frac{C}{6} \left[2 \exp\left(-i\delta_{e_p^-}^{1S_c0}\right) \mathfrak{R}_{e_p^-}^{1S_c0} - 3 \exp\left(-i\delta_{e_p^-}^{1S_c1}\right) \mathfrak{R}_{e_p^-}^{1S_c1} + \exp\left(-i\delta_{e_p^-}^{1S_c2}\right) \mathfrak{R}_{e_p^-}^{1S_c2} \right]$$

$$(7.93c)$$

where (again) C is common factor that will cancel when we substitute (7.93) and (7.90) into (7.92). At this point, the reader is reminded that spin-orbit effects have been neglected (LS-coupling approximation) and that we neglected any configuration interactions when performing the calculations in this example.

When (7.90), (7.93a, 7.93b and 7.93c), and (7.92) are combined, it is clear that an analysis of this pd reaction via angular momentum transfer theory delivers a value for the asymmetry parameter that deviates radically from the (constant) C-Z value of $\beta = 2$. In particular, we see that the radial integrals in (7.93a, 7.93b and 7.93c) ensure that β has a strong dependance on the incident photon energy. In addition, the three values for j_t reflect the fact that there is coupling between the three possible final channels ($L = 0, 1, 2$). In fact, the scattering amplitudes of (7.93a, 7.93b and 7.93c) are linear combinations of the dipole matrix elements for each of the possible final states. For the final state term (^3P), each of the three dipole matrix elements undergoes a change of sign at various (different) photon energies (this is known as reaching a Cooper minimum). As a result, near the Cooper minimum, the influence of $\overline{S}_0(1)$ and $\overline{S}_-(2)$ will have a significant impact on the value of β [90, 95]. Finally, we recover the C-Z value for β when anisotropic interactions are ignored, that is, when the radial dipole matrix elements and phase shifts are independent of the final-state term L. In that case, (7.93b, 7.93c) show that $\overline{S}_0(1) = \overline{S}_-(2) = 0$. It is then from (7.92) that we regain the C-Z value $\beta = 2$ [90, 95]. For similar analyses (and measurements) of different pd (and autoionizing) reactions for other systems using angular momentum transfer theory, the reader is invited to examine [87, 88, 92, 93, 96–106]. ■

Problems

7.1. Starting with (C.70), derive (7.15).

7.2. Derive (7.31b). Use that result to verify (7.35a) and (7.35b).

HINT: Use (B.82)

7.3. Single-photon photodetachment[4]

   The removal of an electron from a neutral atom (or molecule) or from a positively charged atomic (or molecular) species by photon absorption is known as photoionization. The removal of an electron from a negative ion by photon absorption is known as photodetachment. These processes are given different names because a negatively charged ion is qualitatively different from a neutral or a positively charged ion. This is because the structure of a negative ion is fundamentally different from that of a neutral atom or positive ion due to increased screening of the nuclear charge. In neutral atoms and positive ions, bound electrons primarily react to the long-range Coulomb potential of the nuclear core, producing an infinite number of bound states. In contrast, the masking of the (Coulombic) nucleon-electron interaction due to increased screening in negative ions allows for only a limited number of bound states and elevates electron-electron interactions (known as electron correlations) to such a point that they will often dominate the Coulomb forces in these systems, at least for the outer valence electrons. The fundamental difference in structure between a neutral or positive ion on the one hand, and a negative ion on the other, leads to differences in the way that these respective systems interact with photons. For example, the physics of the photoionization process (and thus the photoionization cross-section) will differ from the physics of the photodetachment process (and thus the photodetachment cross-section) in a fundamental way due to these structural differences. We will now prove a theorem for the single-photon photodetachment process which we write as

$$A^- + photon(h\nu) \rightarrow A + e^-$$

In this process, a negative ion A^- absorbs a photon of energy $h\nu$, resulting in the liberation of a photoelectron, leaving behind a neutral core. The photoelectron has a kinetic energy that is equal to the difference between the photon energy $h\nu$ and the binding energy of the electron in the negative ion.

   You will prove the following theorem for the photodetachment cross-section near threshold, that is, under conditions for which the detached electron has a very low kinetic energy, or equivalently, a small momentum, $\vec{k} = \hbar\vec{q}$. The theorem is

---

[4]This problem appears in a slightly different form in [107], pp. 132–133.

$$\sigma_{pd} \sim \left\{ \begin{array}{ll} q^3, & l_0 = 0 \\ q^{2l_0+1}, & l_0 \geq 1 \end{array} \right\}$$

where σ_{pd} is the photodetachment (pd) cross-section and l_0 is the orbital angular momentum of the bound electron (before the pd event).

We start with the basic form of the cross-section in the dipole approximation,

$$\sigma_{pd} \propto q \int |r_{qi}|^2 d\Omega_q = q \int \left| \iint \psi_i\left(\vec{r}\right) \vec{r} \psi\left(\vec{q}\right) r^2 dr d\Omega_r \right|^2 d\Omega_q \tag{1}$$

where $\psi_i\left(\vec{r}\right)$ is the wave function for the initial state (of the bound electron) and $\psi\left(\vec{q}\right)$ is the wave function for the (unbound) photoelectron. The term inside the bracket is the dipole matrix element.

Now we make the following assumptions, which are valid only for negative ions:

For photodetachment, we can assume that the core on which the extra electron resides is spherically symmetrical and that this core remains relatively undisturbed by the pd process. Thus, we will use hydrogenic wave functions for the initial bound state. And so we write, as the initial state of the bound electron,

$$\psi_i\left(\vec{r}\right) = \frac{1}{r} u_{nl_0}(r) Y_{lm}(\theta, \phi) \tag{2}$$

where $u_{nl_0}(r)$ is the radial portion of the (hydrogenic) initial bound state and is normalized as follows

$$\int_0^\infty u_{nl_0}^2 dr = 1 \tag{3}$$

As for the photoelectron, we invoke the standard series expansion in partial waves of angular momentum l for its wave function:

$$\psi\left(\vec{q}\right) = \frac{1}{r} \sum_{l=0}^\infty (2l+1) i^l u_l(q, r) P_l\left(\cos\theta_{\vec{q}\cdot\vec{r}}\right) \tag{4}$$

Now we introduce a direction given by a vector $\vec{s}$ such that the projection of the orbital angular momentum of the bound electron is zero in this direction. Then we can simplify via equation (2.60)

$$Y_{lm}(\theta, \phi) = Y_{l0}(\theta, \phi) = \sqrt{\frac{2l_0 + 1}{4\pi}} P_{l_0}\left(\cos\theta_{\vec{r}\cdot\vec{s}}\right) \tag{5}$$

(a) Evaluate equation (1) to arrive at the following:

$$\sigma_{pd} \propto \frac{q}{(2l_0 + 1)}\left[(l_0 + 1)|\mathfrak{R}_{l_0,l_0+1}|^2 + (l_0)|\mathfrak{R}_{l_0,l_0-1}|^2\right] \tag{6}$$

where the radial portion of the dipole matrix element is given by

$$\mathfrak{R}_{l_0l} \equiv \int dr u_{nl_0}(r) r u_l(q, r) \tag{7}$$

(b) Since the forces involved in the pd event are short range (fall off faster than $1/r^2$), the wave function of the photoelectron can be described by a plane wave in a one-electron approximation. We justify this by observing that photoelectrons produced in a near-threshold pd event reach the far-field regions with very little velocity and so spend most of their time at large r, where the interaction potential is practically zero. Thus, we write

$$\psi\left(\vec{q}\right) = \exp\left(i\vec{q}\cdot\vec{r}\right) = \sum_{l=0}^{\infty}(2l + 1)i^l j_l(qr) P_l\left(\cos\theta_{\vec{q}\cdot\vec{r}}\right) \tag{8}$$

Because the forces involved in the pd event are short ranged, the radial matrix elements $\mathfrak{R}_{l_0,l_0\pm 1}$ will be determined by the behavior of the radial wave functions near the parent neutral, in a region $r\sim 1$ (in atomic units). In this case, the radial portion of the initial-state wave function $u_{nl_0}(r)$ will be on the order of unity, especially if we consider the atom to be in the ground state. Under these conditions, and noting that, near threshold, $q \ll 1$, we can also form an appropriate approximate expression for the final-state radial wave function $u_l(q, r) = r j_l(qr)$ and so finally arrive at the final form of the theorem.

7.4. Another representation for the spherical Bessel functions is given by

$$j_l(\rho) = (-\rho)^l\left(\frac{1}{\rho}\frac{\partial}{\partial\rho}\right)^l\left(\frac{\sin\rho}{\rho}\right) \quad \text{and} \quad n_l(\rho) = (-\rho)^l\left(\frac{1}{\rho}\frac{\partial}{\partial\rho}\right)^l\left(\frac{\cos\rho}{\rho}\right)$$

Prove by induction that these forms satisfy the differential equation for the spherical Bessel functions:

$$\left[\frac{d^2}{d\rho^2} + \frac{2}{\rho}\frac{d}{d\rho} - \frac{l(l + 1)}{\rho^2} + 1\right]R(\rho) = 0; R(\rho) \equiv \begin{cases} j_l(\rho) \\ n_l(\rho) \end{cases}$$

7.5. Inner-shell, s-subshell photoionization for an atomic species A in the LS coupling approximation [90]

(a) Suppose we have an inner-shell photoionization process of the form

$$As^2p^4\left(^3P\right) + \gamma \rightarrow \left[A^+sp^4\left(^2D\right) + e_p^-\right]_{\left(^3P,^3D\right)}$$

Show that the asymmetry parameter has an upper bound of $\beta \leq 1/5$.

(b) Suppose we have an inner-shell photoionization process of the form

$$As^2p^2\left(^3P\right) + \gamma \rightarrow \left[A^+sp^2\left(^2S\right) + e_p^-\right]_{\left(^3P\right)}$$

What is the expected photoelectron angular distribution?

(c) Suppose we have an inner-shell photoionization process of the form

$$As^2p^2\left(^1S\right) + \gamma \rightarrow \left[A^+sp^2\left(^2D\right) + e_p^-\right]_{\left(^1P\right)}$$

Is the asymmetry parameter constant, and if so, what is its value?

7.6. p-subshell photoionization for an atomic species A in the LS coupling approximation [90]

(a) Consider the following photoionization process:

$$As^2p^3\left(^2D\right) + \gamma \rightarrow \left[A^+s^2p^2\left(^1S\right) + e_d^-\right]_{\left(^2D\right)}$$

Find the asymmetry parameter β.

(b) Consider the following photoionization process in which one of the p-electrons is ionized and the other is excited to a bound s-state (the atom may have several other full subshells):

$$Ap^2\left(^3P\right) + \gamma \rightarrow \left[A^+s\left(^2S\right) + e_d^-\right]_{\left(^3P\right)}$$

Find the asymmetry parameter β.

7.7. Consider the following inner-shell photoionization transition in carbon [90]:

$$C2s^2 2p^2\left({}^3P_{2,1,0}\right) + \gamma \rightarrow \left[C^+ 2s2p^2\left({}^2S_{1/2}\right) + e_p^-\right]_{\left({}^1P_1, {}^3P_{2,1,0}\right)}$$

Find the form for the asymmetry parameter β for each of the three listed initial total angular momenta $j_0 = 2, 1, 0$. Do this for both the LS coupling case and the case for which relativistic (spin-orbit) effects are allowed.

7.8. Show that the asymmetry parameter for the photodetachment reaction of (10.1) can be written as

$$\beta = \frac{3\left|\bar{S}_+(1)\right|^2 - 5\left|\bar{S}_0(2)\right|^2 - 3\sqrt{2}\left(\bar{S}_+(1)\bar{S}_-^*(1) + \bar{S}_-(1)\bar{S}_+^*(1)\right)}{3\left[\left|\bar{S}_-(1)\right|^2 + \left|\bar{S}_+(1)\right|^2\right] + 5\left|\bar{S}_0(2)\right|^2}$$

7.9. Consider the following photoionization reaction in sulfur in the LS-coupling approximation [92, 93]:

$$S[Ne]3s^2 3p^4\left({}^3P\right) + \gamma \rightarrow \left[S^+[Ne]3s^2 3p^3\left({}^2D\right) + e_{s,d}^-\right]_{\left({}^3S, {}^3P, {}^3D\right)}$$

Find the asymmetry parameter β. Show that it reduces to equation (4.40) when anisotropic interactions are ignored, that is, when the radial dipole matrix elements and phase shifts are independent of L_{S^+}, S_{S^+}, L.

7.10. Find the asymmetry parameter β for the single-photon photoionization of the alkali metals [96]. Allow for spin-orbit effects. What value for β do you get if you neglect spin-orbit effects? Discuss the reasons for the two different results.

7.11. Consider the following photodetachment reaction in iron in the LS-coupling approximation [47]:

$$Fe^-[Ar]3d^7 4s^2\left({}^4F\right) + \gamma \rightarrow Fe[Ar]3d^6 4s^2\left({}^5D\right) + e_{p,f}^-$$

Find the asymmetry parameter β. Show that it reduces to (4.39b) when anisotropic interactions are ignored, that is, when the radial dipole matrix elements and phase shifts are independent of L_{Fe}, S_{Fe}, L.

7.12. Non-statistical weighting of relative line intensities in LS coupling [47, 108]
Start with (D.15), and associate the operators $T^{(k_1)}$ and $T^{(k_2)}$ with the annihilation operations involved in the photodetachment process of converting a negative ion in the state ${}^{2s+1}L_j$ (total orbital angular momentum L), which is assumed to be well-described by an LS-coupling scheme, into a neutral atom in the state ${}^{2s'+1}L'_{j'}$ (total orbital angular momentum L'). The photodetachment

process annihilates an electron of spin $k_1 = 1/2$ and orbital angular momentum k_2 and promotes it into the continuum via the annihilation operators $T^{(k_1)}$ and $T^{(k_2)}$, respectively (note that the L-operator and the S-operator both satisfy the definitions of a spherical tensor operator). Demonstrate the phenomenon of "non-statistical" line-width narrowing in the photoelectron kinetic energy spectrum and explain its origin. Find any selection rules that are involved. Assume all j-levels of the negative ion are populated in accordance with their degeneracies and that kT is much greater than the fine-structure splitting. Analyze the special case in which an s-electron ($k_2 = 0$) is detached.

Chapter 8
Molecular Photoelectron Angular Distributions

Molecules present different challenges than do atoms when attempting to describe PADs. The noncentral nature of the molecular potential influences the departing photoelectron so that the resulting angular distribution may depend on the portion of the molecular potential the photoelectron samples as it departs. In addition, the molecule may be vibrating and rotating so that the molecular potential is varying in space and time. The bonds between the constituent atoms in the molecule may be flexing, and the angles between the atoms may be changing during the pd process. Indeed, the molecule itself may not survive the pd process, and the photoelectron may be leaving just as some (or all) the atoms (and/or Auger electrons) in the molecular family also take their leave. Nevertheless, there are many similarities between atomic and molecular PADs, and we start our analysis there.

As is commonly done, we express the wave functions of the molecular electrons in terms of atomic wave functions (or at least in terms of functions that are similarly normalized to unity). We continue to stick to the simplified, single-electron, independent-particle model. We can then expand the initial state of the molecule in a series of spherical harmonics in a manner similar to what was done in Chaps. 4 and 7 [c.f. (4.12) and (7.4)] [3, 109, 110]:

$$|i\rangle = \sum_{l'm'} c'_{l'm'} R_{nl'}(r) Y_{l'm'}(\theta', \phi')$$ (8.1)

where (θ', ϕ') are angles measured about an internuclear axis. We must keep in mind that the summation in (8.1) can only include those terms which reflect the symmetry of the molecular potential about that axis [13]. The wave function for the photoelectron can also be represented as it was in Chap. 4 [c.f. (4.15)]:

The original version of this chapter was revised. The correction to this chapter is available at https://doi.org/10.1007/978-3-031-08027-2_11

V. T. Davis, *Introduction to Photoelectron Angular Distributions*, Springer Tracts in Modern Physics 286, https://doi.org/10.1007/978-3-031-08027-2_8

$$|f\rangle \propto 4\pi \sum_{l,m} c_{lm} Y_{lm}^*\left(\widehat{k}\right) Y_{lm}(\widehat{r}) G_{kl}(r) \tag{8.2}$$

As before, $\widehat{k}$ represents the direction of the incident photon and $\widehat{r}$ the direction of the departing photoelectron, both measured in the lab frame from some designated origin (usually inside the molecule). Before we proceed, we must bring the molecular frame into congruence with the lab frame. We do this by describing the orientation of the chosen molecular axis in relation to the fixed lab frame axes with the Euler angles. That means we can transform the angular part of the initial wave function into the laboratory frame via the rotation [c.f. (3.24)]:

$$|l'm'\rangle = \sum_{m''} D_{m''m'}^{l'}(\alpha,\beta,\gamma)|l'm''\rangle \tag{8.3}$$

To find the differential pd cross-section, we can now proceed as we did in Chap. 4. But first, we must realize that (8.3) has made our calculations a function of the orientation of the molecules with respect to the fixed lab frame. We account for the dependence of the differential cross-section on these orientations by averaging in a way that depends upon the design of experiment being used to measure the PADs. If the molecules are randomly oriented (such as in experiments that cross gas jets with photon beams—see Chap. 9), then we are compelled to average over all possible molecular orientations with equal weight, i.e., integrate over the entire range of the Euler angles. Using (8.3) and averaging, we find that (4.22) is consequently multiplied by an additional factor I, where

$$\begin{aligned}
I &= \frac{1}{8\pi^2} \sum_{m''m'''} \int_0^{2\pi} d\alpha \int_0^{\pi} \sin\beta d\beta \int_0^{2\pi} d\gamma D_{m''m'}^{l'*}(\alpha,\beta,\gamma) D_{m'''m'}^{l'}(\alpha,\beta,\gamma) \\
&= \frac{1}{8\pi^2} \frac{8\pi^2}{2l'+1} \sum_{m''m'''} \delta_{m''m'''} = \frac{1}{2l'+1} \sum_{m''} 1 = 1
\end{aligned} \tag{8.4}$$

where we used (2.153). The rest of the analysis unfolds in the same way it did in Chap. 4 with the same ultimate results [3, 109–111].

Let us now proceed to find the form of the differential pd cross-section under the condition that the molecules have a definite (fixed) orientation with respect to the lab axes. We continue to connect the lab frame and the molecular frame via the Euler angles and the rotation operator of (2.92). The orientation of the photon polarization vector and the orientation of the electron detector are characterized by the operators

$$\mathscr{R}_\gamma = \left\{\alpha_\gamma, \beta_\gamma, \gamma_\gamma\right\} \tag{8.5a}$$

and

$$\mathscr{R}_e = \left\{\alpha_e, \beta_e, \gamma_e\right\} \tag{8.5b}$$

which bring the molecular frame into congruence with each of these two frames, respectively. Normally, we combine the two frames into a single, fixed laboratory frame characterized by the rotation operator

$$\mathscr{R} = \{\alpha, \beta, \gamma\} = \mathscr{R}_\gamma^{-1} = \mathscr{R}_e^{-1} \tag{8.6}$$

which carries the lab frame onto the molecular frame. The pd differential cross-section is given by [c.f. (4.9)]

$$\frac{d\sigma}{d\widehat{\mathscr{R}}_e d\widehat{\mathscr{R}}_\gamma} \propto |\langle f|O|i\rangle|^2 \tag{8.7}$$

where (for example) $\widehat{\mathscr{R}}_e$ represents a unit vector in the direction of the indicated rotation.

The operator is the electric dipole operator of (4.10), which we rotate into the molecular frame as follows:

$$O \equiv r\sqrt{\frac{4\pi}{3}} \sum_{m_\gamma} D_{m_\gamma 0}^1 (\mathscr{R}_\gamma) Y_{1m_\gamma}(\widehat{r}) \tag{8.8}$$

Note that this photon is linearly polarized along the z-axis of the photon frame.

The form of the final state is found in (4.15):

$$|f\rangle \equiv \Psi^{(-)}\left(\vec{k}_e, \vec{r}\right) = \sum_{l,m} a_{lm}^{(-)}\left(\widehat{k}_e\right) \psi_{lm}^{(-)}\left(k_e, \vec{r}\right) \tag{8.9}$$

where

$$a_{lm}^{(-)}\left(\widehat{k}_e\right) = 4\pi(i)^l e^{-i\delta_l} Y_{lm}^*\left(\widehat{k}_e\right) \tag{8.10a}$$

$$\psi_{lm}^{(-)}(k_e, r) = Y_{lm}(\widehat{r}) G_{kl}(r) \tag{8.10b}$$

The $(-)$ superscript designates normalization in accordance with the appropriate boundary conditions.

The electron is ejected along the $\widehat{k}_e = \{\theta_e, \phi_e\}$ direction as measured in the frame of the molecule. In anticipation of what is to come, we make the following transformations [c.f. (2.155) and (2.149)]:

$$Y_{l'm'}^*\left(\widehat{k}_e\right) Y_{lm}\left(\widehat{k}_e\right) = (-1)^{m'} \sqrt{\frac{(2l'+1)(2l+1)}{4\pi}} \sum_{K_e} (2K_e + 1)^{-1/2}$$
$$\times \langle l, l'; 0, 0|K_e, 0\rangle \langle l, l'; m, -m'|K_e, M_e\rangle Y_{K_e M_e}\left(\widehat{k}_e\right) \tag{8.11}$$

and

$$D_{m'_\gamma 0}^{1*}(\mathcal{R}_\gamma)D_{m_\gamma 0}^1(\mathcal{R}_\gamma) = (-1)^{m'_\gamma}\sum_{K_\gamma}\langle 1,1;0,0|K_\gamma,0\rangle$$

$$\langle 1,1;m_\gamma,-m'_\gamma|K_\gamma,M_\gamma\rangle D_{M_\gamma 0}^{K_\gamma}(\mathcal{R}_\gamma) \tag{8.12}$$

We substitute (8.8, 8.9, 8.10a, 8.10b, 8.11 and 8.12) into (8.7) and perform the indicated square modulus [109, 110]:

$$\frac{d\sigma}{d\widehat{\mathcal{R}}_\gamma d\widehat{k}_e} = 16\pi^2\sum_{ll'}\sum_{mm'}\sum_{m_\gamma m'_\gamma}(-1)^{m'+m'_\gamma}(i)^{l'-l}e^{-i(\delta_l-\delta_{l'})}\sqrt{\frac{(2l'+1)(2l+1)}{4\pi}}D_{l'm'm'_\gamma}^{(-)\Gamma_0*}D_{lmm_\gamma}^{(-)\Gamma_0}$$

$$\times\sum_{K_e}(2K_e+1)^{-1/2}\langle l,l';0,0|K_e,0\rangle\langle l,l';m,-m'|K_e,M_e\rangle Y_{K_e M_e}\left(\widehat{k}_e\right)$$

$$\times\sum_{K_\gamma}\langle 1,1;0,0|K_\gamma,0\rangle\langle 1,1;m_\gamma,-m'_\gamma|K_\gamma,M_\gamma\rangle D_{M_\gamma,0}^{K_\gamma}(\mathcal{R}_\gamma) \tag{8.13}$$

The dynamics of the pd process are contained in the matrix elements [109, 110]:

$$D_{lmm_\gamma}^{(-)\Gamma_0}(k_e) = \sqrt{\frac{4\pi}{3}}\int d\vec{r}\,\psi_{lm}^{(-)*}\left(k_e,\vec{r}\right)rY_{1m_\gamma}(\widehat{r})\Psi_{\Gamma_0}\left(\vec{r}\right) \tag{8.14}$$

where we have made the identification

$$|i\rangle \equiv \Psi_{\Gamma_0}\left(\vec{r}\right) \tag{8.15}$$

The C-G coefficient(s) on the RHS side of (8.13) restrict the values of K_e and K_γ to the intervals $0 \leq K_e \leq 2l_{max}$ and $0 \leq K_\gamma \leq 2$, respectively. So, for a given orientation $\mathcal{R}_\gamma$ of the photon frame, the differential cross-section has the form [109, 110]

$$\frac{d\sigma}{d\Omega} = \sum_{K_e=0}^{2l_{max}}A_{K_e m_e}Y_{K_e M_e}\left(\widehat{k}_e\right) \tag{8.16}$$

where l_{max} is the largest of the allowed components of the orbital angular momentum of the photoelectron (determined by the various molecular symmetries) and the direction $\widehat{k}_e$ is measured in the frame of the molecule. Equation (8.16) shows the presence of odd harmonics which are not present in the differential cross-section for randomly oriented molecules.[1] Apparently, a fixed-molecule experiment can provide additional information that is otherwise "washed out" by the freely tumbling molecules one encounters in a typical gas-jet experiment [109, 110].

Armed with the general result of (8.16), we can now examine some specific examples.

Example 8.1

Find the differential pd cross-section if the z-axes of the photon and molecule frames coincide.

Setting the angle $\beta_\gamma = 0$, the polarization of the photon now lies along the z-axis of the molecule. In that case, from (2.109) and (2.108),

$$D_{M_\gamma,0}^{K_\gamma}(\alpha_\gamma, 0, \gamma_\gamma) = \delta_{M_\gamma,0} \tag{8.17}$$

giving us [109, 110]

$$\frac{d\sigma(\beta_\gamma = 0)}{d\hat{k}_e} = \frac{d\sigma(\beta_\gamma = 0)}{d\mathcal{R}_\gamma(\beta_\gamma = 0)d\hat{k}_e}$$

$$= 16\pi^2 \sum_{ll'} \sum_{mm'} \sum_{m_\gamma m'_\gamma} (-1)^{m'+m'_\gamma} (i)^{l-l'} e^{-i(\delta_l - \delta_{l'})} \sqrt{\frac{(2l'+1)(2l+1)}{4\pi}} D_{l'm'm'_\gamma}^{(-)\Gamma_0*} D_{lmm_\gamma}^{(-)\Gamma_0}$$

$$\times \sum_{K_e} (2K_e + 1)^{-1/2} \langle l, l'; 0, 0 | K_e, 0 \rangle \langle l, l'; m, -m' | K_e, M_e \rangle Y_{K_e M_e}\left(\hat{k}_e\right)$$

$$\times \sum_{K_\gamma} \langle 1, 1; 0, 0 | K_\gamma, 0 \rangle \langle 1, 1; m_\gamma, -m'_\gamma | K_\gamma, 0 \rangle \tag{8.18}$$

The last C-G coefficient on the RHS of (8.18) ensures that $m_\gamma = m'_\gamma$. For polarization along the molecular axis, we have $m_\gamma = m'_\gamma = 0$.[2]

---

[1] Equations (3.29) and (4.38) are of the form $d\sigma/d\Omega = \sum_{K=0,2} A_K P_K(\cos\theta) = \frac{\sigma_{pd}}{4\pi}[1 + \beta P_2(\cos\theta)]$.

This same low-harmonic dependance is seen if the molecules have random orientations.

[2] A photon polarization perpendicular to the molecular axis would result in a restriction of the values of m_γ, m'_γ to $m_\gamma = m'_\gamma = \pm 1$ [109, 110].

$$\frac{d\sigma(\beta_\gamma = 0)}{d\widehat{k}_e} = \frac{d\sigma(\beta_\gamma = 0)}{d\mathcal{R}_\gamma(\beta_\gamma = 0)d\widehat{k}_e}$$

$$= 16\pi^2 \sum_{ll'} \sum_{mm'} (i)^{l-l'} e^{-i(\delta_l - \delta_{l'})} \sqrt{\frac{(2l'+1)(2l+1)}{4\pi}} D_{l'm'0}^{(-)\Gamma_0*} D_{lm0}^{(-)\Gamma_0}$$

$$\times \sum_{K_e} (2K_e+1)^{-1/2} \langle l, l'; 0, 0 | K_e, 0 \rangle \langle l, l'; m, -m' | K_e, M_e \rangle Y_{K_e M_e}\left(\widehat{k}_e\right)$$

$$\times \underbrace{\sum_{K_\gamma} \langle 1, 1; 0, 0 | K_\gamma, 0 \rangle \langle 1, 1; 0, 0 | K_\gamma, 0 \rangle}_{=1}$$

$$\Rightarrow \frac{d\sigma(\beta_\gamma = 0)}{d\widehat{k}_e} = 16\pi^2 \sum_{ll'} \sum_{mm'} (-1)^{m'} (i)^{l-l'} e^{-i(\delta_l - \delta_{l'})} \sqrt{\frac{(2l'+1)(2l+1)}{4\pi}} D_{l'm'0}^{(-)\Gamma_0*} D_{lm0}^{(-)\Gamma_0}$$

$$\times \sum_{K_e} (2K_e+1)^{-1/2} \langle l, l'; 0, 0 | K_e, 0 \rangle \langle l, l'; m, -m' | K_e, M_e \rangle Y_{K_e M_e}\left(\widehat{k}_e\right)$$

$$(8.19)$$

where we also used (2.130). ∎

Example 8.2

Find the total pd cross-section $\sigma(\beta_\gamma = 0)$ so that it is independent of ejection direction.

The total pd cross-section we want is found by integrating over the ejection directions [109, 110]:

$$\sigma(\beta_\gamma = 0) = \int \left(d\sigma(\beta_\gamma = 0)/d\widehat{k}_e \right) d\widehat{k}_e \qquad (8.20)$$

To find $\sigma(\beta_\gamma = 0)$, we first note from (2.64)

$$\int d\widehat{k}_e Y_{K_e M_e}\left(\widehat{k}_e\right) = \sqrt{4\pi}\delta_{K_e,0}\delta_{M_e,0} \qquad (8.21)$$

So that,

$$\sigma(\beta_\gamma = 0) = 16\pi^2 \sum_{ll'} \sum_{mm'} (-1)^{m'} (i)^{l-l'} e^{-i(\delta_l - \delta_{l'})} \sqrt{(2l'+1)(2l+1)}$$

$$\times \langle l, l'; 0, 0 | 0, 0 \rangle \langle l, l'; m, -m' | 0, 0 \rangle \left| D_{lm0}^{(-)\Gamma_0} \right|^2$$

$$(8.22)$$

where we also used (2.67a). The last C-G coefficient on the RHS of (8.22) collapses the sum over m' to the value $m = m'$. The other C-G coefficient collapses the sum over l' to the value $l' = l$. This gives us [109, 110]

$$
\begin{aligned}
\sigma(\beta_\gamma = 0) &= 16\pi^2 \sum_{lm} \underbrace{(-1)^m (2l+1)\langle l, l'; 0, 0|0, 0\rangle \langle l, l; m, -m|0, 0\rangle}_{=1} \left| D_{lm0}^{(-)\Gamma_0} \right|^2 \\
&= 16\pi^2 \sum_{lm} \left| D_{lm0}^{(-)\Gamma_0} \right|^2
\end{aligned}
$$

$$(8.23)$$

where we also used (2.144b) and (2.146). ∎

Example 8.3

Find the differential pd cross-section for the case in which there is a fixed photoelectron ejection direction relative to the photon-frame z-axis.

Let the direction in question be designated as $\widehat{k}_e^\gamma$. Then we are looking for the differential cross-section:

$$
\frac{d\sigma}{d\widehat{\mathscr{R}}_\gamma d\widehat{k}_e} = \frac{d\sigma}{d\widehat{\mathscr{R}}_\gamma d\widehat{k}_e^\gamma}
$$

$$(8.24)$$

We transform the appropriate portions of (8.13) as follows [c.f. (2.95)]:

$$
Y_{K_e m_e}\left(\widehat{k}_e\right) = \sum_{m_e'} Y_{K_e m_e'}\left(\widehat{k}_e^\gamma\right) D_{m_e' m_e}^{K_e}\left(\mathscr{R}_\gamma^{-1}\right)
$$

$$(8.25a)$$

and [c.f. (2.149)],

$$
D_{m_e' m_e}^{K_e}\left(\mathscr{R}_\gamma^{-1}\right) D_{M_\gamma 0}^{K_\gamma}(\mathscr{R}_\gamma) = \sum_K \langle K_e, K_\gamma; m_e, 0|K, M\rangle \langle K_e, K_\gamma; m_e', M_\gamma|K, M_e'\rangle D_{M_e' M}^{K}(\mathscr{R}_\gamma)
$$

$$(8.25b)$$

giving [109, 110]

$$\frac{d\sigma}{d\hat{\mathscr{R}}_\gamma d\hat{k}_e} = 16\pi^2 \sum_{ll'} \sum_{mm'} \sum_{m_\gamma m'_\gamma} (-1)^{m'+m'_\gamma} (i)^{l-l'} e^{-i(\delta_l - \delta_{l'})} \sqrt{\frac{(2l'+1)(2l+1)}{4\pi}} D^{(-)\Gamma_0 *}_{l'm'm'_\gamma} D^{(-)\Gamma_0}_{lmm_\gamma}$$

$$\times \sum_{K_e, m^\gamma_e} (2K_e+1)^{-1/2} \langle l,l';0,0|K_e,0 \rangle \langle l,l';m,-m'|K_e,m_e \rangle Y_{K_e m^\gamma_e}(\hat{k}^\gamma_e)$$

$$\times \sum_{K_\gamma} \langle 1,1;0,0|K_\gamma,0 \rangle \langle 1,1;m_\gamma,-m'_\gamma|K_\gamma,M_\gamma \rangle$$

$$\times \sum_K \langle K_e,K_\gamma;m_e,0|K,M \rangle \langle K_e,K_\gamma;m^\gamma_e,M_\gamma|K,M^\gamma_e \rangle D^K_{M^\gamma_e M}(\mathscr{R}_\gamma)$$

$$(8.26)$$

Because we already had $0 \le K_e \le 2l_{max}$ and $0 \le K_\gamma \le 2$, we see that here we also have $0 \le K \le 2l_{max} + 2$. ∎

At this point, some additional comments are necessary. First, it is worth remembering that the molecule-frame transition amplitudes of (8.14) are where the dynamics of photon absorption and photoelectron-core interactions are located [109, 110, 112]. It is also important to note that the above analysis did not account for the vibrational/rotational motion of the molecule. That is, the pd event is not instantaneous. If the molecule is undergoing vibrational/rotational motion, then the molecule may transition to a metastable state (including an autoionizing state) of appreciable lifetime during the pd event. If this is the case, then the ratio of the lifetime of the metastable state to the rotational period of the molecule becomes important [113]. Also, if the kinetic energy of the photoelectron is low, then the molecule may rotate through an appreciable angle and expose the slowly-departing photoelectron to a time-changing potential, which will in turn affect the trajectory of the photoelectron and the resulting angular distribution [114] (see Problem 8.6). If, on the other hand, the photoelectron is more energetic, it may transfer appreciable momentum and/or angular momentum to the molecule as it departs [115]. Centrifugal barrier effects in the anisotropic molecular potential (shape resonances) and multi-electron configurational overlap can also affect the observed PAD [116, 117].

In addition, we point out that there are other ways to approach the analysis of molecular PADs. For example, rather than the analysis presented above, one could instead analyze molecular PADs in the context of angular momentum transfer theory [118–120] (see Problem 8.5) or extend the analysis beyond the dipole approximation [121–126]. Other advanced theories can also account for spin-orbit interactions. Specifically, because the dipole operator cannot affect spin, spin-orbit effects must be invoked to explain any changes (during the pd process) in the spin of photoelectrons and the effect of spin polarization on molecular PADs [127]. Theoretical descriptions (and experimental measurements) of molecular PADs have also been offered for molecules with a high degree of symmetry [e.g., diatomic molecules or molecules that possess cylindrical symmetry-see Problems 8.2 and 8.4(e)] [109, 110, 128–143].

And finally, molecular PADs offer a technique to gain insight into the nature of chiral (optically active) molecules using circularly polarized light. The PAD for levorotatory (left-handed) and dextrorotatory (right-handed) molecular stereoisomers behaves (in the dipole approximation) as

$$\frac{d\sigma}{d\Omega} \propto A \pm B\cos\theta + C\cos^2\theta \tag{8.27}$$

where the $\pm$ factor applies to left/right-circularly polarized photons (see Problem 8.7). In (8.27), the angle θ is the angle between the velocity vector of the photoelectron and the direction of the incident photon. This result shows that chiral molecules produce different PADs for the absorption of left- and right-circularly polarized photons and that the leading term (indicating an unequal number of levorotatory and dextrorotatory molecules in the experimental sample) would be of order $2B\cos\theta$ in the dipole approximation [144].

Problems

8.1. Rederive (8.13) for the case of circularly-polarized light. Note that, in the case of circularly polarized light, the photon-frame z-axis is taken as the propagation direction of the incident light.

8.2. For a cylindrically symmetric molecule, we have the restriction $m = m' = m_0$ where m_0 is the initial-state projection of the molecule's angular momentum onto the molecular-frame z-axis. Find $d\sigma(\beta_\gamma = 0)/d\widehat{k}_e$ for a cylindrically symmetric molecule. Also find the integrated cross-section.

8.3. In (8.6), we defined the rotation $\mathscr{R} = \{\alpha, \beta, \gamma\}$. The inverse rotation would then be

$$\mathscr{R}^{-1} = \{-\gamma, -\beta, -\alpha\}$$

(a) Relate the angles in $\mathscr{R} = \{\alpha, \beta, \gamma\}$ to the angles in $\mathscr{R}_\gamma = \{\alpha_\gamma, \beta_\gamma, \gamma_\gamma\}$ under the condition that the laboratory/photon frame z-axis is taken as the direction of the photon polarization in the case of linearly polarized light, or as the direction of photon propagation in the case of circularly polarized light.

(b) Show that: $Y^*_{K_\gamma M_\gamma}(\beta_\gamma, \alpha_\gamma) = (-1)^{M_\gamma} Y_{K_\gamma M_\gamma}(\beta, \gamma)$

HINT: Use (2.54).

(c) Find the relationship between $d\sigma/d\widehat{\mathscr{R}}$ and $d\sigma/d\Omega$, where $d\widehat{\mathscr{R}} = (8\pi^2)^{-1} d\alpha \sin\beta d\beta d\gamma$ and $d\Omega = \sin\beta d\beta d\gamma$.

8.4. Integrated detector angular distributions (IDAD) [109, 110]

The process of measuring PADS involves the relationship between the orientation of the target molecule, the orientation of the electron detector, and the orientation/polarization direction of the photon. If one is interested in isolating the effect of target-molecule orientation on the pd interaction, then one must eliminate all effects due to detector location/orientation. Integrating over detector solid angles (i.e., measuring the total photoelectron current over a "sphere" surrounding the target; a process known as IDAD) is one way to accomplish the desired goal. In IDAD measurements, the molecular PAD is measured as a function of the target orientation only. Such a measurement scheme would be useful in determining the geometry/orientation of a surfaced-adsorbed molecular species or the rotational sublevel population of a molecular beam (for example).

(a) Start with the results of Problem 8.1 and integrate over the ejection directions $d\widehat{k}_e$ in the molecular frame to show that

$$\frac{d\sigma}{d\widehat{\mathscr{R}}_\gamma} = 4\pi\sqrt{4\pi}\sum_{l,m\,m_\gamma m'_\gamma}(-1)^{m'_\gamma-m_p}D^{(-)\Gamma_0*}_{lmm'_\gamma}D^{(-)\Gamma_0}_{lmm_\gamma}$$

$$\times\sum_{K_\gamma}\langle 1,1;m_p,-m_p|K_\gamma,0\rangle\langle 1,1;m_\gamma,-m'_\gamma|K_\gamma,M_\gamma\rangle\sqrt{\frac{4\pi}{2K_\gamma+1}}Y^*_{K_\gamma M_\gamma}(\beta_\gamma,\alpha_\gamma)$$

where $m_p = 0$ for linear polarization along the photon-frame z-axis, and $m_p = \pm 1$ for right/left circular polarization. Are there any restrictions on the values of m_γ, m'_γ? If so, why?

Note: You will need the result of Problem 8.3(b).

(b) The angles β, γ are spherical polar angles measured from the laboratory z-axis and give the orientation of the target in that frame. Make the replacement β, $\gamma \to \theta_T$, ϕ_T and use the result from Problem 8.3(c) to show that we can write the differential cross-section in the form:

$$\frac{d\sigma}{d\Omega_T} = \frac{16\pi^2\sqrt{4\pi}}{3}\sum_{K=0}^{2}\sum_{M=0}^{K}\text{Re}\left[Z^{m_p}_{KM}C^{(K)}_M(\theta_T,\phi_T)\right]$$

where the $C^{(K_\gamma)}_{M_\gamma}$ are the normalized spherical harmonics [c.f. (4.11)]. Find the form of the coefficients $Z^{m_p}_{KM}$.

HINT: Remember that the cross-section is real.

(c) Calculate all the coefficients $Z^{m_p}_{KM}$ explicitly. Which of the $Z^{m_p}_{KM}$ (if any) are explicitly real?

(d) Show that we can write the differential cross-section as

$$\frac{d\sigma}{d\Omega_T} = \frac{16\pi^2\sqrt{4\pi}}{3}\left\{\begin{array}{l} Z_{00}+Z_{10}\cos\theta_T+Z_{20}\frac{1}{2}(3\cos^2\theta_T-1) \\[2mm] \left[\begin{array}{l}\frac{1}{\sqrt{2}}(Z^{\text{Re}}_{11}\cos\phi_T-Z^{\text{Im}}_{11}\sin\phi_T) \\[2mm] +\sqrt{\frac{3}{2}}(Z^{\text{Re}}_{21}\cos\phi_T-Z^{\text{Im}}_{21}\sin\phi_T)\cos\theta_T\end{array}\right]\sin\theta_T \\[4mm] +\sqrt{\frac{3}{8}}(Z^{\text{Re}}_{22}\cos2\phi_T-Z^{\text{Im}}_{22}\sin2\phi_T)\sin^2\theta_T\end{array}\right\}$$

where $Z^{\text{Re}}_{KM}/Z^{\text{Im}}_{KM}$ refers to the real/imaginary part of $Z^{m_p}_{KM}$. How many parameters are needed to completely specify the differential cross-section if linearly polarized light is used?

(e) For a target molecule of cylindrical symmetry, we have the restriction $m = m_\gamma + m_0$ and $m' = m'_\gamma + m_0$. Show that, if linearly polarized light is used, the differential cross-section takes the form

$$\frac{d\sigma}{d\Omega_T} = \frac{\sigma}{4\pi}[1 + \beta_T P_2(\cos\theta_T)]$$

Find σ and β_T. Compare your results to the forms in (3.27) and (3.28).

8.5. Molecular PADs and angular momentum transfer theory [111, 118, 119]

If we express the differential cross-section of (8.7) (in the dipole approximation) in terms of its angular momentum transfer components and integrate over all molecular orientations, then, following the procedures outlined in Chap. 7, we will get

$$\frac{d\sigma}{d\widehat{k}_e} = \sum_{j_t} \frac{d\sigma(j_t)}{d\widehat{k}_e}$$

where

$$\sum_{j_t} \frac{d\sigma(j_t)}{d\widehat{k}_e} \propto \frac{1}{2j_t+1} \sum_{ll'} (i)^{l-l'} e^{-i(\delta_l-\delta_{l'})} \Theta(j_t; 10; ll'; \theta) \sum_{mm'} \sum_{m_\gamma m'_\gamma} D^{(-)j_t*}_{lmm'_\gamma} D^{(-)j_t}_{lmm'_\gamma} \delta_{m-m_\gamma, m'-m'_\gamma}$$

and

$$D^{(-)j_t}_{lmm_\gamma} \equiv (-1)^{m_\gamma} \langle 1, l; -m_\gamma, m | j_t, m-m_\gamma \rangle D^{(-)\Gamma_0}_{lmm_\gamma}$$

(a) Find the normalization integral

$$\int_0^{2\pi} \int_0^{\pi} \sin\theta d\theta d\phi \Theta(j_t; 10; ll'; \theta)$$

(b) Using your result from part (a) and arguing by analogy from the results of Chap. 7 and this chapter, find $\sigma(j_t), \sigma, \beta, \beta(j_t)$.

8.6. Molecular rotations and PADs [114]

Suppose we have a unit vector $\widehat{D}$ that points in the direction of motion of a photoelectron that is about to be ejected from a rotating molecule. As the molecule rotates, this vector changes its orientation in space. Let $\widehat{D}(0)$ describe this unit vector in its initial orientation at time $t = 0$ and $\widehat{D}(t)$ its orientation at some later time t.

According to (8.6), the rotation $\mathcal{R} = \{\alpha_0, \beta_0, \gamma_0\}$ carries the (fixed) lab frame $OXYZ$ onto the molecular frame $Ox_0y_0z_0(M_0)$ at time $t = 0$. A further rotation $d\Omega(t)$ carries this frame onto the molecular frame $Oxyz(M_t)$ as it is at some later time t.

(a) The initial pd probability density is given in (3.19). We want to average the Legendre polynomial $P_l\left[\hat{\varepsilon} \cdot \widehat{D}(t)\right]$ with respect to this probability density over the initial orientations of the molecule:

$$\left\langle P_l\left[\hat{\varepsilon} \cdot \widehat{D}(t)\right]\right\rangle = \int P_l\left[\hat{\varepsilon} \cdot \widehat{D}(t)\right] P_{pd}\left(\alpha_0, \beta_0, \gamma_0\right) d\alpha_0 \sin\beta_0 d\beta_0 d\gamma_0$$

Transform the appropriate quantities into the molecular frame to show that

$$\langle P_l[\hat{\varepsilon} \cdot \widehat{D}(t)]\rangle = \left(\frac{1}{8\pi^2}\right)\left(\frac{4\pi}{2l+1}\right) \int d\alpha_0 \sin\beta_0 d\beta_0 d\gamma_0$$

$$\times \sum_{m, m_2 = -l}^{l} \left(\begin{array}{c} Y_{lm}^*(\hat{\varepsilon}) D_{m_2 m}^l(\mathscr{R}) Y_{lm_2}\left[\widehat{D}_{M_0}(t)\right] \\ \times \left\{ 1 + \frac{8\pi}{5} \sum_{m_1, m_3 = -2}^{2} D_{m_3 m_1}^{2*}(\mathscr{R}) Y_{2m_3}^*\left[\hat{\mu}_{eM_0}(0)\right] Y_{2m_1}(\hat{\varepsilon}) \right\} \end{array} \right)$$

(b) Evaluate the integral(s) to show that

$$\langle P_l[\hat{\varepsilon} \cdot \widehat{D}(t)]\rangle = \frac{8\pi}{25} \sum_{m_2 = -2}^{2} Y_{2m_2}\left[\widehat{D}_{M_0}(t)\right] Y_{2m_2}^*\left[\hat{\mu}_{eM_0}(0)\right]$$

(c) After a time t, the molecule has rotated according to the following transformation via (2.95):

$$Y_{2m_2}\left[\widehat{D}_{M_0}(t)\right] = \sum_{m_1} D_{m_1 m_2}^2[\delta\Omega(t)] Y_{2m_1}\left[\widehat{D}_{M_0}(0)\right]$$

Apply this transformation and then average over the angular momenta and reorientation angles to show that

$$\left\langle P_l\left[\hat{\varepsilon} \cdot \widehat{D}(t)\right]\right\rangle = \frac{8\pi}{5} \sum_{m_1, m_2 = -2}^{2} Y_{2m_1}\left[\widehat{D}_{M_0}(0)\right] Y_{2m_2}^*\left[\hat{\mu}_{eM_0}(0)\right] \left\langle D_{m_1 m_2}^2[\delta\Omega(t)]\right\rangle$$

(d) Apply the reasoning in example 3.3 to argue that this average is the asymmetry parameter β for this molecule. In the event that the pd event is very much faster than the rotational motion of the molecule, show that the asymmetry parameter reduces to

$$\beta = 2P_2(\cos\theta)$$

where θ is the angle between the dipole moment of the molecule and the momentum vector of the photoelectron.

8.7. Chiral molecules [144]

(a) Modify (8.26) to include circular polarization. Average the result over all molecular orientations to show that

$$\frac{d\sigma}{d\hat{k}_e} = 16\pi^2 \left(8\pi^2\right) \sum_{ll'} \sum_{m} \sum_{m_\gamma} (-1)^{m+m_\gamma-m_p} (i)^{l-l'} e^{-i(\delta_l-\delta_{l'})}$$

$$\times \sqrt{\frac{(2l'+1)(2l+1)}{4\pi}} D_{l'mm_\gamma}^{(-)\Gamma_0*} D_{lmm_\gamma}^{(-)\Gamma_0}$$

$$\times \sum_{K_e} \sqrt{2K_e+1} \begin{pmatrix} l & l' & K_e \\ 0 & 0 & 0 \end{pmatrix} \begin{pmatrix} l & l' & K_e \\ m & -m & 0 \end{pmatrix} \begin{pmatrix} 1 & 1 & K_e \\ m_p & -m_p & 0 \end{pmatrix}$$

$$\times \begin{pmatrix} 1 & 1 & K_e \\ m_\gamma & -m_\gamma & 0 \end{pmatrix} P_{K_e}\left(\hat{k}_e^\gamma\right)$$

(b) Use your result from part (a) to prove (8.27).

HINT: For chiral molecules, the matrix elements $D_{lmm_\gamma}^{(-)\Gamma_0}$ are not equivalent for positive and negative values of the azimuthal quantum numbers m, m_γ.

Note: You should also end up proving the Yang theorem of reference [1] while completing this problem. The Yang Theorem states that,

"If only incoming waves of orbital angular momentum L contribute appreciably to the interaction, the angular distribution of the outgoing particles in the center-of-mass system will be a function of even polynomials of $\cos\theta$ with maximum power not higher than $2L$."

Chapter 9
Measuring Photoelectron Angular Distributions in the Laboratory

For the pd process to be possible, the energy of an incident photon of frequency ν and energy $E_\gamma = h\nu$ must exceed the energy (BE) binding the electron to the atom or molecule. This condition may be expressed by the Einstein energy-balance equation [145]:

$$E = h\nu - BE \geq 0 \tag{9.1}$$

where E is the kinetic energy of the ejected electron. A photon of sufficient energy E_γ can access different pd channels, possibly leaving the residual core in an excited state. The single-photon pd process, in which a parent P in an initial state i absorbs a photon γ, resulting in a residual core C in a final state f and a photoelectron e^-, is described by the general bound-free equation:

$$P(i) + \gamma \rightarrow C(f) + e^- \, (if\gamma) \tag{9.2}$$

where, in the rest frame of the parent/core, the kinetic energy of the photoelectron for the dissociation channel $i \rightarrow f$ is [24, 25]

$$E(if\gamma) = E_\gamma - \underbrace{\left(E_f - E_n\right)}_{E_{fi}} = E_\gamma - E_{fi} \tag{9.3}$$

where E_n is the energy of the parent in the initial state and E_f is the energy of the residual core in the final state.

Measuring photodissociation cross-section and asymmetry parameters is only possible if one has a reliable source of known-frequency light. Whether it be light from a laser or light from a synchrotron, or some other source of photons, all photo-spectroscopic measurements rely on these sources, and various techniques are used to make the relevant measurements. Most techniques involve a beam of ions intersecting a collimated photon beam. In such cases, it may be necessary to make

V. T. Davis, *Introduction to Photoelectron Angular Distributions*, Springer Tracts in Modern Physics 286, https://doi.org/10.1007/978-3-031-08027-2_9

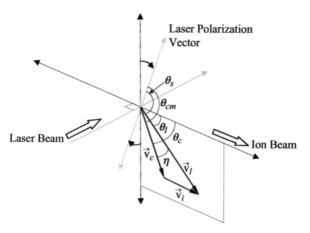

Fig. 9.1 Nonrelativistic kinematic transformation diagram for a typical crossed-beams photodissociation experiment. A fast-moving beam of ions is crossed at right angles with a laser beam. Photoelectrons created at the intersection of the two beams are collected and their kinetic energies measured in a kinetic energy (*KE*) analyzer. The figure shows the relationship between the velocity vectors of a collected photoelectron in the lab frame $\vec{v}_l$, and the CM frame $\vec{v}_c$. $\vec{v}_i$ is the velocity of an ion in the beam for a given ion beam energy E_i, and $\vec{v}_l$ is the vector sum $\vec{v}_l = \vec{v}_c + \vec{v}_i$. $\vec{v}_l$ is the velocity of a photoelectron as seen by a stationary observer in the lab frame. The angle $\eta = \theta_c - \theta_l$ can always be determined after the fact, since $\vec{v}_i$ is a function of the experimental parameters (which are controlled by the experimentalist), and $\vec{v}_c$ is ultimately determined by the Einstein energy balance equation. Also shown is the polarization vector for a linearly polarized laser photon. In this crossed-beams setup, the polarization vector is perpendicular to the laser propagation direction and is thus coplanar with $\vec{v}_l$, $\vec{v}_c$, and $\vec{v}_i$. θ_s is the angle between the polarization vector of the linearly polarized photon and $\vec{v}_l$, and θ_{cm} is the angle between the polarization vector of the linearly polarized photon and $\vec{v}_c$.

kinematic corrections and to account for the effects of the Doppler shift. Photoelectron angular distribution measurements are made in the laboratory, and we must have a way to convert these lab-frame measurements into the parent/center-of-mass (CM)[1] frame to accurately describe the physics. To illustrate this point, we will highlight three common techniques used to measure PADs in the laboratory.

One experimental method that uses a crossed-beams geometry to measure PADs is the laser photodetachment electron spectroscopy (LPES) technique. In the LPES technique, photoelectrons are collected, and their kinetic energies measured with a kinetic energy (KE) analyzer (see, e.g., [46] and problem 9.4). Using the LPES technique, angular distributions are measured one angle at a time, and the kinematic transformation from the lab frame to the CM frame is accomplished as follows (see Fig. 9.1).

---

[1] We assume the parent/core is infinitely heavy, i.e., we neglect the recoil of the core during the pd process.

Let θ_{cm} be the angle measured from the photon polarization vector to the photoelectron velocity vector in the CM frame. Now, consider the following expression (we are working with linearly polarized photons in the dipole approximation):

$$I_{cm}(\theta_{cm}=0) + \left[I_{cm}\left(\theta_{cm}=\frac{\pi}{2}\right) - I_{cm}(\theta_{cm}=0)\right]\sin^2\theta_{cm}$$

$$= \frac{\sigma_{pd}}{4\pi}(1+\beta) + \left[\frac{\sigma_{pd}}{4\pi}\left(1-\frac{\beta}{2}\right) - \frac{\sigma_{pd}}{4\pi}(1+\beta)\right]\sin^2\theta_{cm}$$

$$= \frac{\sigma_{pd}}{4\pi}\left(1+\beta-\frac{3\beta}{2}\sin^2\theta_{cm}\right) = \frac{\sigma_{pd}}{4\pi}\left[1+\beta-\frac{3\beta}{2}\left(1-\cos^2\theta_{cm}\right)\right]$$

$$= \frac{\sigma_{pd}}{4\pi}\left[1+\beta\left(-\frac{1}{2}+\frac{3}{2}\cos^2\theta_{cm}\right)\right] = \frac{\sigma_{pd}}{4\pi}\left[1+\beta\left(\frac{3\cos^2\theta_{cm}-1}{2}\right)\right]$$

$$= \frac{\sigma_{pd}}{4\pi}\left[1+\beta P_2(\cos\theta_{cm})\right] = I_{cm}(\theta_{cm})$$

$$\Rightarrow I_{cm}(\theta_{cm}) = I_{cm}(\theta_{cm}=0) + \left[I_{cm}\left(\theta_{cm}=\frac{\pi}{2}\right) - I_{cm}(\theta_{cm}=0)\right]\sin^2\theta_{cm} \quad (9.4)$$

where we have also used (3.29). From Fig. 9.1:

$$\eta = \theta_{cm} - \theta_s = \theta_c - \theta_l \quad (9.5)$$

where η is the angle pictured in Fig. 9.1 and given by

$$\eta = \theta_c - \theta_l = \sin^{-1}\left(\frac{v_i}{v_c}\sin\theta_l\right) \quad (9.6)$$

Equation (9.5) tells us that when $\theta_s = 0$, then $\theta_{cm} = \eta$, which in turn means

$$I_{lab}(\theta_s=0) = I_{cm}(\theta_{cm}=\eta) \quad (9.7)$$

Combine (9.4) with (9.7):

$$I_{lab}(0) = I_{cm}(0) + \left[I_{cm}\left(\frac{\pi}{2}\right) - I_{cm}(0)\right]\sin^2\eta \quad (9.8)$$

Again, from (9.5), $\theta_s = \frac{\pi}{2} - \eta \Rightarrow \theta_{cm} = \frac{\pi}{2}$, which implies

$$I_{lab}\left(\frac{\pi}{2}-\eta\right) = I_{cm}\left(\frac{\pi}{2}\right) \quad (9.9)$$

Next, we combine (9.3) and (9.6):

$$\eta = \sin^{-1}\left[v_i\left(\frac{2\left(E_\gamma - E_{fi}\right)}{m_e}\right)^{-\frac{1}{2}}\sin\theta_l\right] \tag{9.10}$$

Before we can use (9.10), we must first find a kinematic transformation relating the photoelectron kinetic energy in the ion frame $E(if\gamma) \equiv E_c$, with its corresponding kinetic energy as measured in the lab frame E_l. From Fig. 9.1,

$$\vec{v}_l = \vec{v}_c + \vec{v}_i \Rightarrow \vec{v}_c = \vec{v}_l - \vec{v}_i \Rightarrow |v_c^2| = v_l^2 + v_i^2 - 2|v_l|\,|v_i|\cos\theta_l$$

$$\Rightarrow \underbrace{\frac{1}{2}m_e v_c^2}_{E_c} = \underbrace{\frac{1}{2}m_e v_l^2}_{E_l} + \underbrace{\frac{1}{2}m_e v_i^2}_{\varepsilon\equiv\left(\frac{m_e}{m_i}\right)E_i} - 2v_l v_i\left(\frac{1}{2}m_e\right)\cos\theta_l$$

$$\Rightarrow E_c = E_l + \varepsilon - 2\left[v_l^2 v_i^2\left(\frac{1}{2}m_e\right)^2\right]^{\frac{1}{2}}\cos\theta_l$$

$$\Rightarrow E_c = E_l + \varepsilon - 2\sqrt{\varepsilon E_l}\cos\theta_l \tag{9.11}$$

We can use (9.11) to derive the inverse transformation

$$\left(\sqrt{E_l}\right)^2 + \left(-2\sqrt{\varepsilon}\cos\theta_l\right)\sqrt{E_l} + (\varepsilon - E_c) = 0$$

$$\Rightarrow \sqrt{E_l} = \frac{1}{2}\left(2\sqrt{\varepsilon}\cos\theta_l \pm \sqrt{4\varepsilon\cos^2\theta_l - 4(\varepsilon - E_c)}\right)$$

$$\Rightarrow E_l = \left(\sqrt{\varepsilon}\cos\theta_l \pm \underbrace{\sqrt{\varepsilon\cos^2\theta_l - \varepsilon + E_c}}_{-\varepsilon\sin^2\theta_l}\right)^2 \tag{9.12}$$

$$= \left(\sqrt{\varepsilon}\cos\theta_l \pm \sqrt{E_c - \varepsilon\sin^2\theta_l}\right)^2$$

As an aside, note that $\varepsilon \equiv m_e v_i^2/2 = (m_e/m_i)E_i$ is the kinetic energy of an electron moving with the same speed v_i as an ion in the beam.

Nonrelativistic kinematic shifts can also cause an apparent shift in angular coordinates. For example, the solid angle of acceptance of photoelectrons into the kinetic energy (KE) analyzer changes depending upon one's reference frame. In spherical coordinates, the ratio of the solid angle in the lab frame Ω_l, to the solid angle in the CM frame Ω_c, is

$$\frac{d\Omega_l}{d\Omega_c} = \frac{\sin\theta_l d\theta_l}{\sin\theta_c d\theta_c} \tag{9.13}$$

where we have noted that $d\phi_l = d\phi_c$ due to axial symmetry. We use (9.11) and (9.12) to find

$$dE_c = 2\sqrt{\varepsilon E_l}\sin\theta_l d\theta_l \Rightarrow \sin\theta_l d\theta_l = \frac{dE_c}{2\sqrt{\varepsilon E_l}} \tag{9.14a}$$

$$dE_l = 2\sqrt{\varepsilon E_c}\sin\theta_c d\theta_c \Rightarrow \sin\theta_c d\theta_c = -\frac{dE_l}{2\sqrt{\varepsilon E_c}} \tag{9.14b}$$

Substituting these results into (9.13) gives

$$\left|\frac{d\Omega_l}{d\Omega_c}\right| = \frac{dE_c}{dE_l}\sqrt{\frac{E_c}{E_l}} \tag{9.15}$$

But from (9.11), we see

$$\frac{dE_c}{dE_l} = 1 - \sqrt{\frac{\varepsilon}{E_l}}\cos\theta_l \tag{9.16}$$

which, when substituted into (9.15), gives

$$\left|\frac{d\Omega_l}{d\Omega_c}\right| = \sqrt{\frac{E_c}{E_l}}\left(1 - \sqrt{\frac{\varepsilon}{E_l}}\cos\theta_l\right) \tag{9.17}$$

This equation is used when transforming differential cross-sections from one frame to the other:

$$\left.\frac{d\sigma}{d\Omega}\right|_c = \left.\frac{d\sigma}{d\Omega}\right|_l\left(\frac{d\Omega_l}{d\Omega_c}\right) \tag{9.18}$$

As a reminder, we note that (9.6) gives the relative angle of emission of photoelectrons in the two frames.

As stated earlier, most photo-spectroscopic experiments involve ions moving within a laser field. These ions will experience a shift in the photon frequency ν due to the relativistic Doppler effect [11, 24, 25]:

$$\nu' = \nu\frac{1 - \beta_i\cos\theta}{\sqrt{1 - \beta_i^2}} \tag{9.19}$$

where ν' is the laser frequency as seen in the ion frame, and ν is the laser frequency in the lab frame. Here θ is the angle between the laser beam and the ion beam. The numerator in (9.19) is the nonrelativistic (first order) Doppler effect and is zero for $\theta = \pi/2$. Therefore, to first-order, the geometry of the experiment depicted in Fig. 9.1

will not suffer any Doppler shift. The denominator of (9.19) is a consequence of relativistic time dilation and gives rise to second-order effects regardless of the crossing angle. The shift associated with this term (for $\theta = \pi/2$) is approximated as follows:

$$\Delta\nu_{2nd-order} = \nu' - \nu = \frac{\nu}{\sqrt{1 - \beta_i^2}} - \nu$$

$$= \nu\left[\frac{1}{\sqrt{1 - \beta_i^2}} - 1\right] \approx \nu\left[1 + \frac{\beta_i^2}{2} - 1\right] = \frac{\nu}{2}\beta_i^2 \qquad (9.20)$$

and is second-order in the velocity parameter $\beta_i = v_i/c$ (as expected from the name) and is negligible for nonrelativistic ion beams.

To maximize the photoelectron signal, some experiments invoke a collinear/merged-beams geometry (see, e.g., [146, 147]). One major advantage of a collinear-beam geometry setup is that, if one could measure the relative shift in the absorbed laser frequency when the laser and ion beams are parallel and antiparallel, the uncertainty due to the Doppler effect can be eliminated to all orders of β_i by taking the geometric mean as follows [24, 25, 146]:

$$\nu'_{p,a} = \nu\frac{1 \pm \beta_i}{\sqrt{1 - \beta_i^2}} \Rightarrow \sqrt{\nu'_p\nu'_a} = \sqrt{\left(\nu\frac{1 + \beta_i}{\sqrt{1 - \beta_i^2}}\right)\left(\nu\frac{1 - \beta_i}{\sqrt{1 - \beta_i^2}}\right)} = \sqrt{\nu\nu} = \nu$$

$$(9.21)$$

Example 9.1

Estimate the error associated with ion beam divergence $\Delta\theta$.

For nonrelativistic beams crossed at right angles, the error associated with ion beam divergence $\Delta\theta$ is derived (to first order) as follows [20, 24, 25]:

$$\Delta\nu_{eff} = \nu[1 - \beta_i\cos(\theta + \Delta\theta)] - \nu(1 - \beta_i\cos\theta)$$
$$= \nu\beta_i[\cos\theta - \cos(\theta + \Delta\theta)]$$

In the case that the divergence is from the right angle $\pi/2$, we have

$$\cos\theta = 0; \quad \cos\left(\frac{\pi}{2} + \Delta\theta\right) = \cos\frac{\pi}{2}\cos\Delta\theta - \sin\frac{\pi}{2}\sin\Delta\theta = -\sin\Delta\theta \approx -\Delta\theta$$

for a small divergence. So, we have

$$\Delta \nu_{eff} \approx \nu \beta_i \Delta \theta = \left(\frac{v_i}{c}\right) \nu \Delta \theta = \left(\frac{v_i}{\lambda_L}\right) \Delta \theta; \quad \frac{\nu}{c} = \frac{1}{\lambda_L}$$

Note: For a typical crossed-beams apparatus, we might have $v_i \approx 5 \times 10^5$ m\s, $\Delta \theta \approx 0.02$ rad, and $\lambda_L = 1064$ nm, which gives $\Delta \nu_{eff} \approx 5.9 \mu eV$.

For collinear-beam geometries, the divergence is from $\theta = 0$, which gives [146]

$$\Delta \nu_{eff} = \nu \beta_i [\cos \theta - \cos (\theta + \Delta \theta)] = \nu \beta_i [1 - \cos (0 + \Delta \theta)] \approx \nu \beta_i \left(1 - 1 + \frac{\Delta \theta^2}{2}\right)$$

$$\Rightarrow \Delta \nu_{eff} \approx \nu \beta_i \left(\frac{\Delta \theta^2}{2}\right)$$

∎

Once a pd experiment has been completed, and a photoelectron binding energy (BE) computed, all the quantities in (8.10) are known, allowing us to determine the angle η. Therefore, the quantity $I_{cm}(\pi/2)$ in (9.9) can be computed from a measurement in the lab frame and normalized to one. When this is done, (9.8) becomes

$$I_{lab}(0) = I_{cm}(0) + [1 - I_{cm}(0)] \sin^2 \eta$$

$$\Rightarrow I_{cm}(0) = \frac{I_{lab}(0) - \sin^2 \eta}{1 - \sin^2 \eta} \tag{9.22}$$

The asymmetry parameter is now easily calculated by determining $I_{cm}(\theta_{cm} = 0)$ and $I_{cm}(\theta_{cm} = \pi/2)$ and by using (3.29) [146]

$$\frac{I_{cm}(\theta_{cm} = 0)}{I_{cm}\left(\theta_{cm} = \frac{\pi}{2}\right)} = \frac{\frac{\sigma_{pd}}{4\pi}[1 + \beta P_2(\cos 0)]}{\frac{\sigma_{pd}}{4\pi}[1 + \beta P_2(\cos 90)]} = \frac{(1 + \beta)}{1 - \frac{\beta}{2}} \Rightarrow \beta = \frac{2\left(\frac{I(0)}{I(90)}\right) - 2}{\left(\frac{I(0)}{I(90)}\right) + 2} \tag{9.23}$$

If $I_{cm}(\theta_{cm} = \pi/2)$ is normalized to unity, then (9.23) becomes

$$\beta = \frac{2I_{cm}(0) - 2}{I_{cm}(0) + 2} \tag{9.24}$$

Another crossed-beams (or merged-beams) method now commonly used to measure PADs is the velocity-map imaging (VMI) method. In a VMI spectrometer, a series of three electrostatic plates is used to collect and focus photoelectrons produced in a pd process. The photoelectrons are then projected onto a position-sensitive detector (see Fig. 9.2) [148]. A major advantage of the VMI method over other methods is the near 100% collection efficiency of photoelectrons achieved over the entire 4π solid angle, resulting in the ability to collect a full angular distribution

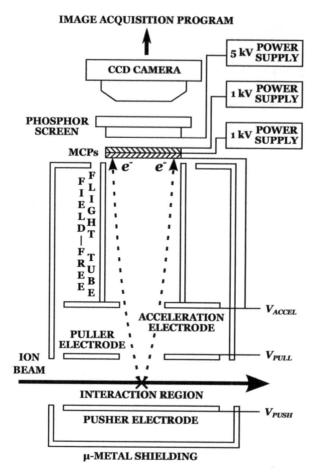

Fig. 9.2 Schematic view of a VMI spectrometer. The trajectories of photoelectrons created in the interaction region via a photodetachment or photoionization process are projected onto a position-sensitive detector by a three-element electrostatic lens and a series of aperture electrodes. In this design, the detector consists of microchannel plates (MCPs), a phosphor screen, and a charge-coupled device (CCD) camera. The VMI spectrometer maps photoelectrons with the same ejection direction and speed onto the same annular region of the position-sensitive detector, regardless of where the photoelectron was initially created in the interaction region. All elements are kept at the appropriate potentials to avoid distortions of the photoelectron trajectories, to focus the image on the detector, and to ensure a near-unity photoelectron collection efficiency. The design of the VMI spectrometer thus preserves the PAD as a two-dimensional projection which is later reconverted back into the original three-dimensional pattern using the appropriate analysis procedures (e.g., see Problem 9.6) [23, 149] Reproduced from [23] with permission from the author

pattern in one scan. Because the VMI technique ensures that photoelectrons with the same velocity strike the detector at the same radius, one can also collect photoelectron (relative) kinetic energy spectra while simultaneously collecting the PAD (see Fig. 9.3).

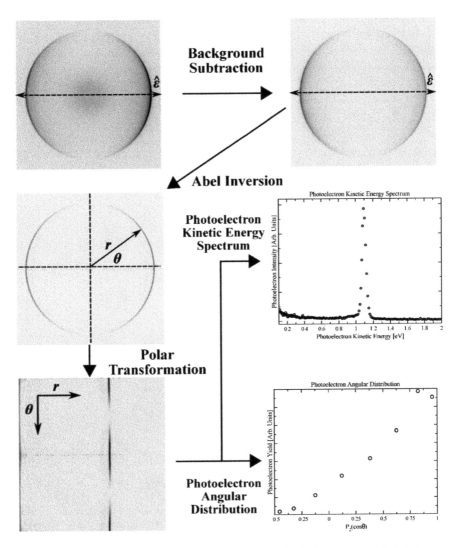

Fig. 9.3 VMI image analysis process (see also Problem 9.6). A photoelectron (relative) kinetic energy spectrum can be extracted from the PAD (middle right figure), and the asymmetry parameter β is calculated via (9.25) and (9.26) (bottom right figure). The VMI images in this figure are from the photodetachment of copper anions using 532 nm linearly polarized laser photons [23]. Reproduced from [23] with permission from the author

The projection method involved in velocity-map imaging requires that the PAD possess azimuthally symmetry, which means the differential cross-section must be in the form of (3.29).[2] If a full angular distribution pattern is collected in one scan via

---

[2]To ensure that the VMI is oriented properly to capture the axially symmetric photoelectron distribution, the polarization vector of the (linearly) polarized laser photon must be oriented horizontally, collinear with the ion velocity vector $\vec{v}_i$. In that case, $\theta_{cm} = \theta_c$ and $\theta_s = \theta_l$ (see Fig. 9.1).

VMI techniques, then one may determine β by producing a linear plot of photoelectron intensity I_{cm} verses $P_2(\cos\theta)$. The asymmetry parameter is then determined by measuring the slope of the resulting line (see Fig. 9.3). However, one still needs to convert the intensity pattern, which is measured in the lab frame, into the true pattern, which is a pattern in the CM frame. Once corrections for photoelectron energies are made, additional corrections are necessary to transform angles and solid angles from the lab frame to the CM frame (see Fig. 9.4).

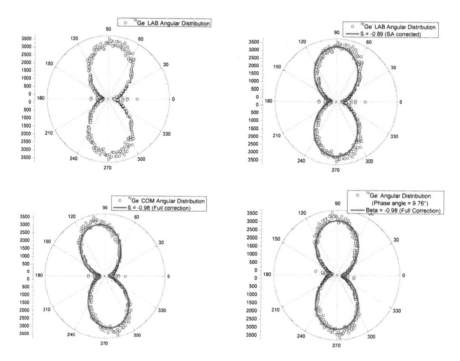

Fig. 9.4 Polar plot of a photoelectron angular distribution for the photodetachment of Ge$^-$. The single-photon photodetachment at a photon wavelength of 532 nm proceeded via the following transitions: Ge$^-$4p^3(^{4}S$_{3/2}$) + $h\nu$(532 nm) → Ge4p^2(^{3}P$_{0,\,1,\,2}$) + e$^-$ (s, d). The radial coordinate in each plot represents (collected) photoelectron relative intensity for one of these transitions. The polarization vector of the linearly polarized laser photons are in the horizontal direction with respect to every diagram. The first diagram (upper left) represents a PAD to which a kinematic correction has not yet been applied. The second diagram (upper right) is the same data as shown in the first diagram, against which the solid angle kinematic correction has now been applied. The third diagram (lower left) is the same as the second diagram, except now plotted against the CM angle. The final diagram (lower right) shows the photoelectron angular distribution fully corrected for both kinematic effects and systematic errors. The small circles represent experimental data. The solid (red) curves indicate the fit to (9.26). Note that the measured asymmetry parameter will be erroneous unless the full kinematic correction is applied [149]

Starting with (3.29) and using the nonrelativistic kinematic transformation from above, we get

$$I_{cm} = \frac{d\sigma_{pd}}{d\Omega_c} = \frac{\sigma_{pd}}{4\pi}[1 + \beta P_2(\cos(\theta_{cm} \pm \eta))]\left[\sqrt{\frac{E_c}{E_l}}\left(1 - \sqrt{\frac{\varepsilon}{E_l}}\cos\theta_l\right)\right] \quad (9.25)$$

where η is determined by (9.10) and the sign in the $\theta_{cm} \pm \eta$ term depends on the quadrant in which the kinematic transformation is made (see Fig. 9.5). Note that the term in the brackets [last term in (9.25)] represents the solid angle kinematic correction.

The form for I_{cm} in (9.25) allows for a straightforward process to fit experimental data with standard curve-fitting routines via the equation

$$I(\alpha) = a[1 + \beta P_2(\cos(\alpha - c))] \quad (9.26)$$

where a, c, and β serve as fitting parameters [41].

Although a kinematic transformation is absolutely necessary to obtain accurate results, one problem that exists with using (9.26) to transform VMI images and PADs from the lab frame to the CM frame is that the photoelectron kinetic energy E_c (and E_l) must be known in advance (or measured during the experiment) for the transformation equation to be useful. Although LPES methods are well-suited to measure absolute photoelectron kinetic energies, other methods (such as VMI techniques) are not. One way to overcome this limitation is to use the displacement of the photoelectron image in the VMI imaging apparatus (for example) to extract the photoelectron kinetic energy (see Fig. 9.5).

In Fig. 9.5, we choose as our origin O_L the center of the photoelectron image as it is viewed in the lab frame. Angular measurements with respect to the z-axis will then correspond to θ_L. Radial measurements from the lab-frame origin, however, correspond to r_C. To see this, we first realize that the origin in the lab frame is shifted from the CM origin by an amount Δz (which can be measured by the experimentalist). Since r_L is defined with respect to the CM origin, it also becomes shifted by Δz when moving to the lab origin. Then by construction, the portion of r_L that intersects the photoelectron distribution is equivalent to r_C, while the difference $r_L - r_C$ is unobserved. Thus, in choosing the center of the photoelectron distribution as our origin, we can measure radii, and hence energies, in the CM frame, while angular measurements are made in the lab frame [consequently, angular measurement must still be frame-corrected via (9.25)].

Our analysis of laboratory techniques for measuring PADs would not be complete without mentioning a recently developed type of reaction microscope called COLTRIMS. Cold Target Recoil Ion Momentum Spectroscopy (COLTRIMS)[3] is an imaging technique in which a beam of collimated cold molecules is bombarded with a beam of high-energy monochromatic photons. A collision between a molecule and a

---

[3]The name "COLTRIMS reaction microscope" is sometimes abbreviated as C-REMI in the literature.

Fig. 9.5 Kinematic effect
on a PADs image collected
by a VMI spectrometer. The
labels in the diagram are
explained in the text [149]

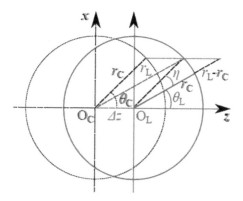

photon may cause the molecule to undergo a "Coulomb explosion," leaving in its wake a number of positive ions ("recoils") and electrons, each of which now has an initial momentum vector pointing in some direction in space. These reaction products are then collected in coincidence for the purpose of reconstructing the orientation of the molecule (at the time of the explosion) with respect to a lab-fixed reference frame. This allows for a PAD from fixed-in-space molecules to be constructed. This type of PAD, called a molecular-frame photoelectron angular distribution, or MFPAD (a photoelectron angular distribution measured in the frame of the molecule) offers a rich source of information about the nature of molecular potentials and photon-molecule interactions. One advantage of the COLTRIMS technique is that, by judicious use of electric and magnetic fields, the COLTRIMS apparatus can collect all particles in the entire 4π solid angle (within limits of detector construction) [150–152].

Another advantage of the COLTRIMS reaction microscope is that it is easily transportable and can be integrated into a synchrotron facility quickly and with little effort (see Figs. 9.6, 9.7 and 9.8). As a result, COLTRIMS devices are ubiquitous in many locations.

Synchrotrons serve as bright sources of collimated, high-energy photons that are needed to penetrate deep into the cores of atoms and molecules. When this light is used to study inner-shell pd processes (e.g., with a COLTRIMS), the core electrons that are ejected can be used to illuminate the molecule from within, allowing one to measure photoemission delays and create detailed maps of the way in which molecular potentials evolve during the pd process (see Chap. 10) [135, 153–158]. COLTRIMS has also been used to study circular dichroism in small molecules [159]. For a basic description of the COLTRIMS device, see App. G.

Finally, it should be mentioned that, although most of the analysis in this chapter was based on linearly polarized incident photons, other polarizations can be used. However, these alternate polarizations do not generally offer any additional dynamical information on the pd process since the differential cross-sections expressed in (3.29) and (4.39) can always be expressed in terms of σ_{pd} and β, regardless of the photon polarization (see examples below) [90].[4]

---

[4]Excepting the aforementioned interaction between circularly polarized light and chiral molecules

Fig. 9.6 Two views of a COLTRIMS attached to beamline U49/2-PGM-1 of the BESSY II synchrotron facility, Helmholtz-Zentrum Berlin, Germany. In the top picture, the COLTRIMS is located to the right of the center of the picture and can be identified by the stacked, circular Helmholtz coils surrounding the apparatus. The bottom picture shows a closer view of the COLTRIMS. The COLTRIMS shown here was designed and constructed by members of the Experimental Atomic and Molecular Physics Group at the Goethe-Universitat, Frankfurt Am Main

Fig. 9.7 Two views of a COLTRIMS device attached to beamline 9.0.1 of the Advanced Light Source (ALS) synchrotron facility, Berkeley National Laboratory, USA. The COLTRIMS itself can be identified by the stacked, circular, copper-sheathed Helmholtz coils surrounding the apparatus. The long "pipe" is the beamline through which collimated light is delivered to the device. The COLTRIMS shown here was designed and constructed by members of the Atomic, Molecular, Optical and Chemical Physics Group at the University of Nevada, Reno

Fig. 9.8 A COLTRIMS attached to beamline SEXTANTS of the SOLEIL synchrotron facility, Saint-Aubin, France. The COLTRIMS itself is located in the left of the picture and can be identified by the stacked, circular Helmholtz coils surrounding the apparatus. The long "pipe" extending the length of the picture is the beamline through which collimated light is delivered to the device. The COLTRIMS shown here was designed and constructed by members of the Experimental Atomic and Molecular Physics Group at the Goethe-Universitat, Frankfurt Am Main

Example 9.2

Find an expression for the differential cross-section when unpolarized/circularly polarized light is used.

Unpolarized/circularly polarized light is equivalent to two incoherent/coherent linearly polarized beams of equal intensity pointing along mutually orthogonal (x and y) axes. In that case, the differential cross-section would be written as [90]

$$
\left[\frac{d\sigma_{pd}}{d\Omega}\right]_{\text{unpol/circ pol}} = \frac{1}{2}\frac{\sigma_{pd}}{4\pi}\left[1 + \beta P_2(\cos\theta_x)\right] + \frac{1}{2}\frac{\sigma_{pd}}{4\pi}\left[1 + \beta P_2(\cos\theta_y)\right]
$$
$$
= \frac{\sigma_{pd}}{4\pi}\left[1 - \frac{1}{2}\beta P_2(\cos\theta_z)\right]
$$
(9.27)

where the z-axis is taken as the beam propagation direction, and we have used the definition of $P_2(\cos\theta)$ found in (4.39c) and the relation [9, 90]

$$
\cos^2(\theta_x) + \cos^2(\theta_y) + \cos^2(\theta_z) = 1
$$
(9.28)

For partially polarized light, the differential cross-section has the same form as in (9.27), except that each component is scaled by a weighting factor describing the fraction of polarized light along each orthogonal axis [90, 160]:

$$\left[\frac{d\sigma_{pd}}{d\Omega}\right]_{\text{par pol}} = \frac{I_x}{I_0}\frac{\sigma_{pd}}{4\pi}[1+\beta P_2(\cos\theta_x)] + \frac{I_y}{I_0}\frac{\sigma_{pd}}{4\pi}[1+\beta P_2(\cos\theta_y)] \quad (9.29)$$

where

$$I_0 = I_x + I_y \quad (9.30)$$

We manipulate portions of the RHS of (9.29) as follows:

$$\frac{I_x}{I_0}[1+\beta P_2(\cos\theta_x)] + \frac{I_y}{I_0}[1+\beta P_2(\cos\theta_y)]$$

$$= \frac{I_x}{I_0} + \frac{I_y}{I_0} + \beta\left[\frac{I_x}{I_0}P_2(\cos\theta_x) + \frac{I_y}{I_0}P_2(\cos\theta_y)\right]$$

$$= 1 + \beta\left[\frac{3}{2}\frac{I_x}{I_0}\cos^2\theta_x - \frac{1}{2}\frac{I_x}{I_0} + \frac{3}{2}\frac{I_y}{I_0}\cos^2\theta_y - \frac{1}{2}\frac{I_y}{I_0}\right]$$

$$= 1 - \frac{\beta}{2}\left[1 - 3\frac{I_x}{I_0}\cos^2\theta_x - 3\frac{I_y}{I_0}\cos^2\theta_y\right]$$

$$= 1 - \frac{\beta}{2}\left[1 - \frac{3}{2}\cos^2\theta_x\left(\frac{2I_x}{I_0}\right) - \frac{3}{2}\cos^2\theta_y\left(\frac{2I_y}{I_0}\right)\right]$$

$$= 1 - \frac{\beta}{2}\left[1 - \frac{3}{2}\cos^2\theta_x\left(1 - \frac{I_y}{I_0} + \frac{I_x}{I_0}\right) - \frac{3}{2}\cos^2\theta_y\left(1 - \frac{I_x}{I_0} + \frac{I_y}{I_0}\right)\right]$$

$$= 1 - \frac{\beta}{4}\left[(3-1) - 3\cos^2\theta_x - 3\cos^2\theta_y - 3\frac{I_x}{I_0}\cos^2\theta_x + 3\frac{I_y}{I_0}\cos^2\theta_x\right.$$
$$\left. + 3\frac{I_x}{I_0}\cos^2\theta_y - 3\frac{I_y}{I_0}\cos^2\theta_y\right]$$

$$= 1 - \frac{\beta}{4}\left[3(1 - \cos^2\theta_x - \cos^2\theta_y) - 1 - 3\frac{I_x - I_y}{I_0}\cos^2\theta_x + 3\frac{I_x - I_y}{I_0}\cos^2\theta_y\right]$$

$$= 1 - \frac{\beta}{4}\left[3\cos^2\theta_z - 1 - 3\frac{I_x - I_y}{I_0}(\cos^2\theta_x - \cos^2\theta_y)\right]$$

$$= 1 - \frac{\beta}{2}\left[P_2(\cos\theta_z) - \frac{3}{2}p(\cos^2\theta_x - \cos^2\theta_y)\right]$$

$$(9.31)$$

where we have used (9.28) and (9.30), and the degree of polarization is given by [90]

$$p = \frac{I_x - I_y}{I_x + I_y} \quad (9.32)$$

Now we can rewrite (9.29) in its final form [90, 146]:

$$\left[\frac{d\sigma_{pd}}{d\Omega}\right]_{\text{par pol}} = \frac{\sigma_{pd}}{4\pi}\left\{1 - \frac{\beta}{2}\left[P_2(\cos\theta_z) - \frac{3}{2}p(\cos^2\theta_x - \cos^2\theta_y)\right]\right\} \qquad (9.33)$$

Equation (9.33) also applies to elliptically polarized light as long as the orthogonal axes are understood to be the major and minor axes of the ellipse which characterizes the incident laser light. ∎

Example 9.3

Repeat the above example and this time include higher-order multipole terms.

To include the higher-order multipole terms, manipulations similar to the ones above can be performed on (5.48). For unpolarized/circularly polarized light, the result is [49, 52]

$$\left[\frac{d\sigma_{pd}}{d\Omega}\right]_{\text{unpol/circ pol}} = \frac{\sigma_{pd}}{4\pi}\left[\left(1 + \frac{\beta}{4}\right) + \left(\delta + \frac{\gamma}{2}\right)\sin\theta\cos\phi \right.$$
$$\left. - \frac{3\beta}{4}\sin^2\theta\cos^2\phi - \frac{\gamma}{2}\sin^3\theta\cos^3\phi\right] \qquad (9.34)$$

Notice that (9.34) is a function only of $\sin\theta\cos\phi$, the cosine of the angle between the photoelectron momentum and the photon propagation vectors.

For partially polarized light, using (9.32), we get [49, 52]

$$\left[\frac{d\sigma_{pd}}{d\Omega}\right]_{\text{par pol}} = \frac{\sigma_{pd}}{4\pi}\left\{
\begin{array}{l}
\left[1 + \frac{\beta}{4} - \frac{3}{4}p\beta + \frac{3}{2}p\beta\cos^2\theta\right] + \left[\delta + \gamma p\cos^2\theta - \gamma\frac{(p-1)}{2}\right] \\
\times\sin\theta\cos\phi \\
+ \left[\frac{3\beta}{4}(p-1)\right]\sin^2\theta\cos^2\phi + \left[\gamma\frac{(p-1)}{2}\right]\sin^3\theta\cos^3\phi
\end{array}
\right\}$$

$$= \frac{\sigma_{pd}}{4\pi}\left\{
\begin{array}{l}
\left[1 + \frac{\beta}{8}(1 + 3p)(3\cos^2\theta - 1)\right] \\
+ \left[\delta + \gamma\cos^2\theta + \gamma\frac{(p-1)}{8}(5\cos^2\theta - 1)\right] \\
\times\sin\theta\cos\phi + \left[\frac{3\beta}{8}(p-1)\right]\sin^2\theta\cos 2\phi + \left[\gamma\frac{(p-1)}{8}\right] \\
\times\sin^3\theta\cos 3\phi
\end{array}
\right\}$$

$$(9.35)$$

∎

Problems

9.1. In most experiments in which atomic or molecular ion beams are used, a method must be found to select the desired atomic or molecular species by mass and/or charge (such a device is known as a momentum analyzer). One selection technique centers around a 90° bending electromagnet. The magnet is so designed as to produce a radial magnetic field in its interior. Show that, to select a particle of mass m_i and charge q_i that has been accelerated by a potential V_i, a magnet of radius R_m (the radius along which the selected ion must travel) must be set to a magnetic field B_{R_m}, whose magnitude is given by

$$B_{R_m} = \sqrt{\frac{2m_i V_i}{q_i R_m^2}}$$

9.2. Another type of momentum analyzer is a Wien filter. In a Wien filter, the magnetic field is supplied by an electromagnet, or by a pair of permanent magnets [161]. The electric field is supplied by a pair of parallel conducting plates separated by a distance d and maintained at a potential difference V. In this type of selector, the magnetic and electrostatic fields ($\vec{B} \& \vec{E}$, respectively) are arranged perpendicularly, as shown in the diagram below:

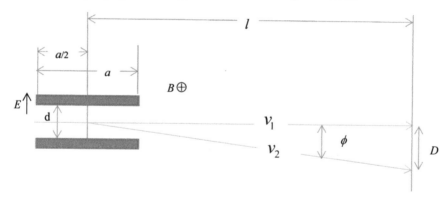

Schematic diagram of a Wien filter

If one were to scan the mass spectrum of an ion beam entering the filter by varying the potential V, only ions with a well-defined mass-to-charge ratio will be transmitted.

(a) Show that, for ions of mass m_i and charge q_i that have been accelerated from rest by a potential V_i, the mass of the transmitted ions is given by

$$m_i = \frac{2q_i V_i B^2 d^2}{V^2}$$

(b) Consider a filter that is set to allow an ion of mass m_1 and charge q_i to pass through undeflected. An ion with the same charge, but of slightly higher mass m_2 ($m_1 \approx m_2$), will experience a (slight) centripetal deflection. Find the angle ϕ (in terms of the relevant given parameters) between the trajectories of m_1 and m_2 for a Wien filter of linear extent a (neglecting end effects). Show that the mass resolution of the Wien filter is

$$\frac{\Delta m_i}{m_i} = \frac{4DV_id}{alV}$$

where D is the length of the post-filter aperture/slit, and l is the distance from the center of the filter to the post-filter aperture.

9.3. Electrostatic beam optics [19, 161]

In experiments designed to measure PADs, ions must be delivered from the ion source to the region in which they will interact with incoming photons. To focus the ion beam as it travels from the source to the interaction region, a series of electrostatic focusing lenses are usually employed.

An Einzel lens consists of hollow conducting tubes in which axially symmetric electrostatic potentials Φ are realized to focus charged particle beams without changing the energy of the beam particles. We will show this as follows. In the paraxial approximation, we can expand this potential in a Taylor series expansion [19]:

$$\Phi(r, z) = \Phi(0, z) + b_2(z)r^2 + b_4(z)r^4 + \cdots \tag{1}$$

where the z-axis is parallel to the axis of the cylindrical tube and is the axis of symmetry of the field, r is the radial coordinate, and ϕ is the azimuthal angle, and where, due to axial symmetry, $\Phi(-r) = \Phi(r)$, only the even terms in the expansion are nonzero. Note also that $\Phi(0, z)$ is the potential along the z-axis. The paraxial approximation tells us that $b_n r^n \ll \Phi(0, z)$, so that terms of order $n > 2$ can be neglected. Physically, the paraxial approximation assumes that the ion beam does not stray very far from the symmetry axis (the z-axis) of the lens.

(a) Show that, in the paraxial approximation and for axially symmetric potentials, the potential at a point close to the symmetry axis can be calculated from solely its values $\Phi(r = 0)$ and $\Phi''(r = 0)$ [19]. Also show that ions which stray from the axis are subject to a restoring force which drives them back to the z-axis and thus focuses the beam.

Note: The paraxial approximation also implies that $\left(\frac{\partial^2 \Phi}{\partial r^2}\right)r^2 \ll \left(\frac{\partial \Phi}{\partial r}\right)r$.

(b) An excellent type of focusing lens is one that is produced by two pairs of axially symmetric hyperbolic electrodes, each pair of which is kept at potentials $\Phi = 0$ and $\Phi = \Phi_0$, respectively. Such a potential has the form [19]

$$\Phi(r, z) = a\left(z^2 - \frac{1}{2}r^2\right)$$

Show that the equipotential surfaces in such a field are hyperboloids centered around the z-axis. Show that the force which drives the ions back to the axis produces harmonic oscillations in the r-component of the ion motion.

9.4. In laser photodetachment electron spectroscopy (LPES) experiments, atomic anions are accelerated into an interaction chamber where electrons ("photoelectrons") are detached from the atoms by laser photons. The photoelectrons are collected, and their kinetic energy is measured by a kinetic-energy-analyzing spectrometer. The spectrometer is in the form of two concentric half-cylinders sharing a common axis (see diagram below). The cylindrical plates are charged in such a way to produce (between the half-cylinders) a radial electrostatic field of the form

$$\vec{E} = \frac{k}{r^2}\hat{r}$$

A photoelectron of mass m, velocity $\vec{v}$, and charge $-q$ enters the region between the plates in a direction perpendicular to the radial direction as shown in the figure. Since the velocity vector has no component along the axis of the half-cylinders, we can safely assume that the photoelectron's subsequent motion lies entirely in the plane as shown in the below diagram.

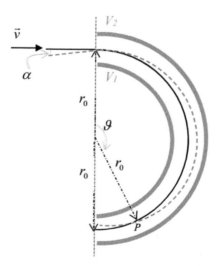

A photoelectron is collected, and its kinetic energy measured via the following procedure. The voltages on the plates are adjusted to allow only those photoelectrons of speed v to successfully traverse the spectrometer by moving in a circular path of radius r_0 (shown in solid black in the diagram) while between the plates (photoelectrons with other speeds will crash into one of the inner walls), where

$$r_0 = \frac{r_{in} + r_{out}}{2}$$

and r_{in} and r_{out} are the radii of the inner and outer cylinders, respectively.

(a) Find an expression relating the potential set across the plates ($\Delta V = V_1 - V_2$) to the pass energy $K_0 = mv^2/2 = eV_{r_0}$ for a photoelectron to successfully traverse the analyzer, where V_{r_0} is the potential that a photoelectron which travels along the circular path of radius r_0 feels.

(b) Find the speed v (in terms of the constants q, k, and m and r_0) that a photoelectron would have if it traveled along the circular path.

(c) Construct the Lagrangian for a photoelectron traveling through the spectrometer and find the resulting differential equation of motion (DEOM) for the radial coordinate. Show also that angular momentum is conserved, and so write your radial DEOM using this conserved quantity. Find r_0 (the radius of a circular orbit shown as a solid black curve in the diagram) in terms of this conserved quantity and other given constants.

(d) Although a set of slits (not shown) is placed at the entrance of the spectrometer to collimate the incoming particles, it is still possible for a photoelectron of speed v to enter at the same distance r_0 from the spectrometer axis, but at a small angle α (known as the spectrometer acceptance angle) from the direction of the original path. In this case, the photoelectron will not travel in a circular path but will instead travel along a different trajectory (shown as a dashed red curve in the diagram). To model this trajectory, expand your radial DEOM to first order about the constant radius r_0, as follows:

$$r \approx r_0 + \eta$$

where η is a small deviation in the radial direction from the radius of the circular orbit. Solve the linearized radial DEOM (remember, you must apply the correct initial conditions for η to do this) and show that the solution for the deviation can be written as

$$\eta = r_0 \sin \alpha \sin \left(\frac{v}{r_0} t \right)$$

and thus find (to first order) the angle ϑ where this new trajectory crosses the trajectory of part (a), (point P in the diagram). Show that your answer for ϑ is independent of α. P is the point at which the spectrometer (to first order) will focus a slightly spreading beam of photoelectrons. This result is known as Barber's Rule.

Note: The spectrometer should be constructed so that the point P lies just outside the cylinders so that the photoelectrons can ultimately be easily collected (and counted) by some other device. That is to say, the cylinders should not span an angle of 180 degrees, but instead some slightly lesser angle.

9.5. Relativistic kinematic transformations for a crossed-beams photodissociation experiment

We begin by noting that (9.11–9.17) were derived based on a (nonrelativistic) Galilean coordinate transformation. A relativistic description requires that velocities be added, or "transformed," in a way that is Lorentz invariant. This requires the standard velocity addition formulas of special relativity, as applied to the geometry of the experimental setup.

Let the z-direction be the direction the ion beam is traveling $\left(\vec{v}_i = v_i \hat{e}_z \right)$. The plane indicated in Fig. 9.1 is the x-z plane, and the velocities are as discussed in the text.

(a) Find the appropriate relativistic velocity transformation equations for the geometry of the experimental setup. Your answer should express the components of $\vec{v}_c$ in terms of $\vec{v}_l$, $\vec{v}_i$, and θ_l.

(b) Use your results from part (a) to show that

$$\left| \vec{v}_c \right| = \frac{\sqrt{v_i^2 + v_l^2 - 2 v_i v_l \cos\theta_l - \frac{v_i^2 v_l^2 \sin^2\theta_l}{c^2}}}{1 - \frac{v_i v_l \cos\theta_l}{c^2}} \quad \text{and} \quad \tan\theta_c = \frac{v_l \sin\theta_l}{(v_l \cos\theta_l - v_i)\gamma_i}$$

where $\gamma_i = [1 - (v_i/c)^2]^{-1/2}$.

(c) Use your results from part (b) to find the kinetic energy of a relativistic photoelectron in the ion (center-of-mass) frame, K_c, and the kinetic energy of a relativistic photoelectron in the lab frame, K_l. Your answer for K_c should be in terms of v_i, v_l, θ_l, γ_i, γ_l, and the appropriate constants. Your answer for K_l should be in terms of v_i, v_c, θ_c, γ_i, γ_c, and the appropriate constants. Note that

$$\gamma_{l,c} \equiv \left[1 - (v_{l,c}/c)^2 \right]^{-1/2}$$

Check to ensure your answers reduce to the correct expressions in the non-relativistic limit.

(d) Prove that the relativistic version of Eq. (9.17) is

$$\left| \frac{d\Omega_l}{d\Omega_c} \right| = \frac{\sin\theta_l \, d\theta_l}{\sin\theta_c \, d\theta_c} = \left(\frac{v_c}{v_l} \right) \left(1 - \frac{v_i \cos\theta_l}{v_l} \right) \left(\frac{\gamma_c \gamma_i}{\gamma_l} \right)$$

Note that this expression reduces to the correct result in the nonrelativistic limit.

9.6. In photoelectron spectroscopy, the inverse Abel transformation is used to analyze data collected using the velocity map imaging (VMI) technique. A 3-D axially symmetric photoelectron distribution is mapped onto a phosphor screen via an electrostatic focusing array. The resulting 2-D image is captured by a CCD camera, and the original 3-D photoelectron distribution is recreated via the inverse Abel transformation [149].

Raw Photoelectron
Image

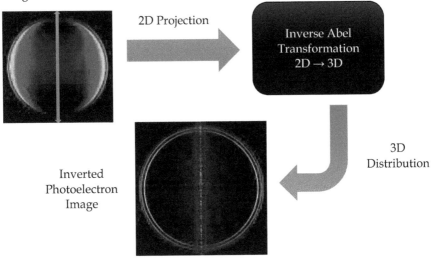

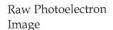

2D Projection

Inverse Abel
Transformation
2D → 3D

3D
Distribution

Inverted
Photoelectron
Image

(a) If the 3-D photoelectron distribution is given by $f(r)$, use the figure below to demonstrate that the Abel transform $g(R)$ of the 3-D photoelectron distribution is given by

$$g(R) = 2 \int_R^\infty \frac{f(r)rdr}{\sqrt{r^2 - R^2}}$$

(b) Show that the inverse Abel transformation is given by

$$f(r) = -\frac{1}{\pi} \int_r^\infty \frac{g'(R)dR}{\sqrt{R^2 - r^2}} \, ; g'(R) \equiv \frac{dg}{dR}$$

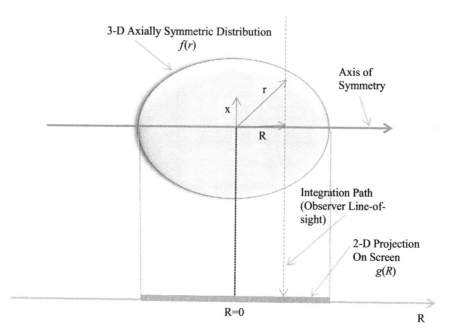

Projection of the axially symmetric 3-D photoelectron angular distribution onto a 2-D screen

9.7. Including a drift region in a COLTRIMS reaction microscope is a way to account for the fact that the photon-ion interaction region (IR) is not truly a point but as a practical matter has a finite size for which one must account. Find the condition under which electrons that have the same initial velocity but originate at different points within the IR will nevertheless arrive at the electron detector at the same time (to first order) as long as they start with a small initial velocity as compared to the velocity they gain while in the acceleration region (see App. G for the basic design elements of the COLTRIMS device).

9.8. Fixed photoelectron collection directions

(a) In the LPES method, the kinetic energy analyzer/spectrometer is located at a fixed angle θ_l. Find θ_l such that the CM asymmetry parameter β_{cm} can be found directly from the asymmetry parameter measured in the lab β_l via

$$\beta_{cm} = \frac{\beta_l}{1 + \left(\frac{v_i}{v_c}\right)^2}$$

(b) For the PEARLS setup of Fig. 10.7, show that the CM asymmetry parameter β_{cm} can be found directly from the asymmetry parameter measured in the lab β_l via [147]

$$\beta_{cm} = \frac{\beta_l}{1 - \left(\frac{v_i}{v_c}\right)^2 \left(1 - \frac{\beta_l}{2}\right)}$$

Chapter 10
Applications of Photoelectron Angular Distribution Measurements

Laboratory PAD measurements are prominently featured in many types of current research. In this chapter, a few applications of PADs are highlighted, and several examples of cutting-edge research involving PADs are briefly described.

Measuring pd reactions of an individual species at multiple photon wavelengths is how one experimentally determines the spectral dependance of pd asymmetry parameters. For such studies, the VMI technique proves to be an efficient technique. VMI images of photoelectron angular distributions made during these studies can also be used to examine pd behavior at threshold and to determine (for example) the range of validity of the Wigner threshold law [c.f. (4.41)] [18].

One common method used to study threshold behavior is the laser pd threshold (LPT) method. LPT studies proceed by fitting total pd cross-sections to various versions of the Wigner threshold law and then inferring the location of the reaction threshold (see, e.g., [162]). But, under certain circumstances, LPT measurements result in an uncertainty as to the location of the pd threshold. No such ambiguity exists when using VMI PAD images. When the threshold is reached, the PAD for that reaction will appear in the image. Using the VMI technique to study pd thresholds can therefore result in a marked improvement in accuracy over other methods.

In addition, VMI images of differential cross-sections preserve information about the contribution of continuum partial waves and their relative phases. All in all, the VMI technique is well-suited to examine every facet of pd cross-sections near threshold.

We examine (as an example) the photodetachment of the sulfur anion via the single-photon reaction

The original version of this chapter was revised. The correction to this chapter is available at
https://doi.org/10.1007/978-3-031-08027-2_11

V. T. Davis, *Introduction to Photoelectron Angular Distributions*, Springer Tracts in
Modern Physics 286, https://doi.org/10.1007/978-3-031-08027-2_10

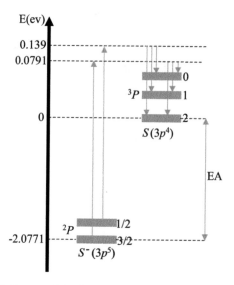

Fig. 10.1 Energy-level diagram of the fine-structure-resolved ground states of S^- and S (not to scale). Shown are the electron affinity EA of sulfur ($EA = 2.0771$ eV) and the six possible photodetachment transitions for a photon of energy/wavelength $\hbar\omega = 2.1562$ eV/$\lambda = 575$ nm. Adapted from [23] and reproduced with permission from the author

$$S^-[Ne]3s^23p^5\left(^2P^0\right) + \gamma \rightarrow S[Ne]3s^23p^4\left(^3P\right) + e^-{}_{s,d} \qquad (10.1)$$

The possible fine-structure-resolved transitions from the S^- ground state to the S ground state are shown in the energy-level diagram of Fig. 10.1. Note that there are six possible photodetachment transitions for a photon of energy $\hbar\omega = 2.1562$ eV. In Fig. 10.2, VMI images of PADs for the photodetachment reaction of (10.1) at various photon wavelengths are shown. When the threshold for a particular fine-structure-resolved photodetachment reaction is reached (and not before), the PAD for that transition will appear in the image.

Using the analysis techniques from Chap. 7, we can write the asymmetry parameter for the photodetachment reaction of (10.1) as (c.f. problem 7.8)

$$\beta = \frac{3\left|\bar{S}_d(1)\right|^2 - 5\left|\bar{S}_d(2)\right|^2 - 3\sqrt{2}\left(\bar{S}_d(1)\bar{S}_s^*(1) + \bar{S}_s(1)\bar{S}_d^*(1)\right)}{3\left[\left|\bar{S}_s(1)\right|^2 + \left|\bar{S}_d(1)\right|^2\right] + 5\left|\bar{S}_d(2)\right|^2} \qquad (10.2)$$

Data taken from Fig. 10.3 can be used to inform theoretical techniques that purport to calculate the photodetachment amplitudes $\left|\bar{S}_\pm(j_t)\right|$ and $\left|\bar{S}_0(j_t)\right|$ and the dependance of pd asymmetry parameters on photon wavelength.

Another advantage of the VMI technique is the relatively high signal-to-noise ratio in the associated photoelectron kinetic energy spectrum. Such "clean" spectra can reveal features that would not otherwise be visible. Consider, for example, the single-photon photodetachment from the ground state of the tin anion. Dipole-

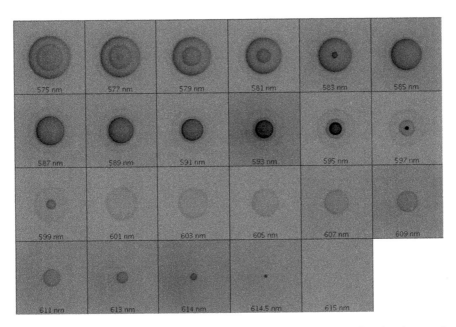

Fig. 10.2 VMI images of photoelectron angular distributions from the photodetachment of S^- using linearly polarized laser photons in the range of 575–615 nm. As the laser wavelength decreases, more transitions from the fine-structure levels of the S^- ground state appear (reading the images from right to left and from bottom to top), and those whose images are already present increase their radii since the photoelectrons acquire more kinetic energy from the increasingly energetic photons. At a photon wavelength of 575 nm, all six possible transitions from the fine-structure levels of the S^- ground state to the fine-structure levels of the S ground state are energetically accessible (upper-left image) [23]. Reproduced from [23] with permission from the author

allowed transitions to the neutral ground state for this reaction are given by the following reactions:

$$Sn^-[Kr]4d^{10}5s^25p^3\left(^4S_{3/2}\right) + \gamma \rightarrow Sn[Kr]5s^25p^2\left(^3P_{0,1,2}\right) + e^-_{s,d} \qquad (10.3)$$

An examination of the VMI PAD image, however, shows four transitions (see Fig. 10.4). An analysis of the photoelectron kinetic energy spectrum prepared from this image reveals that the "extra" feature represents a transition from the anion ground state to the $^1D_2$ excited state in the neutral, which is dipole-forbidden (see Fig. 10.5). In addition, we observe "splittings" between the fine-structure-resolved ^3P_j states that are larger than those predicted in pure LS coupling. Furthermore, the data shows deviations from LS-coupling-predicted transition intensities to the ^3P_j states. Measured ratios for the $^4S_{3/2} \rightarrow {}^3P_{0,\,1,\,2}$ transition intensities are 1:2.1:2.7 rather than the 1:3:5 ratios predicted by the LS-coupling approximation. Similar results are obtained for other group 14 negative ions [23].

This data provides support for recent theoretical predictions of a breakdown in the single-configuration description of the $^1D_2$ excited state. The state is more accurately described as an admixture of states with $^1D_2$ and $^3P_2$ as leading terms [163]. We see

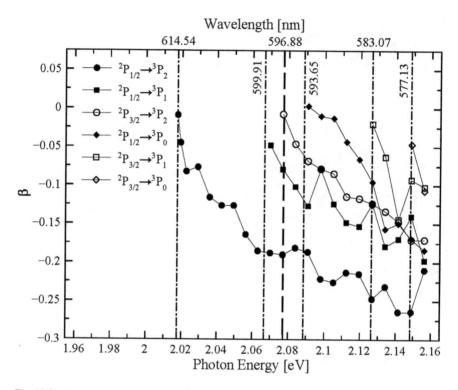

Fig. 10.3 Asymmetry parameters as a function of photon energy for the photodetachment process of (10.1). Each vertical dashed line indicates the threshold energy for the indicated fine-structure-resolved transition. The heavy dashed line indicates the EA for sulfur [23]. Reproduced from [23] with permission from the author

that VMI PADs can be used to experimentally verify theoretical calculations of atomic structures.

Because the photodetachment reaction of (10.3) involves the release of a p-electron, the asymmetry parameters determined from the PADs can be plotted against the Hanstorp model (4.47). Since the Hanstorp model is based on threshold behavior and reflects the short-range behavior of the negative ion potential [20], such a plot serves as a sensitive test of the validity of this model for the given photodetachment reaction (see Fig. 10.6).

We mentioned above that LPT threshold studies were hampered by the fact that the pd threshold is not easy to locate. This is because, in the neighborhood of the threshold, photoelectrons have near-zero kinetic energy, and zero is always difficult to measure. The problem is compounded in the case of p-electron photodetachment by the gradual growth of the cross-section near threshold in accordance with the Wigner threshold law (see, e.g., [162, 164, 165]). Uncertainty in the location of the threshold makes it difficult to employ LPES and LPT methods to determine electron affinities [24, 25]. The problem is further complicated for the transition elements due

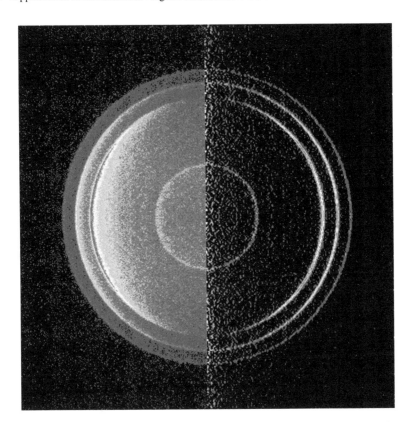

Fig. 10.4 VMI image of the PAD resulting from the single-photon photodetachment from the ground state of the tin anion using 532 nm linearly polarized laser photons. The laser polarization vector lies in the vertical direction. The left half of the image is the raw data. The right half shows the Able-inverted pattern (c.f. problem 9.6). The computer-generated coloring represents the relative number (intensity) of the collected photoelectrons [149]

to complicated electronic structures which require a higher-energy resolution to resolve the fine structures than LPT methods can provide [164, 165]. Observed deviations from the Wigner threshold law further affect the accuracy of LPES and LPT methods (see, e.g., [166]). A variant of the VMI technique called slow-electron velocity-map imaging (SEVI) overcomes these difficulties. Accuracies in values of electronic affinities of one and even two orders of magnitude over previous measurements have been reported using PADs generated via the SEVI method (see, e.g., [164, 165, 167]).

Crossed-beam geometries of the type depicted in Fig. 9.1 are typically used in LPES and VMI experiments and are designed to collect photoelectrons from a well-defined but small interaction region. Photoelectrons can be efficiently collected from these interaction regions (especially in VMI experiments), but the small size of the interaction region means that only a limited number are available. A colinear geometry in which the ion and photons beams are counter-propagating can be

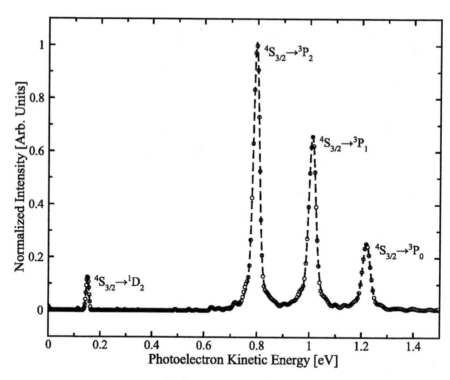

Fig. 10.5 Photoelectron kinetic energy spectrum for the single-photon photodetachment from the ground state of the tin anion using 532 nm linearly polarized laser photons. The small circles represent the experimental data. The dashed curves are Gaussian fits to the data. The three features on the right show the dipole-allowed transitions $^4S_{3/2} \rightarrow {}^3P_{0,\ 1,\ 2}$. The transition on the left, $^4S_{3/2} \rightarrow {}^1D_2$, is dipole-forbidden, providing experimental confirmation of theoretically predicted configuration mixing in the first excited state of neutral tin [23, 149]. Reproduced with permission from the author

used to dramatically increase photoelectron count rates [146, 147, 168]. When employing a colinear/merged-beams geometry in a photoelectron spectroscopy experiment, a method must be developed to collect photoelectrons in the (now) extended linear interaction region. Recently, a technique and a device to collect photoelectrons produced from the interaction of counter-propagating ion and photon beams were developed. The device, called a PhotoElectron Angle-Resolved Linear Spectrometer (PEARLS), consists of two 11-cm-long, shielded, linearly aligned graphite tubes of square cross-section. The PEARLS, within which the ion and photon beams counter-propagate, defines the (field-free) interaction region in which photoelectrons are produced. Photoelectrons produced inside the PEARLS can exit through holes drilled into all four sides of the graphite tubes. The photoelectrons are ultimately collected and counted by 16 channel electron multipliers (CEMs) placed outside the holes along the length of the graphite tubes to ensure collection from the entire extended interaction region (see Fig. 10.7). A filter is placed in the flight path of the photoelectrons between the holes in the graphite and the CEMs to suppress

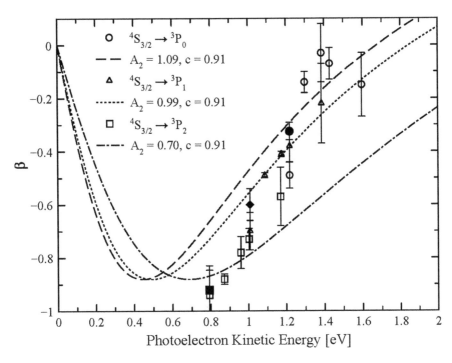

Fig. 10.6 Asymmetry parameters for the photodetachment reaction of (10.3) as a function of photoelectron kinetic energy. Fitting parameters from (4.47) for each fine-structure-resolved transition are indicated in the upper left of the figure. Open symbols are from [41] while filled symbols are from [23]. Reproduced from [23] with permission from the author

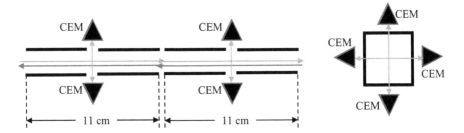

Fig. 10.7 Schematic diagram of the interaction region of the PEARLS device. On the left is a side view. In this view, the path of the counter-propagating ion and photon beams are represented by the horizontal (blue and red) arrows. Photoelectrons created inside the interaction region exit the graphite tubes (ideal paths are represented by vertical, double-headed green arrows) and are ultimately collected and counted in the CEMs (black triangles). On the right is a vertical slice showing a plane of CEM detectors. In the slice view, the photons and ions would be coming into/out of the page. Detectors are placed on all four sides of the graphite tubes to assist in collecting photoelectrons from the entire extended interaction region and to facilitate measuring PADs. This diagram was created from descriptions in [147], DOI 10.1103/PhysRevA.103.033108 and is displayed under a CCA 4.0 International license

low-energy photoelectrons, allowing only transition-specific photoelectrons to be counted [147, 168].

PADs may be measured with the PEARLS either by rotating the linear photon polarization vector with a Fresnel rhomb or, in the case when rotating the photon polarization vector is impractical (e.g., when using a synchrotron photon source), by measuring the photoelectron yields in two orthogonal detectors [c.f. (9.23)] [147]. Although the photoelectron collection efficiency of the PEARLS is considerably less than the near-100% achievable by a VMI spectrometer, the extended linear interaction region makes up for this deficit [168]. PADs from the photodetachment of the P^- anion were measured over a large spectral range using the PEARLS and the data fit to the Hanstorp model [c.f. (4.47)]. The results indicate that the value of the fitting parameter from (4.47) is $c \approx 0.9$ for the photodetachment of a p-electron and that the A_{20} parameter is inversely proportional to the electron affinity of the atomic anion, thus supporting the validity of the Hanstorp model as a special case of the C-Z theory under Wigner threshold assumptions [147].

As mentioned in Chap. 9, the COLTRIMS reaction momentum microscope can measure the momenta of the products of a pd reaction in coincidence with a 4π collection efficiency. Thus, the COLTRIMS is an ideal technique for measuring MFPADs. An analysis of these MFPADs can provide a wealth of information on molecular structure and the dynamics of photon-molecule interactions. For example, because COLTRIMs allows one to determine the complete kinematics of a multibody breakup (e.g., of a molecule), a photoelectron ejected from an inner shell can be used to image the molecular potential, even as the potential varies in time and space. Specifically, a slow-moving core electron acts as a wave that diffracts around the various structures of the molecule, in effect "photographing" the molecule from within. Because the COLTRIMS microscope collects the ejected (photo and Auger) electrons in coincidence with positive fragments, the experimenter is able to reconstruct the photoelectron angular distribution in the body-fixed molecular frame with the resulting MFPAD serving as a sensitive map of the total molecular potential and molecular bond geometry. In this way, three-dimensional images of MFPADs have been constructed that have provided insights into how photoelectrons of various energies are ejected in "Coulomb explosions" during which molecules fragment after absorbing high-energy photons (see Fig. 10.8). Eventually, it may be possible to make "movies" of photon-molecule and chemical reactions using these techniques.

It has been found that low-energy photoelectrons ejected from some molecules emerge primarily along bond axes, allowing the determination of how bonds flex during the breakup process. In heteronuclear diatomic molecules, interference between direct waves and waves reflected from the companion atom in the molecule can show how momentum is distributed among the fragments during molecular breakup. In more complicated molecules, the MFPAD is seen to be the result of the combination of two effects: the propensity of the slow-moving electron to be ejected along the bond axes as a result of the interaction of the departing electron with the molecular potential, and the effect of the dipole-allowed transitions of the reaction which tend to send the electrons in directions determined by (4.38) [150, 157, 158].

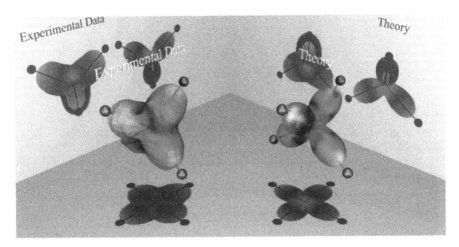

Fig 10.8 Three-dimensional MFPAD of a K-shell photoelectron for the $H^+ + H^+$ + unknown fragmentation channel of methane. On the left is an experimentally determined MFPAD for a photoelectron of energy 4.2 eV above the 1s carbon threshold. On the right is a calculated MFPAD for a photoelectron of energy 4.35 eV integrated over all polarization directions [169]. For more details on the experimental methods and the theoretical calculations, see [157]. Reproduced from [169] with permission from the author

COLTRIMS MFPAD measurements are also used to test the validity of the axial recoil approximation. In the axial recoil approximation, the direction of emission of ion fragments is assumed to give the spatial orientation of the molecule at the time of break-up. This approximation is valid when the duration of the molecular breakup process is faster than the rotational period of the molecule (and the characteristic vibrations in the case of a polyatomic molecule) [157, 158]. In inner-shell photo-ionization, for example, Auger electrons are often emitted after the initial absorption of a high-energy photon and release of an initial photoelectron, usually leading to a Coulomb explosion. For the axial recoil approximation to be valid, the delay between the initial photoionization and breakup of the molecule must be short compared to the rotational period of the molecule. If one can measure pd reaction product momenta in coincidence, one could deduce phase differences from the various momenta vector directions and from those phase differences, determine relative time delays [170]. Using COLTRIMS, a novel technique to measure time-dependent molecular dissociation dynamics at the attosecond level has recently been demonstrated. As stated previously, an ejected electron acts as a wave as it emerges from the molecule. The wave packet of a photoelectron of energy ε can be modeled as a superposition of partial waves as follows [171]:

$$\Psi_\varepsilon(\theta, \phi) = \sum a_{\varepsilon lm} Y_{lm}(\theta, \phi) \qquad (10.4)$$

As it emerges, the electron wave picks up a phase shift (called the Wigner phase) as it interacts with the molecular potential (see Fig. 10.9). By using the COLTRIMS to measure the MFPAD, the Wigner phase can be calculated, and the pd emission

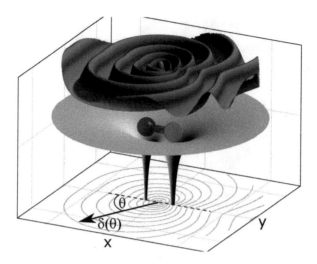

Fig 10.9 An electron wave is emitted from a heteronuclear diatomic molecule. The photoelectron picks up a phase shift as it traverses the molecular potential. This Wigner phase depends on the emission angle with respect to the molecular bond axis due to the anisotropy of the potential. This figure is adapted from [171] under the Creative Commons Attribution 4.0 International License. To view a copy of this license, see http://creativecommons.org/licences/by/4.0

time [known as the Wigner delay $t_w(\varepsilon, \theta, \phi)$] can be determined from this phase shift. Keep in mind that, in the case of a molecular breakup, the delay will depend on which part of the anisotropic molecular potential is sampled by the photoelectron as it departs. The temporal history of the breakup process is determined from the energy- and angle-dependent Wigner phase $\{\arg[\Psi_\varepsilon(\theta, \phi)]\}$ via

$$t_w(\varepsilon, \beta, \theta, \phi) = \hbar \frac{d}{d\varepsilon}\{\arg[\Psi_\varepsilon(\theta, \phi)]\} \qquad (10.5)$$

This derivative is evaluated by direct measurement of the phase shift as a function of the energy of the ionizing photons. The MFPAD is given by (8.16) which, in the context of this experiment, reads as

$$\frac{d\sigma_\varepsilon}{d\Omega} \sim \left| \sum_{lmk} (-1)^l D^l_{k0}(\beta_\gamma) \langle \psi^-_{\varepsilon lm} | d^l_k | \Psi_{\Gamma_0} \rangle Y_{lm}(\theta, \phi) \right|^2 \qquad (10.6)$$

where the $D^l_{k0}(\beta_\gamma)$ are the matrix elements of the rotation operator (for a diatomic molecule, all the Euler angles other than β_γ are irrelevant) and d^l_k is the dipole operator for the absorption of a photon of polarization k that results in the emission of a photoelectron of partial wave quantum numbers l and m. Comparing (10.5) and (10.6) makes the connection between the MFPAD and the Wigner delay clear. This technique was put onto practice at the BESSY II synchrotron facility, Helmholtz-

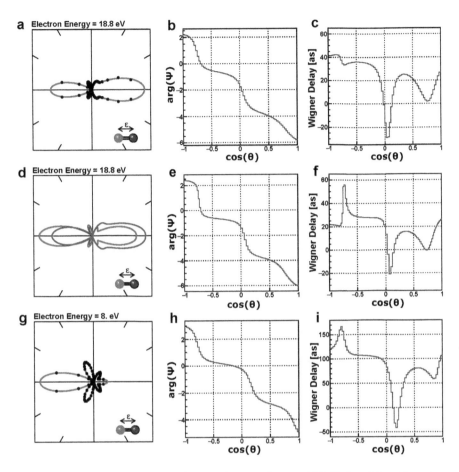

Fig 10.10 (**a**) Measured MFPADs of a core electron of energy 18.8eV ejected from the carbon atom of a CO molecule following absorption of linearly polarized light [the carbon atom is represented by the (black) spheres on the right and the oxygen atom by the (red) spheres on the left in the bottom right corner of diagrams (**a**), (**d**), and (**g**)]. The molecule is oriented horizontally as is the polarization vector of the photons. (**b** and **c**) The extracted phase arg[$\Psi_e(\theta)$] and the Wigner delay, respectively. (**d–f**) Results from theoretical modeling(s) of the same photoemission process. (**g–i**) Same as (**a–c**), except for a photoelectron of energy 0.8 eV. The angle θ is measured from the horizontal in a counterclockwise direction. This figure is taken from [171] under the Creative Commons Attribution 4.0 International License. To view a copy of this license, see http:// creativecommons.org/licences/by/4.0

Zentrum Berlin, in a novel experiment using the CO molecule (see Figs. 10.10 and 10.11). For more details, see [171].

The CO molecule features prominently in another novel COLTRIMS experiment in which circularly polarized (CP) light was used to measure high-energy MFPADs that in turn were used to determine the lengths of the bonds in the molecule. In the heteronuclear diatomic molecule CO, if the MFPAD is confined to the polarization

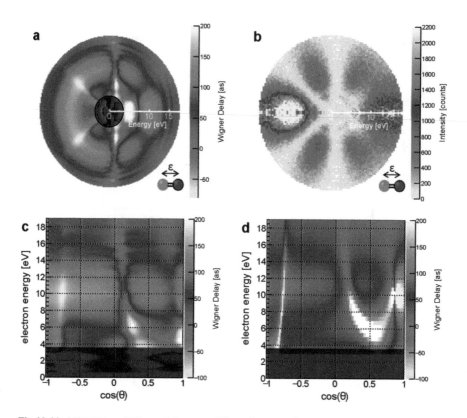

Fig 10.11 MFPADs and Wigner delay maps. The carbon atom is represented by the (black) sphere on the right and the oxygen atom by the (red) sphere on the left in the bottom right corner of diagrams (**a**) and (**b**). The molecule is oriented horizontally as is the polarization vector of the photons. (**a**) Wigner delay map which shows the Wigner delay as a function of electron emission angle and electron kinetic energy. The electron kinetic energy is shown as a function of the distance from the center of the image. The Wigner delay is encoded in the color scale. (**b**) Polar map of the MFPAD. The electron kinetic energy is again shown as a function of the distance from the center of the image, while the relative emission intensity is encoded in the color scale. Note the distinct matching radial features ("spokes") in (**a**) and (**b**). It seems that there is a correlation in the Wigner delay along emission angles that are minima in the MFPADs. (**c**) The same as (**a**) but shown in a different mapping representation. (**d**) Theoretical modeling of the Wigner delay map shown in (**c**). This figure is taken from [171] under the Creative Commons Attribution 4.0 International License. To view a copy of this license, see http://creativecommons.org/licences/by/4.0

plane and if the photoelectron energies are high enough so that the single-scattering approximation is valid, then the observed MFPADs will be the result of the interference between direct and scattered photoelectron waves originating from the atom that emits the photoelectron and the companion atom which scatters the wave, respectively. A characteristic of these CP-MFPADs is a prominent, forward-scattered peak in the direction of the scattering atom (e.g., see Fig. G.5). This peak is not located directly on the molecular axis but is tilted in the direction of the rotation of the CP light (upward or downward depending on whether the light is right- or left-handed circularly polarized). The tilt angle is also seen to decrease with increasing photoelectron kinetic energy. This fact allows the bond length to be directly calculated from the tilt angle. For details on the experiment and the theory used to calculate the bond lengths, see [154].

Correction to: Introduction to Photoelectron Angular Distributions

V. T. Davis

Correction to:
V. T. Davis, *Introduction to Photoelectron Angular Distributions***, Springer Tracts in Modern Physics 286,**
https://doi.org/10.1007/978-3-031-08027-2

The original version of this book was inadvertently published with few errors in Chapters 2, 4, 6, 7, 8, and 10. This has now been updated in the mentioned chapters in this version.

The updated original version of these chapters can be found at
https://doi.org/10.1007/978-3-031-08027-2_2
https://doi.org/10.1007/978-3-031-08027-2_4
https://doi.org/10.1007/978-3-031-08027-2_6
https://doi.org/10.1007/978-3-031-08027-2_7
https://doi.org/10.1007/978-3-031-08027-2_8
https://doi.org/10.1007/978-3-031-08027-2_10

Appendixes

Appendix A: Proof of Equation (2.68) and Evaluation of the Integral $\int_{-1}^{1} x^n P_n(x)dx$[1]

Equation (2.54) gives an expression for the spherical harmonics, but the spherical harmonics can also be expressed in terms of other well-known functions:

$$Y_{lm}(\theta, \phi) = \sqrt{\frac{2l+1}{4\pi}\frac{(l-m)!}{(l+m)!}}P_l^m(\cos\theta)e^{im\phi} \tag{A.1}$$

where $P_l^m(\cos\theta)$ are the associated Legendre functions,

$$P_l^m(\cos\theta) = (-1)^m \sin{}^m\theta \frac{d^m}{d(\cos\theta)^m}P_l(\cos\theta) \tag{A.2}$$

and $P_l(\cos\theta)$ are the Legendre polynomials, given by Rodrigues' formula

$$P_l(\cos\theta) = \frac{1}{l!2^l}\frac{d^l}{d(\cos\theta)^l}(\sin{}^{2l}\theta) \tag{A.3}$$

The Legendre polynomials can also be found via the following generating function:

---

[1] Primary references for App. A: [9, 172, 173]

$$G(x,\mu) \equiv \frac{1}{\sqrt{1 - 2x\mu + x^2}} = \sum_{l=0}^{\infty} x^l P_l(\mu) \tag{A.4}$$

where for ease of writing, we have renamed $\cos\theta \to \mu$. We begin by differentiating the LHS of the generating function with respect to x:

$$\frac{\partial G}{\partial x} = -\frac{(x-\mu)}{(1-2x\mu+x^2)^{3/2}} = -\frac{(x-\mu)}{(1-2x\mu+x^2)}G \Rightarrow (1-2x\mu+x^2)\frac{\partial G}{\partial x}$$
$$= (\mu - x)G \tag{A.5}$$

Combine (A.5) with the RHS of (A.4):

$$\Rightarrow (1 - 2x\mu + x^2) \sum_{l=0}^{\infty} lx^{l-1}P_l(\mu) = (\mu - x) \sum_{l=0}^{\infty} x^l P_l(\mu)$$

$$\Rightarrow \sum_{l=0}^{\infty} (l+1)x^{l-1}P_l(\mu) - \mu \sum_{l=0}^{\infty} (2l+1)x^l P_l(\mu) + \sum_{l=0}^{\infty} lx^{l-1}P_l(\mu) = 0 \tag{A.6}$$

The coefficient of each power of x must be separately equal to zero:

$$lP_{l-1}(\mu) - (2l+1)\mu P_l(\mu) + (l+1)P_{l+1}(\mu) = 0; l \geq 1 \tag{A.7}$$

Now we go back and differentiate the LHS of the generating function with respect to μ:

$$\frac{\partial G}{\partial \mu} = \frac{x}{(1-2x\mu+x^2)^{3/2}} = \frac{x}{(1-2x\mu+x^2)}G \Rightarrow (1-2x\mu+x^2)\frac{\partial G}{\partial \mu} = xG \tag{A.8}$$

Combine (A.8) with the RHS of (A.4):

$$(1 - 2x\mu + x^2) \sum_{l=0}^{\infty} x^l P_l'(\mu) = x \sum_{l=0}^{\infty} x^l P_l(\mu)$$

$$\Rightarrow \sum_{l=0}^{\infty} x^{l+2}P_l'(\mu) - \sum_{l=0}^{\infty} x^{l+1}[P_l(\mu) + 2\mu P_l'(\mu)] + \sum_{l=0}^{\infty} x^l P_l'(\mu) = 0 \tag{A.9}$$

Again, set the coefficients of like powers of x to zero:

$$P_l(\mu) = P_{l-1}'(\mu) - 2\mu P_l'(\mu) + P_{l+1}'(\mu); l \geq 1 \tag{A.10}$$

Differentiate (A.7):

$$lP'_{l-1}(\mu) - (2l+1)P_l(\mu) - (2l+1)\mu P'_l(\mu) + (l+1)P'_{l+1}(\mu) = 0; l \geq 1 \quad (A.11)$$

Use (A.11) to eliminate $P'_{l+1}(\mu)$ from (A.10):

$$
\begin{aligned}
P_l(\mu) &= P'_{l-1}(\mu) - 2\mu P'_l(\mu) - \frac{lP'_{l-1}(\mu) - (2l+1)P_l(\mu) - (2l+1)\mu P'_l(\mu)}{(l+1)} \\
&= \frac{1}{l+1}P'_{l-1}(\mu) - \frac{\mu}{l+1}P'_l(\mu) - \frac{2l+1}{l+1}P_l(\mu) \\
&\Rightarrow lP_l(\mu) = \mu P'_l(\mu) - P'_{l-1}(\mu); l \geq 1
\end{aligned}
$$

$$(A.12)$$

Differentiate both sides of (A.12) $m - 1$ times:

$$l\frac{d^{m-1}}{d\mu^{m-1}}P_l(\mu) = \frac{d^{m-1}}{d\mu^{m-1}}\left[\mu P'_l(\mu)\right] - \frac{d^{m-1}}{d\mu^{m-1}}P'_{l-1}(\mu) \quad (A.13)$$

The first term on the RHS is evaluated using the Leibnitz rule:

$$(f \cdot g)^{(n)} = \sum_{k=0}^{n} \frac{n!}{k!(n-k)!}f^{(k)}g^{(n-k)} \quad (A.14)$$

$$
\begin{aligned}
\Rightarrow \frac{d^{m-1}}{d\mu^{m-1}}\left[\mu P'_l(\mu)\right] &= \sum_{k=0}^{m-1} \frac{(m-1)!}{k!(m-1-k)!}\mu^{(k)}\left[P'_l(\mu)\right]^{(m-1-k)} \\
&= \mu P'_l(\mu)^{(m-1)} + (m-1)\left[P'_l(\mu)\right]^{(m-2)} + 0 + 0 + \cdots
\end{aligned}
$$

$$(A.15)$$

Combine (A.13) and (A.15):

$$
\begin{aligned}
\Rightarrow l\frac{d^{m-1}}{d\mu^{m-1}}P_l(\mu) &= \mu\frac{d^m}{d\mu^m}P_l(\mu) + (m-1)\frac{d^{m-1}}{d\mu^{m-1}}P_l(\mu) - \frac{d^m}{d\mu^m}P_{l-1}(\mu) \\
\Rightarrow (l-m+1)\frac{d^{m-1}}{d\mu^{m-1}}P_l(\mu) &= \mu\frac{d^m}{d\mu^m}P_l(\mu) - \frac{d^m}{d\mu^m}P_{l-1}(\mu)
\end{aligned}
$$

$$(A.16)$$

Multiply both sides of (A.16) by $(-1)^m(1-\mu^2)^{m/2}$ and use (A.2):

$$
\begin{aligned}
&\Rightarrow (l-m+1)(-1)^m\left(1-\mu^2\right)^{m/2}\frac{d^{m-1}}{d\mu^{m-1}}P_l(\mu) = \mu P_l^m(\mu) - P_{l-1}^m(\mu) \\
&\Rightarrow (l-m+1)\frac{(-1)^{-1}\left(1-\mu^2\right)^{-1/2}}{(-1)^{-1}\left(1-\mu^2\right)^{-1/2}}(-1)^m\left(1-\mu^2\right)^{m/2}\frac{d^{m-1}}{d\mu^{m-1}}P_l(\mu) \quad (A.17) \\
&= \mu P_l^m(\mu) - P_{l-1}^m(\mu) \\
&\Rightarrow -(l-m+1)\sqrt{1-\mu^2}P_l^{m-1}(\mu) = \mu P_l^m(\mu) - P_{l-1}^m(\mu)
\end{aligned}
$$

We differentiate (A.7) m times:

$$l\frac{d^m}{d\mu^m}P_{l-1}(\mu) - (2l+1)\left[\mu\frac{d^m}{d\mu^m}P_l(\mu) + m\frac{d^{m-1}}{d\mu^{m-1}}P_l(\mu)\right] + (l+1)\frac{d^m}{d\mu^m}P_{l+1}(\mu) = 0$$

$$(A.18)$$

Note that we used (A.14) to differentiate the middle term on the LHS of (A.7).

Multiply both sides of equation (A.18) by $(-1)^m(1-\mu^2)^{m/2}$ and use (A.2):

$$lP_{l-1}^m(\mu) - (2l+1)\left[\mu P_l^m(\mu) - m\sqrt{1-\mu^2}P_l^{m-1}(\mu)\right] + (l+1)P_{l+1}^m(\mu) = 0 \quad (A.19)$$

Use (A.17) to eliminate $\mu P_l^m(\mu)$ in (A.19)

$$lP_{l-1}^m(\mu) - (2l+1)\left[\begin{matrix}P_{l-1}^m(\mu) - (l-m+1)\sqrt{1-\mu^2}P_l^{m-1}(\mu) \\ -m\sqrt{1-\mu^2}P_l^{m-1}(\mu)\end{matrix}\right]$$
$$+(l+1)P_{l+1}^m(\mu) = 0$$
$$\Rightarrow (-l-1)P_{l-1}^m(\mu) + (2l+1)(l+1)\sqrt{1-\mu^2}P_l^{m-1}(\mu) \quad (A.20)$$
$$+(l+1)P_{l+1}^m(\mu) = 0$$
$$\Rightarrow (2l+1)\sqrt{1-\mu^2}P_l^{m-1}(\mu) = P_{l-1}^m(\mu) - P_{l+1}^m(\mu)$$

Combine (A.17) and (A.20) to eliminate $\sqrt{1-\mu^2}P_l^{m-1}(\mu)$:

$$\Rightarrow \mu P_l^m(\mu) = \frac{(l-m+1)}{(2l+1)}P_{l+1}^m(\mu) + \frac{(l+m)}{(2l+1)}P_{l-1}^m(\mu) \quad (A.21)$$

We use (A.1) to make the following identifications:

$$P_l^m(\cos\theta) = Y_{lm}(\theta,\phi)e^{-im\phi}\sqrt{\frac{4\pi}{2l+1}\frac{(l+m)!}{(l-m)!}} \quad (A.22)$$

$$P_{l+1}^m(\cos\theta) = Y_{l+1,m}(\theta,\phi)e^{-im\phi}\sqrt{\frac{4\pi}{2l+3}\frac{(l+m+1)!}{(l-m+1)!}} \quad (A.23)$$

$$P_{l-1}^m(\cos\theta) = Y_{l-1,m}(\theta,\phi)e^{-im\phi}\sqrt{\frac{4\pi}{2l-1}\frac{(l+m-1)!}{(l-m-1)!}} \quad (A.24)$$

Combine (A.21), (A.22), (A.23), and (A.24)

$$\cos\theta Y_{lm}(\theta,\phi)\sqrt{\frac{(l+m)!}{(2l+1)(l-m)!}} =$$

$$\frac{(l-m+1)}{(2l+1)}\sqrt{\frac{(l+m+1)!}{(2l+3)(l-m+1)!}}Y_{l+1,m}(\theta,\phi)$$

$$+\frac{(l+m)}{(2l+1)}\sqrt{\frac{(l+m-1)!}{(2l-1)(l-m-1)!}}Y_{l-1,m}(\theta,\phi)$$

$$\Rightarrow \cos\theta Y_{lm}(\theta,\phi) = \frac{(l-m+1)}{(2l+1)}\sqrt{\frac{(l+m+1)!(2l+1)(l-m)!}{(2l+3)(l-m+1)!(l+m)!}}Y_{l+1,m}(\theta,\phi)$$

$$+\frac{(l+m)}{(2l+1)}\sqrt{\frac{(l+m-1)!(2l+1)(l-m)!}{(2l-1)(l-m-1)!(l+m)!}}Y_{l-1,m}(\theta,\phi)$$

$$\Rightarrow \cos\theta Y_{lm}(\theta,\phi) = \sqrt{\frac{(l+m+1)(l-m+1)}{(2l+1)(2l+3)}}Y_{l+1,m}(\theta,\phi)$$

$$+\sqrt{\frac{(l-m)(l+m)}{(2l-1)(2l+1)}}Y_{l-1,m}(\theta,\phi)$$

$$(A.25)$$

(A.25) is the same as (2.68), so the recurrence relation is proved.

In a similar fashion, one may also prove the following [9] (but see problem 4.9 for an alternate method):

$$\sin\theta e^{i\phi}Y_{lm}(\theta,\phi) = -\sqrt{\frac{(l+m+1)(l+m+2)}{(2l+1)(2l+3)}}Y_{l+1,m+1}(\theta,\phi)$$

$$+\sqrt{\frac{(l-m)(l-m-1)}{(2l-1)(2l+1)}}Y_{l-1,m+1}(\theta,\phi) \qquad (A.26a)$$

$$\sin\theta e^{-i\phi}Y_{lm}(\theta,\phi) = \sqrt{\frac{(l-m+1)(l-m+2)}{(2l+1)(2l+3)}}Y_{l+1,m-1}(\theta,\phi)$$

$$-\sqrt{\frac{(l+m)(l+m-1)}{(2l-1)(2l+1)}}Y_{l-1,m-1}(\theta,\phi) \qquad (A.26b)$$

So far, we have not used Rodrigues' formula (A.3) for anything. That does not mean the formula is not extremely useful, as we now demonstrate. We will use Rodrigues' formula to evaluate an integral that we will need in App. F. The integral is

$$I_n \equiv \int_{-1}^{1} x^n P_n(x)dx = \frac{1}{2^n n!} \int_{-1}^{1} x^n \frac{d^n}{dx^n} \left(x^2 - 1\right)^n dx \qquad (A.27)$$

where we have substituted (A.3) for $P_n(x)$. Integrating by parts,

$$I_n = \frac{1}{2^n n!} \left[x^n \left(\frac{d}{dx}\right)^{n-1} \left(x^2 - 1\right)^n \right]_{-1}^{1}$$

$$- \frac{1}{2^n n!} \int_{-1}^{1} n x^{n-1} \left(\frac{d}{dx}\right)^{n-1} \left(x^2 - 1\right)^n dx \qquad (A.28)$$

The integrated term vanishes. We integrate by parts a second time, noting that the integrated term will again vanish:

$$I_n = \frac{1}{2^n n!} \int_{-1}^{1} n(n-1) x^{n-2} \left(\frac{d}{dx}\right)^{n-2} \left(x^2 - 1\right)^n dx \qquad (A.29)$$

We continue to integrate by parts until the differentiation under the integral is removed:

$$\Rightarrow I_n = \frac{(-1)^n}{2^n n!} \int_{-1}^{1} n! \left(x^2 - 1\right)^n dx = \frac{1}{2^n} \int_{-1}^{1} \left(1 - x^2\right)^n dx \qquad (A.30)$$

The remaining integral is part of a class of integrals called Beta integrals whose solutions are well-known [9]:

$$\int_{-1}^{1} \left(1 - x^2\right)^n dx = \frac{\Gamma\left(\frac{1}{2}\right)\Gamma(n+1)}{\Gamma\left(n+\frac{3}{2}\right)} = \frac{\sqrt{\pi} n!}{\sqrt{\pi} \left[(2n+1)!!/2^{n+1}\right]} = \frac{2^{n+1} n!}{(2n+1)!!} \qquad (A.31)$$

Giving us finally,

$$I_n = \int_{-1}^{1} x^n P_n(x)dx = \frac{1}{2^n} \frac{2^{n+1} n!}{(2n+1)!!} = \frac{2n!}{(2n+1)!!} = \frac{2^{n+1} (n!)^2}{(2n+1)!} \qquad (A.32)$$

Appendix B: The Racah Formula for the Clebsch-Gordan Coefficients

This appendix outlines the derivation of the Racah formulas for the C-G coefficients. It follows the steps first given by Racah in his seminal papers [2].

Clebsch-Gordan Coefficient Recursion Relation

We start our derivation of (2.138) by picking up where the main text leaves off after (2.137). Equations (2.136) and (2.137) are the recursion relations for the C-G coefficients in m, m_1, m_2. But we will also need the C-G coefficient recursion relation in j, which is more difficult to derive. Nevertheless, we now derive it as follows.

The W-E theorem, applied to a vector operator (spherical tensor of rank 1), i.e., a tensor operator of the form given in (2.176) results in[2]

$$\langle j', m' | A_q^{(1)} | j, m \rangle = C_{j,m;\,1,q}^{j'\,m'} \langle j' \| A^{(1)} \| j \rangle \rightleftharpoons C_{j,m;\,1,q}^{j'\,m'} \langle j' \| \vec{A} \| j \rangle \tag{B.1}$$

where (2.183) was also used. If $\vec{A} = \vec{J}$, then,

$$\langle j', m' | J_q | j, m \rangle = C_{j,m;\,1,q}^{j'\,m'} \langle j' \| \vec{J} \| j \rangle \tag{B.2}$$

Before we apply the W-E theorem to the scalar product $\vec{J} \cdot \vec{A}$, we first we note that

$$\vec{J} \cdot \vec{A} = J_0 A_0 - J_{+1} A_{-1} - J_{-1} A_{+1} \tag{B.3}$$

And

$$J_0 | j, m \rangle = \hbar m | j, m \rangle; \quad J_{\pm 1} | j, m \rangle = \frac{\hbar}{2} \sqrt{j(j+1) - m(m \pm 1)} | j, m \pm 1 \rangle \tag{B.4}$$

which gives

---

[2] Different sources define the reduced matrix elements slightly differently; hence, the exact form of the W-E theorem varies from source to source [see also (2.180) and (2.182)]. The exact form of the reduced matrix elements will not affect the arguments in this section.

$$\langle j, m | \vec{J} \cdot \vec{A} | j, m \rangle = \hbar m \langle j, m | A_0 | j, m \rangle - \frac{\hbar}{2} \sqrt{j(j+1) - m(m+1)} \langle j, m+1 | A_{+1} | j, m \rangle$$
$$- \frac{\hbar}{2} \sqrt{j(j+1) - m(m-1)} \langle j, m-1 | A_{-1} | j, m \rangle$$

$$(B.5)$$

Appying the W-E theorem:

$$\langle j, m | A_0 | j, m \rangle = C^{j,m}_{j,m;\,1,0} \langle j' \| \vec{A} \| j \rangle \tag{B.6a}$$

$$\langle j, m+1 | A_{+1} | j, m \rangle = C^{j,m+1}_{j,m;\,1,1} \langle j' \| \vec{A} \| j \rangle \tag{B.6b}$$

$$\langle j, m-1 | A_{-1} | j, m \rangle = C^{j,m-1}_{j,m;\,1,-1} \langle j' \| \vec{A} \| j \rangle \tag{B.6c}$$

Combining (B.5) and (B.6),

$$\langle j, m | \vec{J} \cdot \vec{A} | j, m \rangle = \left[\begin{array}{l} \hbar m C^{j,m}_{j,m;\,1,0} - \dfrac{\hbar}{2} C^{j,m+1}_{j,m;\,1,1} \sqrt{j(j+1) - m(m+1)} \\ - \dfrac{\hbar}{2} C^{j,m-1}_{j,m;\,1,-1} \sqrt{j(j+1) - m(m-1)} \end{array} \right] \langle j \| \vec{A} \| j \rangle$$

$$(B.7)$$

If $\vec{A} = \vec{J}$, then,

$$\langle j, m | J^2 | j, m \rangle = \left[\begin{array}{l} \hbar m C^{j,m}_{j,m;\,1,0} - \dfrac{\hbar}{2} C^{j,m+1}_{j,m;\,1,1} \sqrt{j(j+1) - m(m+1)} \\ - \dfrac{\hbar}{2} C^{j,m-1}_{j,m;\,1,-1} \sqrt{j(j+1) - m(m-1)} \end{array} \right] \langle j \| \vec{J} \| j \rangle \tag{B.8}$$

From (B.2) and (B.1) we have (for $j' = j$):

$$\frac{\langle j, m' | A_q | j, m \rangle}{\langle j, m' | J_q | j, m \rangle} = \frac{\langle j \| \vec{A} \| j \rangle}{\langle j \| \vec{J} \| j \rangle} \tag{B.9}$$

From (B.7) and (B.8) we have,

$$\frac{\langle j, m | \vec{J} \cdot \vec{A} | j, m \rangle}{\langle j, m | J^2 | j, m \rangle} = \frac{\langle j \| \vec{A} \| j \rangle}{\langle j \| \vec{J} \| j \rangle} \Rightarrow \frac{\langle j, m | \vec{J} \cdot \vec{A} | j, m \rangle}{\hbar^2 j(j+1)} = \frac{\langle j \| \vec{A} \| j \rangle}{\langle j \| \vec{J} \| j \rangle} \tag{B.10}$$

Combining (B.9) and (B.10) gives

$$\langle j, m' | A_q | j, m \rangle = \frac{\langle j, m | \vec{J} \cdot \vec{A} | j, m \rangle}{\hbar^2 j(j+1)} \langle j, m' | J_q | j, m \rangle \qquad (B.11)$$

Eq. (B.11) is called the projection theorem. Now note,

$$\vec{J} = \vec{J}_1 + \vec{J}_2 \Rightarrow J^2 = \left(\vec{J}_1 + \vec{J}_2\right)^2 = J_1^2 + J_2^2 + 2\vec{J}_1 \cdot \vec{J}_2 \Rightarrow \vec{J}_1 \cdot \vec{J}_2$$

$$= J^2 - \frac{J_1^2 - J_2^2}{2} \qquad (B.12)$$

and also that

$$\vec{J} \cdot \vec{J}_1 = \left(\vec{J}_1 + \vec{J}_2\right) \cdot \vec{J}_1 = \vec{J}_1 \cdot \vec{J}_2 + J_1^2 = J^2 - \frac{J_1^2 - J_2^2}{2} + J_1^2$$

$$= \frac{J^2 + J_1^2 - J_2^2}{2} \qquad (B.13)$$

Let $\vec{A} = \vec{J}_1$, $q = 0$, $m = m'$ in (B.11), and combine with (B.13) to get

$$\langle j, m | J_{1z} | j, m \rangle = \frac{\hbar^2 [j(j+1) + j_1(j_1+1) - j_2(j_2+1)]}{\hbar^2 2j(j+1)} \hbar m$$

$$= \frac{[j(j+1) + j_1(j_1+1) - j_2(j_2+1)]}{2j(j+1)} \hbar m \qquad (B.14)$$

Recall that we are working with vector operators that obey the following commutator relations:

$$[J_x, J_{1x}] = [J_y, J_{1y}] = [J_z, J_{1z}] = 0 \qquad (B.15a)$$

$$[J_x, J_{1y}] = i\hbar J_{1z}; \quad [J_y, J_{1z}] = i\hbar J_{1x}; [J_z, J_{1x}] = i\hbar J_{1y} \qquad (B.15b)$$

$$[J_x, J_{1z}] = -i\hbar J_{1y}; \quad [J_y, J_{1x}] = -i\hbar J_{1z}; \quad [J_z, J_{1y}] = -i\hbar J_{1x} \qquad (B.15c)$$

And

$$\left[\vec{J} \cdot \vec{J}_1, J^2\right] = [J_-, J_{1-}] = 0 \qquad (B.16a)$$

$$[J_z, J_{1-}] = -\hbar J_{1-} \qquad (B.16b)$$

These vector observables have the form $\vec{A} = A_x \hat{e}_x + A_y \hat{e}_y + A_z \hat{e}_z$, where A_x, A_y, A_z are scalar observables that obey the following multiplication rules:

$$\vec{A} \cdot \vec{B} = A_x B_x + A_y B_y + A_z B_z \tag{B.17a}$$

$$\left(\vec{A} \times \vec{B}\right)_x = A_y B_z - A_z B_y, \text{ etc.} \tag{B.17b}$$

Using (B.15) and (B.17), one can show[3]

$$\vec{A} \times \vec{B} = -\vec{B} \times \vec{A} + \left[\vec{A}, \vec{B}\right] \tag{B.18}$$

and thus,

$$\vec{J}_1 \times \vec{J} = -\vec{J} \times \vec{J}_1 + 2i\hbar\vec{J}_1 \tag{B.19}$$

all of which allow us to deduce,

$$
\begin{aligned}
\left[J^2, \vec{J}_1\right] &= \vec{J} \cdot \left[\vec{J}, \vec{J}_1\right] - \left[\vec{J}_1, \vec{J}\right] \cdot \vec{J} \\
&= -i\hbar \left(\vec{J} \cdot \vec{J}_1 \times \mathfrak{J} - \vec{J}_1 \times \mathfrak{J} \cdot \vec{J}\right) \\
&= -i\hbar \left(\vec{J} \times \vec{J}_1 - \vec{J}_1 \times \vec{J}\right) \\
&= -2i\hbar \left(\vec{J} \times \vec{J}_1 - i\hbar\vec{J}_1\right)
\end{aligned}
\tag{B.20}
$$

where $\mathfrak{J}$ is the unit dyadic and where we have used the vector identity

$$\left[\vec{A} \cdot \vec{B}, \vec{C}\right] = \vec{A} \cdot \left[\vec{B}, \vec{C}\right] - \left[\vec{C}, \vec{A}\right] \cdot \vec{B} \tag{B.21}$$

From (B.20) we have,

$$
\begin{aligned}
\left[J^2, \left[J^2, \vec{J}_1\right]\right] &= -2i\hbar\left[J^2, \left(\vec{J} \times \vec{J}_1 - i\hbar\vec{J}_1\right)\right] \\
&= -2i\hbar\left\{\vec{J} \times \left[J^2, \vec{J}_1\right] - i\hbar\left[J^2, \vec{J}_1\right]\right\} \\
&= -2i\hbar\left\{-2i\hbar\vec{J} \times \left(\vec{J} \times \vec{J}_1 - i\hbar\vec{J}_1\right) - i\hbar\left(J^2\vec{J}_1 - \vec{J}_1 J^2\right)\right\} \\
&= 2\hbar^2 \left(J^2\vec{J}_1 + \vec{J}_1 J^2\right) - 4\hbar^2\vec{J}\left(\vec{J} \cdot \vec{J}_1\right)
\end{aligned}
\tag{B.22}
$$

where we have used

---

[3] $\left[\vec{A}, \vec{B}\right] \equiv [A_x, B_x]ii + [A_x, B_y]ij + \dots$ That is, the commutator connecting the components of $\vec{A}$ and $\vec{B}$ transform like the components of a dyadic. Thus (B.15) can be written as $\left[\vec{J}, \vec{J}_1\right] = -i\hbar\vec{J}_1 \times \mathfrak{J}$ where $\mathfrak{J}$ is the unit dyadic $= ii + jj + kk$.

$$\vec{A} \times \left(\vec{B} \times \vec{C} \right) = \left(\vec{B} \cdot \vec{A} \right) \vec{C} - \vec{A} \cdot \left(\vec{B} \cdot \vec{C} \right) - \left[\vec{B}, \vec{A} \right] \cdot \vec{C} \tag{B.23}$$

to expand $\vec{J} \times \left(\vec{J} \times \vec{J}_1 \right)$. But now also,

$$\left[J^2, \left[J^2, \vec{J}_1 \right] \right] = \left[J^2, \left(J^2 \vec{J}_1 - \vec{J}_1 J^2 \right) \right] = J^4 \vec{J}_1 - 2 J^2 \vec{J}_1 J^2 + \vec{J}_1 J^4 \tag{B.24}$$

Equations (B.22) and (B.24) together give

$$J^4 \vec{J}_1 - 2 J^2 \vec{J}_1 J^2 + \vec{J}_1 J^4 = 2\hbar^2 \left(J^2 \vec{J}_1 + \vec{J}_1 J^2 \right) - 4\hbar^2 \vec{J} \left(\vec{J} \cdot \vec{J}_1 \right) \tag{B.25}$$

Now we take the matrix element of (B.25) between $\langle j, m|$ and $|j', m' \rangle$, noting that the last term on the RHS of that equation will vanish by (B.16a):

$$\hbar^4 \left[j^2(j+1)^2 - 2j(j+1)j'(j'+1) + j'^2(j'+1)^2 \right] \langle j, m|\vec{J}_1|j', m' \rangle$$
$$= 2\hbar^4 \left[j(j+1) + j'(j'+1) \right] \langle j, m|\vec{J}_1|j', m' \rangle \tag{B.26}$$

Two useful relations are

$$\left[j^2(j+1)^2 - 2j(j+1)j'(j'+1) + j'^2(j'+1)^2 \right]$$
$$= [j(j+1) - j'(j'+1)]^2 = (j+j'+1)^2(j-j')^2 \tag{B.27}$$

and

$$2[j(j+1) + j'(j'+1)] = (j+j'+1) + (j-j')^2 - 1 \tag{B.28}$$

Combining (B.26, B.27, and B.28),

$$\left[(j+j'+1)^2(j-j')^2 - (j+j'+1) - (j-j')^2 + 1 \right] \langle j, m|\vec{J}_1|j', m' \rangle = 0 \tag{B.29}$$

or

$$\left[(j+j'+1)^2 - 1 \right] \left[(j-j')^2 - 1 \right] \langle j, m|\vec{J}_1|j', m' \rangle = 0 \tag{B.30}$$

If the matrix element is nonzero, then one of the two terms in the brackets must vanish. The first vanishes if $j' = j = 0$. The second vanishes if $j' - j = \pm 1$. Therefore, for a nonvanishing matrix component, $\langle j, m|\vec{J}_1|j', m' \rangle$, we must have

$$j' = j, j \pm 1 \tag{B.31}$$

In other words, the only nonvanishing matrix elements of the vector operator $\vec{J}_1$ in the $|j, m\rangle$ basis are those for which (B.31) is satisfied.

From (B.16b) we can say,

$$m\hbar\langle j, m|J_{1-}|j', m'\rangle - \langle j, m|J_{1-}|j', m'\rangle m'\hbar = -\hbar\langle j, m|J_{1-}|j', m'\rangle$$

$$\Rightarrow (m - m' + 1)\langle j, m|J_{1-}|j', m'\rangle = 0 \tag{B.32}$$

Similarly,

$$(m - m' - 1)\langle j, m|J_{1+}|j', m'\rangle = 0 \tag{B.33}$$

Therefore, the only nonvanishing matrix elements of the operators J_{1x} and J_{1y} in the $|j, m\rangle$ basis are those for which $m' = m \pm 1$. Finally, since $[J_{1z}, J_z] = 0$, the only nonvanishing matrix elements for J_{1z} in the $|j, m\rangle$ basis are those for which $m' = m$.

Again, from (B.16b) we can say

$$
\begin{aligned}
&\langle j, m - 1|J_- J_{1-}|j', m + 1\rangle = \langle j, m - 1|J_{1-} J_-|j', m + 1\rangle \\
&\Rightarrow \sum_{j'', m''} [\langle j, m - 1|J_-|j'', m''\rangle \langle j'', m''|J_{1-}|j', m + 1\rangle \\
&= \langle j, m - 1|J_{1-}|j'', m''\rangle \langle j'', m''|J_-|j', m + 1\rangle] \\
&\Rightarrow \sum_{j''} [\langle j, m - 1|J_-|j'', m\rangle \langle j'', m|J_{1-}|j', m + 1\rangle \\
&= \langle j, m - 1|J_{1-}|j'', m\rangle \langle j'', m|J_-|j', m + 1\rangle] \\
&\Rightarrow \sum_{j''} \left[\sqrt{(j'' + m)(j'' - m + 1)}\, \delta_{jj''} \langle j'', m|J_{1-}|j', m + 1\rangle \right. \\
&\left. = \langle j, m - 1|J_{1-}|j'', m\rangle \sqrt{(j' - m)(j' + m + 1)}\, \delta_{j'j''} \right] \\
&\Rightarrow \sqrt{(j + m)(j - m + 1)}\langle j, m|J_{1-}|j', m + 1\rangle \\
&= \langle j, m - 1|J_{1-}|j', m\rangle \sqrt{(j' - m)(j' + m + 1)}
\end{aligned} \tag{B.34}
$$

where the sum over m'' was collapsed using (B.32) and (B.33) and where (2.132) was also used.

Now recall from (B.31) that, for the matrix elements in (B.34), we can have $j' = j$; $j \pm 1$ only. For $j' = j$, (B.34) gives

$$\langle j, m | J_{1-} | j, m+1 \rangle = \langle j : \vec{J}_1 : j \rangle \sqrt{(j-m)(j+m+1)} \qquad \text{(B.35)}$$

where we define

$$\langle j : \vec{J}_1 : j \rangle \equiv \frac{\langle j, m-1 | J_{1-} | j, m \rangle}{\sqrt{(j+m)(j-m+1)}} \qquad \text{(B.36)}$$

For $j' = j - 1$, (B.34) gives

$$\sqrt{(j+m)(j-m+1)} \langle j, m | J_{1-} | j-1, m+1 \rangle$$
$$= \langle j, m-1 | J_{1-} | j-1, m \rangle \sqrt{(j-m-1)(j+m)} \qquad \text{(B.37)}$$

Multiplying through by $\sqrt{\dfrac{(j-m)}{(j+m)}}$,

$$\langle j, m | J_{1-} | j-1, m+1 \rangle = \langle j : J_1 : j-1 \rangle \sqrt{(j-m-1)(j-m)} \qquad \text{(B.38)}$$

where

$$\langle j : \vec{J}_1 : j-1 \rangle \equiv \frac{\langle j, m-1 | J_{1-} | j-1, m \rangle}{\sqrt{(j-m)(j-m+1)}} \qquad \text{(B.39)}$$

Similarly, for $j' = j + 1$, (B.34) gives

$$\langle j, m | J_{1-} | j+1, m+1 \rangle = \langle j : \vec{J}_1 : j+1 \rangle \sqrt{(j+m+1)(j+m+2)} \qquad \text{(B.40)}$$

where

$$\langle j : \vec{J}_1 : j+1 \rangle \equiv \frac{\langle j, m-1 | J_{1-} | j+1, m \rangle}{\sqrt{(j+m)(j+m+1)}} \qquad \text{(B.41)}$$

The reason for the symbology $\langle j : J_1 : j \rangle$ etc. is because the ratios they represent are independent of m by construction, which can be noted for $\langle j : J_1 : j \rangle$ (for example) by examining the last line in (B.34) for $j' = j$.

Now noting that

$$[J_+, J_{1-}] = 2\hbar J_{1z} \qquad \text{(B.42)}$$

we can say

$$\langle j,m|J_+J_{1-}|j',m\rangle - \langle j,m|J_{1-}J_+|j',m\rangle = 2\hbar\langle j,m|J_{1z}|j',m\rangle$$

$$\Rightarrow \sum_{j'',m''}[\langle j,m|J_+|j'',m''\rangle\langle j'',m''|J_{1-}|j',m\rangle - \langle j,m|J_{1-}|j'',m''\rangle\langle j'',m''|J_+|j',m\rangle]$$

$$= 2\hbar\langle j,m|J_{1z}|j',m\rangle$$

$$\Rightarrow \sum_{j''}\begin{bmatrix} \langle j,m|J_+|j'',m-1\rangle\langle j'',m-1|J_{1-}|j',m\rangle - \\ \langle j,m|J_{1-}|j'',m+1\rangle\langle j'',m+1|J_+|j',m\rangle \end{bmatrix} = 2\hbar\langle j,m|J_{1z}|j',m\rangle$$

$$\Rightarrow \sum_{j''}\begin{bmatrix} \hbar\sqrt{(j''+m)(j''-m+1)}\,\delta_{jj''}\langle j'',m-1|J_{1-}|j',m\rangle \\ -\langle j,m|J_{1-}|j'',m+1\rangle\hbar\sqrt{(j'-m)(j'+m+1)}\,\delta_{j'j''} \end{bmatrix} = 2\hbar\langle j,m|J_{1z}|j',m\rangle$$

$$\Rightarrow \sqrt{(j+m)(j-m+1)}\langle j,m-1|J_{1-}|j',m\rangle$$

$$-\langle j,m|J_{1-}|j',m+1\rangle\sqrt{(j'-m)(j'+m+1)} = 2\langle j,m|J_{1z}|j',m\rangle$$

$$(\text{B.43})$$

where the sum over m'' was collapsed using (B.32) and (B.33). Specifically, for the first term on the LHS, we had $m'' = m - 1$ and for the second term on the LHS, we had $m'' = m + 1$. Recalling again, from (B.31) that, for the matrix elements in (B.43), we can have $j' = j; j \pm 1$ only, for $j' = j$, (B.43) gives

$$\sqrt{(j+m)(j-m+1)}\overbrace{\left(\langle j:\vec{J}_1:j\rangle\sqrt{(j+m)(j-m+1)}\right)}^{\langle j,m-1|J_{1-}|j,m\rangle}$$

$$-\overbrace{\left(\langle j:\vec{J}_1:j\rangle\sqrt{(j-m)(j+m+1)}\right)}^{\langle j,m|J_{1-}|j,m+1\rangle \text{ for } m\to m+1}\sqrt{(j-m)(j+m+1)} = 2\langle j,m|J_{1z}|j,m\rangle$$

$$\Rightarrow m\langle j:\vec{J}_1:j\rangle = \langle j,m|J_{1z}|j,m\rangle$$

$$(\text{B.44})$$

where we have used (B.36) and the fact that $\langle j:\vec{J}_1:j\rangle$ is independent of m. As an interesting aside, we could now find the relation between $\langle j:\vec{J}_1:j\rangle$ and the reduced matrix element $\langle j'\|\vec{J}_1\|j\rangle$. Since both are independent of the angular orientation of the system, it is probable that they are proportional to one another. We can now make that demonstration. From (B.2), we have

$$\langle j,m|J_{1z}|j,m\rangle = C^{jm}_{j,m;1,0}\langle j'\|\vec{J}_1\|j\rangle \tag{B.45}$$

Combine this with the result from (B.44):

$$\langle j:\vec{J}_1:j\rangle = \frac{C^{jm}_{j,m;1,0}\langle j\|\vec{J}_1\|j\rangle}{m} \tag{B.46}$$

But from the recursion relations in (2.135) and (2.136), it is easy to deduce that

$$C^{jm}_{j,m;1,0} = \frac{m}{\sqrt{j(j+1)}} \tag{B.47}$$

Combining (B.46) and (B.47) gives us the result

$$\langle j\|\vec{J}_1\|j\rangle = \sqrt{j(j+1)}\langle j : \vec{J}_1 : j\rangle \tag{B.48}$$

Getting back to the main argument, we now go back to (B.43) for the case $j' = j - 1$. Using (B.38) and (B.39),

$$\sqrt{(j+m)(j-m+1)}\langle j, m-1|J_{1-}|j-1, m\rangle - \langle j, m|J_{1-}|j-1, m+1\rangle$$
$$\times\sqrt{(j-m-1)(j+m)} = 2\langle j, m|J_{1z}|j-1, m\rangle$$
$$\Rightarrow \sqrt{(j+m)(j-m+1)}(\langle j : \vec{J}_1 : j-1\rangle\sqrt{(j-m)(j-m+1)})$$
$$-\left(\langle j : \vec{J}_1 : j-1\rangle\sqrt{(j-m)(j-m-1)}\right)\sqrt{(j+m)(j-m-1)} \tag{B.49}$$
$$= 2\langle j, m|J_{1z}|j-1, m\rangle$$
$$\Rightarrow \sqrt{j^2 - m^2}\langle j : \vec{J}_1 : j-1\rangle = \langle j, m|J_{1z}|j-1, m\rangle$$

Similarly, for $j' = j + 1$, using (B.40) and (B.41), (B.43) gives

$$\sqrt{(j+1)^2 - m^2}\langle j : \vec{J}_1 : j+1\rangle = \langle j, m|J_{1z}|j+1, m\rangle \tag{B.50}$$

As another aside, we note that in an analysis similar to the one that produced (B.48), we could use the results of (B.49) and (B.50) to demonstrate

$$\langle j\|\vec{J}_1\|j-1\rangle = \sqrt{j(2j-1)}\langle j : \vec{J}_1 : j-1\rangle \tag{B.51a}$$

$$\langle j\|\vec{J}_1\|j+1\rangle = -\sqrt{(j+1)(2j+3)}\langle j : \vec{J}_1 : j-1\rangle \tag{B.51b}$$

Consider the matrix elements of $[J_{1-}, J_{1z}] = \hbar J_{1-}$ between the states $\langle j, m|$ and $|j, m+1\rangle$,

$$\langle j, m|J_{1-}J_{1z}|j, m+1\rangle - \langle j, m|J_{1z}J_{1-}|j, m+1\rangle = \hbar\langle j, m|J_{1-}|j, m+1\rangle$$

$$\Rightarrow \sum_{j', m'} \left[\begin{array}{c} \underbrace{\langle j, m|J_{1-}|j', m'\rangle}\ \underbrace{\langle j', m'|J_{1z}|j, m+1\rangle} \\ {\scriptstyle \Rightarrow m'=m+1 \text{ from (B.32)}} \\ -\langle j, m|J_{1z}|j', m'\rangle\underbrace{\langle j', m'|J_{1-}|j, m+1\rangle} \\ {\scriptstyle \Rightarrow m'=m \text{ from (B.32)}} \end{array} \right] = \hbar\langle j, m|J_{1-}|j, m+1\rangle$$

$$\Rightarrow \sum_{j'} \left[\begin{array}{c} \langle j, m|J_{1-}|j', m+1\rangle\langle j', m+1|J_{1z}|j, m+1\rangle \\ -\langle j, m|J_{1z}|j', m\rangle\langle j', m|J_{1-}|j, m+1\rangle \end{array} \right] = \hbar\langle j, m|J_{1-}|j, m+1\rangle$$

$$\tag{B.52}$$

As per (B.31), the sum in (B.52) is over the interval $j' = j; j \pm 1$. We evaluate the sum and use (B.35), (B.38), (B.40), (B.44), (B.49), and (B.50).

$$\sqrt{(j-m)(j+m+1)}(m+1)\langle j : \vec{J}_1 : j\rangle - \langle j : \vec{J}_1 : j\rangle m\sqrt{(j-m)(j+m+1)}$$
$$+\langle j : \vec{J}_1 : j-1\rangle\sqrt{(j-m)(j-m-1)}\langle j-1 : \vec{J}_1 : j\rangle\sqrt{j^2 - (m+1)^2}$$
$$\overbrace{-\langle j : \vec{J}_1 : j-1\rangle\sqrt{j^2 - m^2}\ \langle j-1 : \vec{J}_1 : j\rangle}^{*}\sqrt{(j+m)(j+m+1)}$$
$$+(-1)\langle j : \vec{J}_1 : j+1\rangle\sqrt{(j+m+1)(j+m+2)}$$
$$\times\langle j+1 : \vec{J}_1 : j\rangle\sqrt{(j+1)^2 - (m+1)^2} - (-1)\sqrt{(j+1)^2 - m^2}$$
$$\times\langle j : \vec{J}_1 : j+1\rangle\ \overbrace{\langle j+1 : \vec{J}_1 : j\rangle\sqrt{(j-m+1)(j-m)}}^{*}$$
$$= \hbar\langle j : \vec{J}_1 : j\rangle\sqrt{(j-m)(j+m+1)} \tag{B.53}$$

where, for the * terms in (B.53), we have also used the fact that we can say, from (B.40) with $j \to j - 1$,

$$\langle j-1, m|J_{1-}|j, m+1\rangle = \langle j-1 : \vec{J}_1 : j\rangle\sqrt{(j+m+1)(j+m)} \tag{B.54}$$

and from (B.38) with $j \to j + 1$,

$$\langle j+1, m|J_{1-}|j, m+1\rangle = \langle j+1 : \vec{J}_1 : j\rangle\sqrt{(j-m+1)(j-m)} \tag{B.55}$$

Doing the algebra in (B.53),

$$\langle j : \vec{J}_1 : j\rangle^2 - \langle j : \vec{J}_1 : j+1\rangle^2(2j+3) + \langle j : \vec{J}_1 : j-1\rangle^2(2j-1) = \hbar\langle j : \vec{J}_1 : j\rangle \tag{B.56}$$

If we combine (B.14) and (B.44), we have

$$m\langle j : \vec{J}_1 : j\rangle = \frac{[j(j+1) + j_1(j_1+1) - j_2(j_2+1)]}{2j(j+1)}\hbar m$$
$$\Rightarrow \langle j : \vec{J}_1 : j\rangle = \frac{[j(j+1) + j_1(j_1+1) - j_2(j_2+1)]}{2j(j+1)}\hbar \tag{B.57}$$

which further gives

$$\hbar \langle j : \vec{J}_1 : j \rangle - \langle j : \vec{J}_1 : j \rangle^2$$

$$= \hbar^2 \frac{[j(j+1)+j_1(j_1+1)-j_2(j_2+1)]}{2j(j+1)} - \hbar^2 \frac{[j(j+1)+j_1(j_1+1)-j_2(j_2+1)]^2}{(2j)^2(j+1)^2}$$

$$= \frac{\hbar^2 [j(j+1)+j_1(j_1+1)-j_2(j_2+1)]}{2j(j+1)} \left[1 - \frac{[j(j+1)+j_1(j_1+1)-j_2(j_2+1)]}{2j(j+1)} \right]$$

$$= \frac{\hbar^2 [j(j+1)+j_1(j_1+1)-j_2(j_2+1)]}{[2j(j+1)]^2} \{ 2j(j+1) - [j(j+1)+j_1(j_1+1)-j_2(j_2+1)] \}$$

$$= \frac{\hbar^2 \left[j(j+1) + \overbrace{j_1(j_1+1)-j_2(j_2+1)}^{p} \right]}{[2j(j+1)]^2} \left[j(j+1) \overbrace{-j_1(j_1+1)+j_2(j_2+1)}^{-p} \right]$$

$$\tag{B.58}$$

where we defined

$$p \equiv j_1(j_1+1) - j_2(j_2+1) \tag{B.59}$$

And now we combine (B.58) and (B.56):

$$\langle j : \vec{J}_1 : j-1 \rangle^2 (2j-1) - \langle j : \vec{J}_1 : j+1 \rangle^2 (2j+3) = \frac{\hbar^2 [j(j+1)]^2 - p^2}{[2j(j+1)]^2} \tag{B.60}$$

Consider:

$$J_1^2 = J_{1z}^2 + \frac{1}{2} (J_{1+}J_{1-} + J_{1-}J_{1+}) \tag{B.61}$$

which leads to

$$\langle j,m|J_1^2|j,m \rangle = \langle j,m|J_{1z}^2|j,m \rangle + \frac{1}{2} [\langle j,m|J_{1+}J_{1-}|j,m \rangle + \langle j,m|J_{1-}J_{1+}|j,m \rangle] \tag{B.62}$$

We also have

$$\begin{aligned} J_1^2|j,m \rangle &= J_1^2 \sum_{m_1,m_2} C_{j_1,m_1;j_2,m_2}^{j,m} |j_1,m_1 \rangle |j_2,m_2 \rangle \\ &= \hbar^2 j_1(j_1+1) \sum_{m_1,m_2} C_{j_1,m_1;j_2,m_2}^{j,m} |j_1,m_1 \rangle |j_2,m_2 \rangle \\ &= \hbar^2 j_1(j_1+1)|j,m \rangle \end{aligned} \tag{B.63}$$

Combine (B.62) and (B.63):

$$
\hbar^2 j_1 (j_1 + 1)
$$

$$
= \sum_{j',m'} \left\{ \langle j,m|J_{1z}|j',m' \rangle \langle j',m'|J_{1z}|j,m \rangle + \frac{1}{2} \left[\begin{array}{c} \langle j,m|J_{1+}|j',m' \rangle \langle j',m'|J_{1-}|j,m \rangle + \\ \langle j,m|J_{1-}|j',m' \rangle \langle j',m'|J_{1+}|j,m \rangle \end{array} \right] \right\}
$$

$$
= \sum_{j'} \left\{ \langle j,m|J_{1z}|j',m \rangle \langle j',m|J_{1z}|j,m \rangle + \frac{1}{2} \left[\begin{array}{c} \langle j,m|J_{1+}|j',m-1 \rangle \langle j',m-1|J_{1-}|j,m \rangle + \\ \langle j,m|J_{1-}|j',m+1 \rangle \langle j',m+1|J_{1+}|j,m \rangle \end{array} \right] \right\}
$$

$$
= \sum_{j'} \left\{ |\langle j,m|J_{1z}|j',m \rangle|^2 + \frac{1}{2} \left[|\langle j',m-1|J_{1-}|j,m \rangle|^2 + |\langle j,m|J_{1-}|j',m+1 \rangle|^2 \right] \right\}
$$

$$
\overset{\text{from (B.44)}}{= m^2 \langle j : \vec{J}_1 : j \rangle^2} + \frac{1}{2} \left[\overset{\text{from (B.36)}}{\langle j : \vec{J}_1 : j \rangle^2 (j-m+1)(j+m)} + \overset{\text{from (B.35)}}{\langle j : \vec{J}_1 : j \rangle^2 (j+m+1)(j-m)} \right]
$$

$$
\overset{\text{from (B.49)}}{+ (j^2 - m^2) \langle j : \vec{J}_1 : j-1 \rangle^2}
$$

$$
+ \frac{1}{2} \left[\overset{*}{\langle j-1 : \vec{J}_1 : j \rangle^2 (j+m)(j+m-1)} + \overset{\text{from (B.38)}}{\langle j : \vec{J}_1 : j-1 \rangle^2 (j-m)(j-m-1)} \right]
$$

$$
\overset{\text{from (B.50)}}{+ \left[(j+1)^2 - m^2 \right] \langle j : \vec{J}_1 : j+1 \rangle^2} + \frac{1}{2} \left[\begin{array}{c} \overset{*}{\langle j+1 : \vec{J}_1 : j \rangle^2 (j-m+1)(J-m+2)} \\ \overset{\text{from (B.39)}}{+ \langle j : \vec{J}_1 : j+1 \rangle^2 (j+m+1)(j+m+2)} \end{array} \right]
$$

$$
\text{(B.64)}
$$

where we used (B.31) and (B.33), and (as indicated) (B.35), (B.36), (B.38), (B.39), (B.44), (B.49), and (B.50), and where, for the * terms in (B.64), we also used the fact that we can say, from (B.41) with $j \to j - 1$,

$$
\langle j-1, m-1|J_{1-}|j,m \rangle = \langle j-1 : \vec{J}_1 : j \rangle \sqrt{(j+m-1)(j+m)} \tag{B.65}
$$

and from (B.39) with $j \to j + 1$,

$$
\langle j+1, m-1|J_{1-}|j,m \rangle = \langle j+1 : \vec{J}_1 : j \rangle \sqrt{(j-m+1)(j-m+2)} \tag{B.66}
$$

Combining terms in (B.64) gives

$$\hbar^2 j_1(j_1 + 1) = \langle j \vdots J_1 \vdots j - 1 \rangle^2 j(2j - 1)$$
$$+ \langle j \vdots J_1 \vdots j - 1 \rangle^2 (j + 1)(2j + 3) + \langle j \vdots J_1 \vdots j \rangle^2 j(j - 1) \quad \text{(B.67)}$$

where we have used the fact that

$$\langle j \vdots \vec{J}_1 \vdots j \pm 1 \rangle^2 = \langle j \pm 1 \vdots \vec{J}_1 \vdots j \rangle^2 \quad \text{(B.68)}$$

Using (B.57) and

$$q \equiv j_1(j_1 + 1) + j_2(j_2 + 1) \quad \text{(B.69)}$$

we rewrite (B.67) as follows:

$$\langle j \vdots \vec{J}_1 \vdots j - 1 \rangle^2 j(2j - 1) + \langle j \vdots \vec{J}_1 \vdots j + 1 \rangle^2 (j + 1)(2j + 3)$$
$$= \frac{-p^2 + 2qj(j + 1) - j^2(j + 1)^2 \hbar^2}{4j(j + 1)} \quad \text{(B.70)}$$

We can solve (B.70) and (B.60) together for $\langle j \vdots \vec{J}_1 \vdots j - 1 \rangle^2$ and $\langle j \vdots \vec{J}_1 \vdots j + 1 \rangle^2$:

$$\langle j \vdots \vec{J}_1 \vdots j - 1 \rangle^2 [j(2j - 1) + (2j - 1)(j + 1)]$$
$$= \hbar^2 \left\{ \frac{[j(j + 1)]^2 - p^2}{4j^2(j + 1)} + \frac{-p^2 + 2qj(j + 1) - j^2(j + 1)^2}{4j(j + 1)} \right\}$$
$$\Rightarrow \langle j \vdots \vec{J}_1 \vdots j - 1 \rangle^2 (2j - 1)(2j + 1)$$
$$= \hbar^2 \left\{ \frac{[j(j + 1)]^2 - p^2 - p^2 j + 2qj^2(j + 1) - j^3(j + 1)^2}{4j^2(j + 1)} \right\} \quad \text{(B.71)}$$
$$= \hbar^2 \left\{ \frac{j^2(j + 1)^2 - p^2(1 + j) + 2qj^2(j + 1) - j^3(j + 1)^2}{4j^2(j + 1)} \right\}$$
$$= \hbar^2 \frac{-j^4 + (2q + 1)j^2 - p^2}{4j^2}$$

Substituting,

$$2q + 1 = (j_1 - j_2)^2 + (j_2 + j_1 + 1)^2; \quad p^2 = (j_1 - j_2)^2 (j_2 + j_1 + 1)^2 \quad \text{(B.72)}$$

gives

$$\langle j \vdots \vec{J}_1 \vdots j - 1 \rangle^2 = \frac{\left[j^2 - (j_1 - j_2)^2 \right] \left[(j_1 + j_2 + 1)^2 - j^2 \right] \hbar^2}{4j^2(4j^2 - 1)} \quad \text{(B.73)}$$

Taking the square root,

$$\langle j : \vec{J}_1 : j-1 \rangle = \langle j-1 : \vec{J}_1 : j \rangle$$

$$= \pm \hbar \sqrt{\frac{(j-j_1-j_2)(j+j_1-j_2)(j_1+j+j_2+1)(j_1-j+j_2+1)}{4j^2(2j-1)(2j+1)}} \qquad (B.74)$$

By using (B.68), we see that we can get $\langle j : \vec{J}_1 : j+1 \rangle$ by substituting $j \rightarrow j+1$ into (B.73) and then taking the square root,

$$\langle j : \vec{J}_1 : j+1 \rangle = \pm \hbar \sqrt{\frac{(j-j_1+j_2+1)(j+j_1-j_2+1)(j+j_1+j_2+2)(j_1-j+j_2)}{4(j+1)^2(2j+1)(2j+3)}}$$

$$(B.75)$$

Note finally that we can resolve a lingering phase ambiguity by letting

$$\langle j : \vec{J}_1 : j-1 \rangle = -\langle j : \vec{J}_2 : j-1 \rangle \qquad (B.76)$$

As a final aside, we realize that we could now use (B.48), (B.51), (B.57), and (B.74) to compute the reduced matrix elements $\langle j \| \vec{J}_1 \| j \rangle$ and $\langle j \| \vec{J}_1 \| j \pm 1 \rangle$.

We finally have all we need to construct the last recursion relation we need to derive Racah's formula. To construct this recursion relation, consider the following:

$$J_{1z}|j,m\rangle = \sum_{j',m'} \langle j',m'|J_{1z}|j,m\rangle |j',m'\rangle$$

$$= \sum_{m_1} \sum_{j',m'} \langle j',m'|J_{1z}|j,m\rangle C^{j',m'}_{j_1,m_1;j_2,m_2=m'-m_1} |j_1,m_1\rangle |j_2, m'-m_1\rangle \qquad (B.77)$$

But also,

$$J_{1z}|j,m\rangle = \sum_{m_1} m_1 C^{j,m}_{j_1,m_1;j_2,m_2=m-m_1} |j_1,m_1\rangle |j_2, m-m_1\rangle \qquad (B.78)$$

Combine (B.77) and (B.78):

$$m_1 C^{j,m}_{j_1,m_1;j_2,m_2=m-m_1} = \sum_{j',m'} \langle j',m'|J_{1z}|j,m\rangle C^{j',m'}_{j_1,m_1;j_2,m_2=m'-m_1} \qquad (B.79)$$

But the comment after (B.33) $\Rightarrow m' = m$ on the RHS of (B.79)

$$m_1 C^{j,m}_{j_1,m_1;j_2,m_2} = \sum_{j'=j,j\pm 1} \langle j',m|J_{1z}|j,m\rangle C^{j',m}_{j_1,m_1;j_2,m_2}$$

$$= m\langle j \vdots \vec{J}_1 \vdots j\rangle C^{j,m}_{j_1,m_1;j_2,m_2} + \overbrace{\sqrt{j^2-m^2}}^{\sqrt{(j+m)(j-m)}} C^{j-1,m}_{j_1,m_1;j_2,m_2} \langle j \vdots \vec{J}_1 \vdots j-1\rangle$$

$$+ \overbrace{\sqrt{(j+1)^2-m^2}}^{\sqrt{(j-m+1)(j+m+1)}} C^{j+1,m}_{j_1,m_1;j_2,m_2} \langle j \vdots J_1 \vdots j+1\rangle$$

$$\text{(B.80)}$$

where we have used (B.44), (B.49), and (B.50). So now we have

$$\sqrt{(j+m)(j-m)}\langle j \vdots \vec{J}_1 \vdots j-1\rangle C^{j-1,m}_{j_1,m_1;j_2,m_2}$$
$$= [m_1 - m\langle j \vdots \vec{J}_1 \vdots j\rangle]C^{j,m}_{j_1,m_1;j_2,m_2} \qquad \text{(B.81)}$$
$$-\sqrt{(j-m+1)(j+m+1)}\langle j \vdots \vec{J}_1 \vdots j+1\rangle C^{j+1,m}_{j_1,m_1;j_2,m_2}$$

Putting in the values for $\langle j \vdots \vec{J}_1 \vdots j' = j, j\pm 1\rangle$ from (B.74), (B.57), and (B.75) we get

$$
\begin{aligned}
&\sqrt{\frac{(j+m)(j-m)(j-j_1+j_2)(j+j_1-j_2)(j+j_1+j_2+1)(j_1-j+j_2+1)}{4j^2(2j-1)(2j+1)}}C^{j-1,m}_{j_1,m_1;j_2,m_2}\\
&= \left[m_1 - m\frac{j_1(j_1+1)-j_2(j_2+1)+j(j+1)}{2j(j+1)}\right]C^{j,m}_{j_1,m_1;j_2,m_2}\\
&\quad -\sqrt{\frac{\left[(j+1)^2-m^2\right](j-j_1+j_2+1)(j+j_1-j_2+1)(j+j_1+j_2+2)(j_1-j+j_2)}{4(j+1)^2(2j+1)(2j+3)}}C^{j+1,m}_{j_1,m_1;j_2,m_2}
\end{aligned}
$$

$$\text{(B.82)}$$

which is the recursion relation for the C-G coefficients in j that we will need.[4]

The Racah Formula for the Clebsch-Gordon Coefficients

We begin the derivation of the Racah formula for the C-G coefficients by defining (after Racah),

---

[4]The derivation of Eq. (B.82) followed closely the one presented in chapter III of [174].

$$f(j_1, m_1; j_2, m_2; j, m) \equiv \frac{C^{j,m}_{j_1, m_1; j_2, m_2} \sqrt{(j_1 - m_1)!(j_2 - m_2)!(j - m)!}}{(-1)^{j_1 - m_1} \sqrt{(j_1 + m_1)!(j_2 + m_2)!(j + m)!}} \qquad (B.83)$$

which implies

$$f(j_1, m_1; j_2, m_2; j, m - 1) = \frac{C^{j,m-1}_{j_1, m_1; j_2, m_2} \sqrt{(j_1 - m_1)!(j_2 - m_2)!(j - m + 1)!}}{(-1)^{j_1 - m_1} \sqrt{(j_1 + m_1)!(j_2 + m_2)!(j + m - 1)!}} \quad (B.84)$$

Now we use the lower sign of (2.136):

$$f(j_1, m_1; j_2, m_2; j, m - 1) = \frac{\sqrt{(j_1 - m_1)!(j_2 - m_2)!(j - m + 1)!}}{(-1)^{j_1 - m_1} \sqrt{(j_1 + m_1)!(j_2 + m_2)!(j + m - 1)!}}$$

$$\times \left(\frac{1}{\sqrt{(j + m)(j - m + 1)}} \right)$$

$$\times \left[\sqrt{(j_1 + m_1 + 1)(j_1 - m_1)} C^{j,m}_{j_1, m_1 + 1; j_2, m_2} + \sqrt{(j_2 + m_2 + 1)(j_2 - m_2)} C^{j,m}_{j_1, m_1; j_2, m_2 + 1} \right]$$

$$= \frac{(j_1 + m_1 + 1)(j_1 - m_1)}{(-1)}$$

$$\times \underbrace{\left[\frac{\sqrt{(j_1 - m_1 - 1)!(j_2 - m_2)!(j - m + 1)!} \; C^{j,m}_{j_1, m_1 + 1; j_2, m_2}}{(-1)^{j_1 - m_1 - 1} \sqrt{(j_1 + m_1 + 1)!(j_2 + m_2)!(j + m)!(j - m + 1)}} \right]}_{f(j_1, m_1 + 1; j_2, m_2; j, m)}$$

$$+ (j_2 + m_2 + 1)(j_2 - m_2)$$

$$\times \underbrace{\left[\frac{\sqrt{(j_1 - m_1)!(j_2 - m_2 - 1)!(j - m + 1)!} \; C^{j,m}_{j_1, m_1; j_2, m_2 + 1}}{(-1)^{j_1 - m_1} \sqrt{(j_1 + m_1)!(j_2 + m_2 + 1)!(j + m - 1)!(j + m)(j - m + 1)}} \right]}_{f(j_1, m_1; j_2, m_2 + 1; j, m)}$$

$$\qquad (B.85)$$

which gives

$$f(j_1, m_1; j_2, m_2; j, m - 1) = (j_2 + m_2 + 1)(j_2 - m_2) f(j_1, m_1; j_2, m_2 + 1; j, m)$$
$$- (j_1 + m_1 + 1)(j_1 - m_1) f(j_1, m_1 + 1; j_2, m_2; j, m)$$

$$\qquad (B.86)$$

Equation (B.83) implies

$$f(j_1, m_1; j_2, m_2; j, m + 1) = \frac{C^{j,m+1}_{j_1, m_1; j_2, m_2} \sqrt{(j_1 - m_1)!(j_2 - m_2)!(j - m - 1)!}}{(-1)^{j_1 - m_1} \sqrt{(j_1 + m_1)!(j_2 + m_2)!(j + m + 1)!}} \quad (B.87)$$

and using the upper sign of (2.136), we have

$$f(j_1,m_1;j_2,m_2;j,m+1) = \frac{\sqrt{(j_1-m_1)!(j_2-m_2)!(j-m-1)!}}{(-1)^{j_1-m_1}\sqrt{(j_1+m_1)!(j_2+m_2)!(j+m+1)!}}\left(\frac{1}{\sqrt{(j-m)(j+m+1)}}\right)$$

$$\times\left[\sqrt{(j_1-m_1+1)(j_1+m_1)}\,C^{j,m}_{j_1,m_1-1;j_2,m_2}+\sqrt{(j_2-m_2+1)(j_2+m_2)}\,C^{j,m}_{j_1,m_1;j_2,m_2-1}\right]$$

$$=\frac{1}{(j+m+1)(j-m)}$$

$$\times\left\{\begin{array}{c}\left[\dfrac{\sqrt{(j_1-m_1)!(j_1-m_1+1)(j_2-m_2)!(j-m)(j-m-1)!(j+m+1)(j_1+m_1)}\,C^{j,m}_{j_1,m_1-1;j_2,m_2}}{(-1)(-1)^{j_1-m_1-1}\sqrt{(j_2+m_2)!(j+m+1)!(j_1+m_1)!}}\right]\\[4mm]+\left[\dfrac{\sqrt{(j_1-m_1)!(j_2-m_2)!(j_2-m_2+1)(j-m)(j-m-1)!(j+m+1)(j_2+m_2)}\,C^{j,m}_{j_1,m_1;j_2,m_2-1}}{(-1)^{j_1-m_1}\sqrt{(j_1+m_1)!(j+m+1)!(j_2+m_2)!}}\right]\end{array}\right\}$$

$$=\frac{1}{(j+m+1)(j-m)}\left\{\begin{array}{c}\underbrace{\left[\dfrac{\sqrt{(j_1-m_1+1)!(j_2-m_2)!(j-m)!}\,C^{j,m}_{j_1,m_1-1;j_2,m_2}}{(-1)(-1)^{j_1-m_1-1}\sqrt{(j_2+m_2)!(j+m)!(j_1+m_1-1)!}}\right]}_{-f(j_1,m_1-1;j_2,m_2;j,m)}\\[6mm]+\underbrace{\left[\dfrac{\sqrt{(j_1-m_1)!(j_2-m_2+1)!(j-m)!}\,C^{j,m}_{j_1,m_1;j_2,m_2-1}}{(-1)^{j_1-m_1}\sqrt{(j_1+m_1)!(j+m)!(j_2+m_2-1)!}}\right]}_{-f(j_1,m_1;j_2,m_2-1;j,m)}\end{array}\right\}$$

$$\text{(B.88)}$$

which gives

$$\begin{aligned}(j+m+1)(j-m)f(j_1,m_1;j_2,m_2;j,m+1)\\=f(j_1,m_1-1;j_2,m_2;j,m)-f(j_1,m_1;j_2,m_2-1;j,m)\end{aligned}\qquad\text{(B.89)}$$

Letting $m=j$ in (B.89),

$$f(j_1,m_1;j_2,m_2-1;j,j)=f(j_1,m_1-1;j_2,m_2;j,j)\qquad\text{(B.90)}$$

The form of (B.90) implies that $f(j_1,m_1;j_2,m_2;j,j)$ is independent of m_1 and m_2, so we write

$$f(j_1,m_1;j_2,m_2;j,j)\equiv A_j\qquad\text{(B.91)}$$

From (B.91) and (B.86), we get

$$f(j_1,m_1;j_2,m_2;j,j-1)=[(j_2+m_2+1)(j_2-m_2)-(j_1+m_1+1)(j_1-m_1)]A_j\qquad\text{(B.92)}$$

Now let $m=j-1$ in (B.86). This gives

$$f(j_1, m_1; j_2, m_2; j, j - 2) = (j_2 + m_2 + 1)(j_2 - m_2)f(j_1, m_1; j_2, m_2 + 1; j, j - 1)$$
$$- (j_1 + m_1 + 1)(j_1 - m_1)f(j_1, m_1 + 1; j_2, m_2; j, j - 1)$$

$$(B.93)$$

But from (B.92),

$$f(j_1, m_1; j_2, m_2 + 1; j, j - 1) = [(j_2 + m_2 + 2)(j_2 - m_2 - 1) - (j_1 + m_1 + 1)(j_1 - m_1)]A_j$$

$$(B.94)$$

and

$$f(j_1, m_1 + 1; j_2, m_2; j, j - 1) = [(j_2 + m_2 + 1)(j_2 - m_2)$$
$$- (j_1 + m_1 + 2)(j_1 - m_1 - 1)]A_j \qquad (B.95)$$

Combining (B.93), (B.94), and (B.95),

$$f(j_1, m_1; j_2, m_2; j, j - 2)$$
$$= (j_2 + m_2 + 1)(j_2 - m_2)[(j_2 + m_2 + 2)(j_2 - m_2 - 1)$$
$$- (j_1 + m_1 + 1)(j_1 - m_1)]A_j - (j_1 + m_1 + 1)(j_1 - m_1)[(j_2 + m_2 + 1)(j_2 - m_2)$$
$$- (j_1 + m_1 + 2)(j_1 - m_1 - 1)]A_j$$
$$= \begin{bmatrix} (j_2 + m_2 + 1)(j_2 + m_2 + 2)(j_2 - m_2)(j_2 - m_2 - 1) \\ -2(j_2 + m_2 + 1)(j_2 - m_2)(j_1 + m_1 + 1)(j_1 - m_1) \\ +(j_1 + m_1 + 1)(j_1 + m_1 + 2)(j_1 - m_1)(j_1 - m_1 - 1) \end{bmatrix} A_j$$

$$(B.96)$$

The general formula seems to be:

$$f(j_1, m_1; j_2, m_2; j, j - u)$$
$$= A_j \sum_{t=0}^{u} (-1)^t \binom{u}{t} \frac{(j_1 + m_1 + t)!(j_1 - m_1)!(j_2 + m_2 + u - t)!(j_2 - m_2)!}{(j_1 + m_1)!(j_1 - m_1 - t)!(j_2 + m_2)!(j_2 - m_2 - u + t)!}$$

$$(B.97)$$

As a check, we note that (B.97) solves (B.95), (B.96) and (B.86)-try it and see!
Now remembering that

$$m = m_1 + m_2 \Rightarrow j - u = m_1 + m_2 \Rightarrow u = j - m \qquad (B.98)$$

and then substituting (B.97) into (B.83), and solving for $C_{j_1, m_1; j_2, m_2}^{j, m}$,

$$C^{j,m}_{j_1,m_1;j_2,m_2} = \delta_{m_1+m_2,m}A_j \left[\frac{(j_1-m_1)!(j_2-m_2)!(j-m)!(j+m)!}{(j_1+m_1)!(j_2+m_2)!}\right]^{\frac{1}{2}} \sum_t (-1)^{j_1-m_1+t}$$

$$\times \left[\frac{(j_1+m_1+t)!(j+j_2-m_1-t)!}{t!(j-m-t)!(j_1-m_1-t)!(j_2-j+m_1-t)!}\right]$$

$$\text{(B.99)}$$

This gives us the dependence of $C^{j,m}_{j_1,m_1;j_2,m_2}$ on m_1 and m_2. We can obtain the dependence of A_j on j from (B.82) by first using (B.99) to calculate $C^{j+1,j}_{j_1,m_1;j_2,m_2}$:

$$C^{j+1,j}_{j_1,m_1;j_2,m_2} = \delta_{m_1+m_2,j}A_{j+1}\left[\frac{(j_1-m_1)!(j_2-m_2)!\overbrace{(j+1-j)!}^{=1}\overbrace{(j+1+j)!}^{=(2j+1)!}}{(j_1+m_1)!(j_2+m_2)!}\right]^{\frac{1}{2}}$$

$$\times \sum_t (-1)^{j_1-m_1+t}\left[\frac{(j_1+m_1+t)!(j+1+j_2-m_1-t)!}{t!\underbrace{(j+1-j-t)!}_{=(1-t)!\Rightarrow t=0,1\,\text{only}}(j_1-m_1-t)!(j_2-j-1+m_1-t)!}\right]$$

$$\Rightarrow C^{j+1,j}_{j_1,m_1;j_2,m_2} = \delta_{m_1+m_2,j}A_{j+1}\left[\frac{(j_1-m_1)!(j_2-m_2)!(2j+1)!}{(j_1+m_1)!(j_2+m_2)!}\right]^{\frac{1}{2}}$$

$$\times \left[\frac{(-1)^{j_1-m_1}(j_1+m_1)!(j+1+j_2-m_1)!}{(j_1-m_1)!(j_2-j-1+m_1)!}\right.$$

$$\left.+\frac{(-1)^{j_1-m_1+1}(j_1+m_1+1)!(j+j_2-m_1)!}{(j_1-m_1-1)!(j_2-j+m_1)!}\right]$$

$$= \delta_{m_1+m_2,j}A_{j+1}(-1)^{j_1-m_1}\left[\frac{(j_1-m_1)!(j_2-m_2)!(2j+1)!}{(j_1+m_1)!(j_2+m_2)!}\right]^{\frac{1}{2}}$$

$$\times \frac{(j_1+m_1)!(j+j_2-m_1)!}{(j_1-m_1)!(j_2-j+m_1)!}$$

$$\times \left[(j+1+j_2-m_1)(j_2-j+m_1)-(j_1+m_1+1)(j_1-m_1)\right] \quad \text{(B.100)}$$

But $j = m = m_1 + m_2$,

$$C^{j+1,j}_{j_1,m_1;j_2,m_2} = \delta_{m_1+m_2,j}(-1)^{j_1-m_1}A_{j+1}\left[\frac{(j_1-m_1)!(j_2-m_2)!(2j+1)!}{(j_1+m_1)!(j_2+m_2)!}\right]^{\frac{1}{2}}$$

$$\times\left[\frac{(j_1+m_1)!(m_2+j_2)!}{(j_1-m_1)!(j_2-m_2)!}\right]\left[\begin{array}{c}(j+1+j_2-m_1)(j_2-j+m_1)\\ -(j_1+m_1+1)(j_1-m_1)\end{array}\right]$$

$$= \delta_{m_1+m_2,j}(-1)^{j_1-m_1}A_{j+1}\left[\frac{(j_1+m_1)!(j_2+m_2)!(2j+1)!}{(j_1-m_1)!(j_2-m_2)!}\right]^{\frac{1}{2}}$$

$$\times\left[\begin{array}{c}(j+1+j_2-m_1)(j_2-j+m_1)\\ -(j_1+m_1+1)(j_1-m_1)\end{array}\right]$$

(B.101)

Consider the last term on the RHS of (B.101):

$$(j+1+j_2-m_1)(j_2-j+m_1) - (j_1+m_1+1)(j_1-m_1)$$
$$= (2m_1 j + 2m_1) - (j_2+j-j_2^2+j^2+j_1)$$
$$= 2(j+1)m_1 - [j_1(j_1+1) - j_2(j_2+1) + j(j+1)]$$
$$= 2(j+1)[m_1 - j\langle j \vdots J_1 \vdots j\rangle]$$

(B.102)

where we have used (B.57). Now combining (B.101) and (B.102) gives

$$C^{j+1,j}_{j_1,m_1;j_2,m_2} = \delta_{m_1+m_2,j}(-1)^{j_1-m_1}A_{j+1}\left[\frac{(j_1+m_1)!(j_2+m_2)!(2j+1)!}{(j_1-m_1)!(j_2-m_2)!}\right]^{\frac{1}{2}}2(j+1)$$

$$\times\left[m_1 - \overset{=m}{\overbrace{j}}\langle j \vdots J_1 \vdots j\rangle\right]$$

(B.103)

The LHS of (B.82) vanishes for $m = j$,

$$0 = [m_1 - m\langle j \vdots J_1 \vdots j\rangle]C^{j,j}_{j_1,m_1;j_2,m_2} - \sqrt{(2j+1)}\langle j \vdots J_1 \vdots j+1\rangle C^{j+1,j}_{j_1,m_1;j_2,m_2}\quad\text{(B.104)}$$

From (B.83) we have

$$C^{j,j}_{j_1,m_1;j_2,m_2} = (-1)^{j_1-m_1}\underbrace{f(j_1,m_1;j_2,m_2;j,j)}_{=A_j\text{ from (B.91)}}\frac{\sqrt{(j_1+m_1)!(j_2+m_2)!(2j)!}}{\sqrt{(j_1-m_1)!(j_2-m_2)!}}\quad\text{(B.105)}$$

where we have also used (B.91), as indicated.

We can use (B.105) to substitute for $C^{j,j}_{j_1,m_1;j_2,m_2}$ and (B.103) to substitute for $C^{j+1,j}_{j_1,m_1;j_2,m_2}$ in (B.104):

$$0 = [m_1 - m\langle j \vdots J_1 \vdots j\rangle](-1)^{j_1-m_1} A_j \frac{\sqrt{(j_1+m_1)!(j_2+m_2)!(2j)!}}{\sqrt{(j_1-m_1)!(j_2-m_2)!}}$$

$$- \sqrt{(2j+1)}\langle j \vdots J_1 \vdots j+1\rangle (-1)^{j_1-m_1} A_{j+1} \tag{B.106}$$

$$\times \left[\frac{(j_1+m_1)!(j_2+m_2)!(2j+1)!}{(j_1-m_1)!(j_2-m_2)!}\right]^{\frac{1}{2}} 2(j+1)\left[m_1 - \overset{=m}{j}\langle j \vdots J_1 \vdots j\rangle\right]$$

Eliminating all the common factors leaves

$$0 = A_j - 2(j+1)(2j+1)\langle j \vdots J_1 \vdots j+1\rangle A_{j+1} \tag{B.107}$$

Using the expression for $\langle j \vdots J_1 \vdots j+1\rangle$ from (B.75) in (B.107), we have

$$\frac{A_j}{A_{j+1}} = \sqrt{\frac{(j_1+j_2+j+2)(j_1+j_2-j)(j_1-j_2+j+1)(j+j_2-j_1+1)(2j+1)}{(2j+3)}}$$

$$\tag{B.108}$$

After a little thought, the recursion relation in (B.108) is seen to be solved by

$$A_j = B\sqrt{\frac{(2j+1)(j_1+j_2-j)!}{(j_1+j_2+j+1)!(j_1-j_2+j)!(j+j_2-j_1)!}} \tag{B.109}$$

where B must be a constant independent of j. To check, we first compute A_{j+1} from (B.109):

$$A_{j+1} = B\sqrt{\frac{(2j+3)(j_1+j_2-j-1)!}{(j_1+j_2+j+2)!(j_1-j_2+j+1)!(j+1+j_2-j_1)!}} \tag{B.110}$$

Now we form the ratio A_j/A_{j+1} using (B.109) and (B.110):

$$\frac{A_j}{A_{j+1}} = \sqrt{\frac{(2j+1)(j_1+j_2-j)!}{(j_1+j_2+j+1)!(j_1-j_2+j)!(j+j_2-j_1)!}}$$

$$\times \sqrt{\frac{(j_1+j_2+j+2)!(j_1-j_2+j+1)!(j+1+j_2-j_1)!}{(2j+3)(j_1+j_2-j-1)!}}$$

$$= \sqrt{\frac{(j_1+j_2+j+2)(j_1+j_2-j)(j_1-j_2+j+1)(j+j_2-j_1+1)(2j+1)}{(2j+3)}}$$

$$\tag{B.111}$$

which matches the result of (B.108).

We must now find the value of B. We start by noting that (B.110) is true for any j, so we let $j = j_1 + j_2$:

$$A_{j=j_1+j_2} = B\sqrt{\frac{(2j+1)}{(2j+1)!(2j_1)!(2j_2)!}} \tag{B.112}$$

(B.83) and (B.91) imply

$$C^{j_1+j_2,j_1+j_2}_{j_1,j_1;j_2,j_2} = A_{j=j_1+j_2}\sqrt{(2j_1)!(2j_2)!(2j)!} \tag{B.113}$$

Combining (B.112) and (B.113),

$$B\sqrt{\frac{(2j+1)}{(2j+1)!(2j_1)!(2j_2)!}} = \frac{C^{j_1+j_2,j_1+j_2}_{j_1,j_1;j_2,j_2}}{\sqrt{(2j_1)!(2j_2)!(2j)!}} \tag{B.114}$$

But $C^{j_1+j_2,j_1+j_2}_{j_1,j_1;j_2,j_2} = 1$ by (2.124), which leaves us with

$$B = 1 \tag{B.115}$$

Combining (B.99), (B.109), and (B.115), we get

$$C^{j,m}_{j_1,m_1;j_2,m_2} = \delta_{m_1+m_2,m}\left[\frac{(2j+1)(j_1+j_2-j)!(j_1-m_1)!(j_2-m_2)!(j-m)!(j+m)!}{(j_1+j_2+j+1)!(j_1-j_2+j)!(j+j_2-j_1)!(j_1+m_1)!(j_2+m_2)!}\right]^{\frac{1}{2}}$$
$$\times\sum_t(-1)^{j_1-m_1+t}\left[\frac{(j_1+m_1+t)!(j+j_2-m_1-t)!}{t!(j-m-t)!(j_1-m_1-t)!(j_2-j+m_1-t)!}\right] \tag{B.116}$$

This is the formula for the C-G coefficients as it was developed by Wigner (via group-theoretical methods). It can be put into another more symmetric and, therefore, more useful form. To do so, we first recall the binomial theorem:

$$(x+y)^n = \sum_k\binom{n}{k}x^k y^{n-k}; \quad \binom{n}{k} \equiv \frac{n!}{k!(n-k)!} \tag{B.117}$$

This theorem implies

$$(x+y)^{n+m} = \sum_r\binom{n+m}{r}x^r y^{n+m-r} = y^{n+m}\sum_r\binom{n+m}{r}\left(\frac{x}{y}\right)^r \tag{B.118}$$

Using this result, and the identity

$$(x+y)^{n+m} = (x+y)^n(x+y)^m \tag{B.119}$$

we get

$$\sum_r \binom{n+m}{r}\left(\frac{x}{y}\right)^r = \left[\sum_s \binom{n}{s}\left(\frac{x}{y}\right)^s\right]\left[\sum_j \binom{m}{j}\left(\frac{x}{y}\right)^j\right] \qquad (B.120)$$

Equating like powers of x/y on both sides of (B.120) [being very careful not to double count $(x + y)^n(x + y)^m$ and $(x + y)^m(x + y)^n$],

$$\binom{n+m}{r} = \sum_{s=0}^{r}\binom{n}{s}\binom{m}{j=r-s} \Rightarrow \frac{(n+m)!}{r!(n+m-r)!}$$

$$= \sum_s \frac{n!}{s!(n-s)!}\cdot\frac{m!}{(r-s)!(m-r+s)!} \qquad (B.121)$$

Now let $n = a - b,\, m = b,\, r = a - c$:

$$\frac{a!}{(a-c)!c!} = \sum_s \frac{(a-b)!b!}{s!(a-b-s)!(a-c-s)!(b-a+c+s)!}$$

$$\Rightarrow \frac{a!}{b!c!} = \sum_s \frac{(a-b)!(a-c)!}{s!(a-b-s)!(a-c-s)!(b-a+c+s)!} \qquad (B.122)$$

This result is known as the binomial addition theorem. We now apply (B.122) to the last term on the RHS of (B.116) with the following identifications:

$$a \equiv (j+j_2 - m_1 - t); \quad b \equiv (j - m - t); \quad c \equiv (j_1 - m_1 - t); \quad s \to u \qquad (B.123)$$

Making these substitutions, remembering that $m = m_1 + m_2$, and performing all the algebra gives, for the last term on the RHS of (B.116),

$$\sum_t (-1)^{j_1 - m_1 + t}\left[\frac{(j_1 + m_1 + t)!(j+j_2 - m_1 - t)!}{t!(j-m-t)!(j_1 - m_1 - t)!(j_2 - j + m_1 + t)!}\right]$$

$$= \sum_{t,u} (-1)^{j_1 - m_1 + t}\left[\frac{(j_1 + m_1 + t)!(j_2 + m_2)!(j+j_2 - j_1)!}{t!(j_2 - j + m_1 + t)!u!(j_2 + m_2 - u)!(j+j_2 - j_1 - u)!(j_1 - j_2 - m - t + u)!}\right]$$

$$(B.124)$$

Consider the binomial coefficient $\binom{a}{i}$. How do we handle the case $a < 0$?

$$\binom{a}{i} = \frac{a!}{i!(a-i)!} = \frac{a(a-1)(a-2)\cdots(a-i+1)}{i!}$$

$$= (-1)^i \frac{(-a)(1-a)(2-a)\cdots(i-1-a)}{i!}$$

$$= (-1)^i \frac{(-a)(1-a)(2-a)\cdots(i-1-a)}{i!} \cdot \frac{(-a-1)(-a-2)\cdots(1)}{(-a-1)(-a-2)\cdots(1)}$$

$$= (-1)^i \frac{(i-1-a)!}{i!(-a-1)!} = (-1)^i \binom{i-1-a}{i} \tag{B.125}$$

Let us apply this result to the LHS of (B.121) with $a \to n+m$ and $i \to r$. In this case, (B.125) would give

$$\binom{n+m}{r} = (-1)^r \binom{r-1-n-m}{r} \tag{B.126}$$

Similarly, we can apply the results of (B.125) to the first term on the RHS of (B.121) with $a \to n$ and $i \to s$. Here (B.125) gives

$$\binom{n}{s} = (-1)^s \binom{s-1-n}{s} \tag{B.127}$$

Combining (B.121), (B.126), and (B.127)

$$(-1)^r \binom{r-1-n-m}{r} = \sum_{s=0}^{r} (-1)^s \binom{s-1-n}{s} \binom{m}{r-s} \tag{B.128}$$

In (B.128), we are assuming m is positive, but n and $n+m$ are both negative. Now we let $m \to b$ (to avoid confusion later), and $n \to r - c - 1$ in (B.128). This substitution gives

$$(-1)^r \binom{c-b}{r} = \sum_{s=0}^{r} (-1)^s \binom{s-r+c}{s} \binom{b}{r-s}$$

$$\Rightarrow (-1)^r \frac{(c-b)!}{r!(c-m-r)!} = \sum_{s=0}^{r} (-1)^s \frac{(s-r+c)!}{s!(c-r)!} \cdot \frac{b!}{(r-s)!(b-r+s)!} \tag{B.129}$$

$$\Rightarrow (-1)^r \frac{(c-b)!(c-r)!}{r!b!(c-b-r)!} = \sum_{s=0}^{r} (-1)^s \frac{(s-r+c)!}{s!(r-s)!(b-r+s)!}$$

We apply the results of (B.129) to the last line of (B.124) with the following substitutions:

$$s \rightarrow t$$
$$c \rightarrow 2j_1 - j_2 + u - m_2$$
$$b \rightarrow j_1 - j + u - m_2 \tag{B.130}$$
$$r \rightarrow j_1 - j_2 + u - m$$

Keep in mind that only the terms that are affected by the summation over t in (B.124) will be affected by the application of (B.129). The terms in the last line of (B.124) that are independent of the summation will remain unchanged. The whole idea here is to collapse the summation over t by the application of (B.129) to (B.124), to wit,

$$\sum_t (-1)^{j_1 - m_1 + t} \left[\frac{(j_1 + m_1 + t)!}{t!(j_2 - j + m_1 + t)!(j_1 - j_2 - m - t + u)!} \right]$$
$$= (-1)^{j_2 - m_2 - u} \frac{(j_1 + m_1)!(j + j_1 - j_2)!}{(j_1 - j_2 - m + u)!(j_1 - j - m_2 + u)!(j + m - u)!} \tag{B.131}$$

Let us now recap what we have done with the last term on the RHS of (B.116). This would include the results of (B.124) and (B.131):

$$\sum_t (-1)^{j_1 - m_1 + t} \left[\frac{(j_1 + m_1 + t)!(j + j_2 - m_1 - t)!}{t!(j - m - t)!(j_1 - m_1 - t)!(j_2 - j + m_1 + t)!} \right]$$
$$= \sum_{t,u} (-1)^{j_1 - m_1 + t}$$
$$\times \left[\frac{(j_1 + m_1 + t)!(j_2 + m_2)!(j + j_2 - j_1)!}{t!(j_2 - j + m_1 + t)!u!(j_2 + m_2 - u)!(j + j_2 - j_1 - u)!(j_1 - j_2 - m - t + u)!} \right]$$
$$= \sum_u (-1)^{j_2 + m_2 - u}$$
$$\times \left[\frac{(j_1 + m_1)!(j + j_1 - j_2)!(j_2 + m_2)!(j + j_2 - j_1)!}{(j_1 - j_2 - m + u)!(j_1 - j - m_2 + u)!(j + m - u)!u!(j_2 + m_2 - u)!(j + j_2 - j_1 - u)!} \right] \tag{B.132}$$

To finish, we let $z \equiv j_2 + m_2 - u$ and substitute the results of (B.132) into (B.116):

$$C_{j_1, m_1; j_2, m_2}^{j,m} = \delta_{m_1 + m_2, m} \left[\frac{(2j + 1)(j_1 + j_2 - j)!(j_1 - j_2 + j)!(j + j_2 - j_1)!}{(j_1 + j_2 + j + 1)!} \right]^{\frac{1}{2}}$$
$$\times \sum_z (-1)^z \frac{\sqrt{(j_1 + m_1)!(j_1 - m_1)!(j_2 + m_2)!(j_2 - m_2)!(j - m)!(j + m)!}}{z!(j_1 + j_2 - j - z)!(j_1 - m_1 - z)!(j_2 + m_2 - z)!(j - j_2 + m_1 + z)!(j - j_1 - m_2 + z)!} \tag{B.133}$$

This matches (2.138) from the main text.[5]

―――――――――
[5] For variations on the formula (B.133) for the C-G coefficients, and for more formulas involving the C-G coefficients, see [10].

Appendix C: The 6-*j* Symbols and the Racah Formula (2.172)

This appendix outlines the derivation of the formula for the Racah coefficients and how those coefficients relate to the 6-*j* symbols. It follows the steps first given by Racah in his seminal papers [2]. Before we begin, we must develop some more algebraic tools. First, we start with (B.125) and allow $i \to r - s$, $a \to m$,

$$\binom{m}{r-s} = (-1)^{r-s}\binom{r-s-m-1}{r-s} \tag{C.1}$$

As an interesting aside, we can substitute this result into (B.121):

$$\binom{n+m}{r} = \sum_{s=0}^{r}\binom{n}{s}(-1)^{r-s}\binom{r-s-m-1}{r-s} \Rightarrow (-1)^{-r}\binom{n+m}{r}$$

$$= \sum_{s=0}^{r}\binom{n}{s}(-1)^{s}\binom{r-s-m-1}{r-s} \tag{C.2}$$

for $m < 0$; s, r= integer. Now use (B.126) to substitute for $\binom{n+m}{r}$:

$$\binom{r-n-m-1}{r} = \sum_{s=0}^{r}\binom{n}{s}(-1)^{s}\binom{r-s-m-1}{r-s} \tag{C.3}$$

Let $m \to r - t' - 1$ in (C.3):

$$\binom{t'-n}{r} = \sum_{s=0}^{r}\binom{n}{s}(-1)^{s}\binom{t'-s}{r-s}$$

$$\Rightarrow \sum_{s=0}^{r}(-1)^{s}\frac{n!}{s!(n-s)!}\frac{(t'-s)!}{(r-s)!(t'-r)!} = \frac{(t'-n)!}{r!(t'-n-r)!} \tag{C.4}$$

$$\Rightarrow \sum_{s=0}^{r}(-1)^{s}\frac{(t'-s)!}{s!(n-s)!(r-s)!} = \frac{(t'-n)!(t'-r)!}{n!r!(t'-n-r)!}$$

This is another version of (B.129). [(B.129) is one of the relationships we will need later.] But for now, we continue by combining (B.121), (B.126), (B.127), and (C.1):

$$(-1)^{r}\binom{r-1-n-m}{r} = \sum_{s=0}^{r}(-1)^{s}\binom{s-1-n}{s}(-1)^{r-s}\binom{r-s-1-m}{r-s}$$

$$\Rightarrow \frac{(r-1-n-m)!}{r!(-1-n-m)!} = \sum_{s=0}^{r} \frac{(s-1-n)!}{s!(-1-n)!} \frac{(r-s-1-m)!}{(r-s)!(-1-m)!} \tag{C.5}$$

In (C.5), set,

$$\begin{aligned}
s &= c + \sigma \\
-n &= a - c - 1 \\
-m &= b - r + c + 1 = b - d - 1 \\
r &= c + d
\end{aligned} \tag{C.6}$$

which gives

$$\frac{(a+b+1)!}{(c+d)!(a+b-c-c+1)!} = \sum_{\sigma} \frac{(\sigma+a)!}{(c+\sigma)!(a-c)!} \frac{(b-\sigma)!}{(d-\sigma)!(b-d)!}$$

$$\Rightarrow \frac{(a+b+1)!(a-c)!(b-d)!}{(c+d)!(a+b-c-d+1)!} = \sum_{\sigma} \frac{(\sigma+a)!}{(c+\sigma)!} \frac{(b-\sigma)!}{(d-\sigma)!} \tag{C.7}$$

This equation, along with (B.122) and (B.129), are the three relations that we will use to assist us in developing an algebraic expression for the 6-j symbols. Therefore, let us rewrite those three equations and relabel as appropriate to avoid confusion later:

$$\frac{a'!}{b'!c'!} = \sum_{s} \frac{(a'-b')!(a'-c')!}{s!(a'-b'-s')!(a'-c'-s')!(b'-a'+c'+s')!} \tag{C.8}$$

$$(-1)^{r} \frac{(c'-b')!(c'-r')!}{r'!b'!(c'-b'-r')!} = \sum_{s=0}^{r} (-1)^{s} \frac{(s-r'+c')!}{s!(r'-s)!(b'-r'+s)!} \tag{C.9}$$

$$\frac{(a'+b'+1)!(a'-c')!(b'-d')!}{(c'+d')!(a'+b'-c'-d'+1)!} = \sum_{\sigma} \frac{(\sigma+a')!}{(c'+\sigma)!} \frac{(b'-\sigma)!}{(d'-\sigma)!} \tag{C.10}$$

We begin by defining the quantity (after Racah [2, 4]) (note the slightly different presentation of the Kronecker delta):

$$v(abc; \alpha\beta\gamma) \equiv \delta(\alpha+\beta+\gamma, 0)$$

$$\times \sum_{z} \frac{(-1)^{c-a+z}[(a+\alpha)!(a-\alpha)!(b+\beta)!(b-\beta)!(c+\gamma)!(c-\gamma)!]^{\frac{1}{2}}}{z!(a+b-c-z)!(a-\alpha-z)!(b+\beta-z)!(c-b+\alpha+z)!(c-a-\beta+z)!} \tag{C.11}$$

and use this definition to create the following sum:

$$\sum_{\alpha,\beta,\gamma,\delta,\phi} (-1)^{f+\phi} v(abe;\alpha,\beta,-e)v(acf;-\alpha,\gamma,\phi)v(bdf;-\beta,\delta,-\phi)v(cde;\gamma,\delta,-e)$$

$$= \delta(\alpha+\beta-e,0)\delta(-\alpha+\gamma+\phi,0)\delta(-\beta+\delta-\phi,0)\delta(\alpha+\delta-e,0)$$

$$\times \sum_{\substack{\alpha,\beta,\gamma,\delta,\phi \\ z,z',z'',z'''}} \frac{\left[\begin{array}{l}(-1)^{3f+4e+\phi+z+z'+z''+z'''}(a+\alpha)!(a-\alpha)!(b+\beta)!(b-\beta)!2e! \\ \times(c+\gamma)!(c-\gamma)!(f+\phi)!(f-\phi)!(d+\delta)!(d-\delta)!\end{array}\right]}{\left[\begin{array}{l}z!(a+b-e-z)!(a-\alpha-z)!(b+\beta-z)!(e-b+\alpha+z)!(e-\alpha-\beta+z)! \\ z'!(a+c-f-z')!(a+\alpha-z')!(c+\gamma-z')!(f-c-\alpha+z')!(f-a-\gamma+z')! \\ z''!(b+d-f-z'')!(b+\beta-z'')!(d+\delta-z'')!(f-d-\beta+z'')!(f-b-\delta+z'')! \\ z'''!\left(c+d-e-z'''\right)!\left(c-\gamma-z'''\right)!\left(d+\delta-z'''\right)!\left(e-d+\gamma+z'''\right)!\left(e-c-\delta+z'''\right)!\end{array}\right]}$$

$$\tag{C.12}$$

The Kronecker δs in this sum imply

$$\beta = e-\alpha; \quad \phi = \alpha-\gamma; \quad \delta = e-\gamma \tag{C.13}$$

and collapse the sums over β, ϕ, and, δ, leaving us with

$$\sum_{\alpha,\beta,\gamma,\delta,\phi} (-1)^{f+\phi} v(abe;\alpha,\beta,-e)v(acf;-\alpha,\gamma,\phi)v(bdf;-\beta,\delta,-\phi)v(cde;\gamma,\delta,-e)$$

$$= \sum_{\substack{\alpha,\gamma,z \\ z',z'',z'''}} \frac{(-1)^{\begin{array}{l}3f+4e+\alpha-\gamma \\ +z+z'+z''+z'''\end{array}} \left[\begin{array}{l}(a+\alpha)!(a-\alpha)!(b+e-\alpha)!(b-e+\alpha)!2e!(c+\gamma)! \\ (c-\gamma)!(f+\alpha-\gamma)!(f-\alpha+\gamma)!(d+e-\gamma)!(d-e+\gamma)!\end{array}\right]}{\left[\begin{array}{l}z!(a+b-e-z)!(a-\alpha-z)!(b+e-\alpha-z)!(e-b+\alpha+z)!(-\alpha+\alpha+z)! \\ z'!(a+c-f-z')!(a+\alpha-z')!(c+\gamma-z')!(f-c-\alpha+z')!(f-a-\gamma+z')! \\ z''!(b+d-f-z'')!(b-\alpha+e-z'')!(d+e-\gamma-z'')!(f-d+\alpha-e+z'')! \\ \quad (f-b-e+\gamma+z'')! \\ z'''!(c+d-e-z''')!(c-\gamma-z''')!(d+e-\gamma-z''')!(e-d+\gamma+z''')! \\ \quad (-c+\gamma+z''')!\end{array}\right]}$$

$$\tag{C.14}$$

Now isolate just the sum over z, let $z \to -\alpha+q+a$, and consider the term

$$\sum_z \frac{(-1)^{z+\alpha}(a+\alpha)!(b-e+\alpha)!}{z!(a+b-e-z)!(a-\alpha-z)!(b+e-\alpha-z)!(e-b+\alpha+z)!(-\alpha+\alpha+z)!}$$

$$= \sum_q \frac{(-1)^{a+q}(a+\alpha)!(b-e+\alpha)!}{(a-\alpha+q)!(b-e+\alpha-q)!(-q)!(b+e-\alpha-q)!(e-b+\alpha+q)!(q)!}$$

$$\tag{C.15}$$

The comment preceding (2.143) still being applicable in this case, we see that the sum over q in (C.15) collapses to one value ($q = 0$) due to the presence of both $(q)!$ and $(-q)!$ in the denominator, which leaves us with

$$\sum_z \frac{(-1)^{z+a}(a+a)!(b-e+a)!}{z!(a+b-e-z)!(a-a-z)!(b+e-a-z)!(e-b+a+z)!(-a+a+z)!}$$
$$= \frac{(-1)^a}{(b+e-a)!(e-b+a)!}$$

$$(\text{C.16})$$

Similarly, let $z''' \to -\gamma + c + n$, and then consider the following term and collapse the sum as follows:

$$\sum_{z'''} \frac{(-1)^{-\gamma+z'''}(c-\gamma)!(d-e+\gamma)!}{z'''!(c+d-e-z''')!(c-\gamma-z''')!(d+e-\gamma-z''')!(e-d+\gamma+z''')!(-c+\gamma+z''')!}$$
$$= \sum_n \frac{(-1)^{-2\gamma+n+c}(c-\gamma)!(d-e+\gamma)!}{(c-\gamma+n)(d-e+\gamma-n)!(-n)!(d+e-c-n)!(e-d+c+n)!(n)!}$$
$$= \frac{(-1)^c}{(d+e-c)!(e-d+c)!}$$

$$(\text{C.17})$$

Substituting the results of (C.16) and (C.17) into (C.14):

$$\sum_{a,\beta,\gamma,\delta,\phi} (-1)^{f+\phi} v(abe;a,\beta,-e)v(acf;-a,\gamma,\phi)v(bdf;-\beta,\delta,-\phi)v(cde;\gamma,\delta,-e)$$

$$= \sum_{a,\gamma,t,u} \frac{(-1)^{a-c-f+t+u}\left[\begin{array}{c}(a+a)!(b+e-a)!2e!(c+\gamma)! \\ (f+a-\gamma)!(f-a+\gamma)!(d+e-\gamma)!\end{array}\right]}{\left[\begin{array}{l}(e+a-b)!(e+b-a)! \\ t!(a+c-f-t)!(a+a-t)!(c+\gamma-t)!(f-c-a+t)!(f-a-\gamma+t)! \\ u!(b+d-f-u)!(b-a+e-u)!(d+e-\gamma-u)!(f-d+a-e+u)!(f-b-e+\gamma+u)! \\ (e+c-d)!(e+d-c)!\end{array}\right]}$$

$$(\text{C.18})$$

where we saw that $(-1)^{4f + 4e - 2c - 2\gamma} = 1$ and relabeled $z' \to t$; $z'' \to u$.

Let us extract a fragment from (C.18) as follows:

$$F = \frac{(f+\alpha-\gamma)!(f-\alpha+\gamma)!}{\left[\begin{array}{l}(a+\alpha-t)!(c+\gamma-t)!(f-c-\alpha+t)!(f-a-\gamma+t)! \\ (b-\alpha+e-u)!(d+e-\gamma-u)!(f-d+\alpha-e+u)!(f-b-e+\gamma+u)!\end{array}\right]}$$

$$(C.19)$$

and operate on a portion of it with (C.8) with the following identifications:

$$\begin{aligned} a' &\to (f+\alpha-\gamma) \\ b' &\to (f-a-\gamma+t) \\ c' &\to (f-d+\alpha-e+u) \end{aligned}$$

$$(C.20)$$

which gives

$$\frac{(f+\alpha-\gamma)!}{(f-a-\gamma+t)!(f-d+\alpha-e+u)!}$$
$$= \sum_{v} \frac{(a+\alpha-t)!(d+e-\gamma-u)!}{v!(a+\alpha-t-v)!(d+e-\gamma-u-v)!(-a+t+f-d-e+u+v)!}$$

$$(C.21)$$

Substituting back into (C.19),

$$F = \frac{(f+\alpha-\gamma)!}{\left[\begin{array}{l}(a+\alpha-t)!(c+\gamma-t)!(f-c-\alpha-t)! \\ (b-\alpha+e-u)!(d+e-\gamma-u)!(f-b-e+\gamma+u)!\end{array}\right]}$$
$$\times \sum_{v} \frac{(a+\alpha-t)!(d+e-\gamma-u)!}{v!(a+\alpha-t-v)!(d+e-\gamma-u-v)!(-a+t+f-d-e+u+v)!}$$
$$= \frac{(f+\alpha-\gamma)!}{\left[\begin{array}{l}(c+\gamma-t)!(f-c-\alpha+t)! \\ (b-\alpha+e-u)!(f-b-e+\gamma+u)!\end{array}\right]}$$
$$\times \sum_{v} \frac{1}{\left[\begin{array}{l}v!(a+\alpha-t-v)!(d+e-\gamma-u-v)! \\ (-a+t+f-d-e+u+v)!\end{array}\right]}$$

$$(C.22)$$

Similarly, we operate on another portion of F with (C.8) and the following identifications:

$$\begin{aligned} a' &\to (f-\alpha-\gamma) \\ b' &\to (f-c-\alpha+t) \\ c' &\to (f-b+\gamma-e+u) \end{aligned}$$

$$(C.23)$$

which gives

$$\frac{(f-\alpha+\gamma)!}{(f-c-\alpha+t)!(f-b+\gamma-e+u)!}$$
$$=\sum_{w}\frac{(b+e-u-\alpha)!(\gamma+c-t)!}{w!(b+e-u-\alpha-w)!(\gamma+c-t-w)!(-c+t+f-b-e+u+w)!}$$

(C.24)

Substituting back into (C.22),

$$F=\frac{1}{(c+\gamma-t)!(b-\alpha+e-u)!}$$
$$\sum_{v,w}\frac{1}{\left[\begin{array}{c}v!(a+\alpha-t-v)!(c-t-\gamma-v)!\\(-a+t+f-d-e+u+v)!\end{array}\right]}\frac{(b+e-u-\alpha)!(\gamma+c-t)!}{\left[\begin{array}{c}w!(b+e-u-\alpha-w)!(\gamma+c-t-w)!\\(-c+t+f-b-e+u+w)!\end{array}\right]}$$
$$=\sum_{v,w}\frac{1}{\left[\begin{array}{c}v!(a+\alpha-t-v)!(d+e-\gamma--u-v)!(-a+t+f-d-e+u+v)!\\w!(b+e-u-\alpha-w)!(\gamma+c-t-w)!(-c+t+f-b-e+u+w)!\end{array}\right]}$$

(C.25)

Let $w\rightarrow v''-t-w''$ and sum over w'':

$$F=\sum_{v,w''}\frac{1}{\left[\begin{array}{c}v!(a+\alpha-t-v)!(d+e-\gamma-u-v)!(-a+t+f-d-e+u+v)!(v''-t-w'')!\\(b+e-u-\alpha-v''+t+w'')!(\gamma+c-v''+w'')!(-c+f-b-e+u+v''+w'')!\end{array}\right]}$$

(C.26)

Let $v\rightarrow v''-c-\gamma$ and sum over v'':

$$F=\sum_{v'',w''}\frac{1}{\left[\begin{array}{c}(v''-c-\gamma)!(a+\alpha-t-v''+c+\gamma)!(d+e-u-v''+c)!\\(-a+t+f-d-e+u+v''-c-\gamma)!(v''-t-w'')!(b+e-u-\alpha-v''+t+w'')!\\(\gamma+c-v''+w'')!(-c+f-b-e+u+v''+w'')!\end{array}\right]}$$

(C.27)

Let $(v''-c)\rightarrow d+e-u-v'''+c+\gamma$ and sum over v''':

$$F=\sum_{v''',w''}\frac{1}{\left[\begin{array}{c}(c+d+e-u-v''')!(a-c-d-e+\alpha-t+u+v''')!(v'''-c-\gamma)!\\(f-a+c+t-v''')!(v'''-t-w'')!(b+e-u-\alpha-v'''+t+w'')!\\(\gamma+c-v'''+w'')!(-c+f-b-e+u+v'''+w'')!\end{array}\right]}$$

(C.28)

Substituting the results of (C.28) into (C.18) gives (after dropping the primes)

$$\sum_{\alpha,\beta,\gamma,\delta,\phi} (-1)^{f+\phi} v(abe;\alpha,\beta,-e)v(acf;-\alpha,\gamma,\phi)v(bdf;-\beta,\delta,-\phi)v(cde;\gamma,\delta,-e)$$

$$= \sum_{\alpha,\gamma,t,u,v,w} \frac{(-1)^{a-c-f+t+u}(a+\alpha)!(b+e-\alpha)!2e!(c+\gamma)!(d+e-\gamma)!}{\begin{bmatrix} (e+a-b)!(e+b-a)!(e+c-d)!(e+d-c)!t!(a+c-f-t)!u!(b+d-f-u)! \\ (f-a+c+t-v)!(c+d+e-u-v)!(a-c-d-e+\alpha-t+u+v)!(v-c-\gamma)! \\ (v-t-w)!(-c+f-b-e+u+v+w)!(b+e-u-\alpha-v+t+w)!(\gamma+c-v+w)! \end{bmatrix}}$$

$$(C.29)$$

Now apply (C.10) with the following identifications, in the following manner:

$$\sigma \rightarrow \alpha$$
$$a' \rightarrow a$$
$$b' \rightarrow b+e$$
$$c' \rightarrow a-c-d-e-t+u+v$$
$$d' \rightarrow b+e+t-u-v+w$$

$$(C.30)$$

so that,

$$\sum_{\alpha} \frac{(a+\alpha)!(b+e-\alpha)!}{(a-c-d-e+\alpha-t+u+v)!(b+e-u-\alpha-v+t+w)!}$$
$$= \frac{(a+b+e+1)!(c+d+e+t-u-v)!(-t+u+v-w)!}{(a-c-d+b+w)!(e+c+d-w+1)!}$$

$$(C.31)$$

Similarly, we let

$$\sigma \rightarrow \gamma$$
$$a' \rightarrow c$$
$$b' \rightarrow d+e$$
$$c' \rightarrow c-v+w$$
$$d' \rightarrow v-c$$

$$(C.32)$$

to get

$$\sum_{\gamma} \frac{(c+\gamma)!(d+e-\gamma)!}{(v-c-\gamma)!(c+\gamma-v+w)!} = \frac{(c+d+e+1)!(v-w)!(d+e-v+c)!}{w!(e+d-w+c+1)!}$$

$$(C.33)$$

Now we have

$$\sum_{\alpha,\beta,\gamma,\delta,\phi} (-1)^{f+\phi} v(abe;\alpha,\beta,-e) v(acf;-\alpha,\gamma,\phi) v(bdf;-\beta,\delta,-\phi) v(cde;\gamma,\delta,-e)$$

$$= \sum_{t,u,v,w} \frac{(-1)^{a-c-f+t+u} \left[\begin{array}{c} 2e!(a+b+e+1)!(c+d+e+t-u-v)!(-t+u+v-w)! \\ (c+d+e+1)!(v-w)!(d+e-v+c)! \end{array} \right]}{\left[\begin{array}{c} (e+a-b)!(e+b-a)!(e+c-d)!(e+d-c)!w!(a-c-d+b+w)! \\ (e+c+d-w+1)!^2 \\ (-c+f-b-e+u+v+w)!u!(v-t-w)!(a+c-f-t)! \\ (f-a+c+t-v)!t!(c+d+e-u-v)!(b+d-f-u)! \end{array} \right]}$$

(C.34)

Let us detach another fragment, this time from (C.34):

$$F' = \frac{(c+d+e+t-u-v)!(u+v-t-w)!}{u!(v-t-w)!(a+c-f-t)!t!(c+d+e-u-v)!(b+d-f-u)!}$$

(C.35)

and apply (C.8) with the following identifications:

$$a' \rightarrow c+d+e+t-u-v$$
$$b' \rightarrow c+d+e-u-v$$
$$c' \rightarrow b+d-f-u$$

(C.36)

to get

$$\frac{(c+d+e+t-u-v)!}{(c+d+e-u-v)!(b+d-f-u)!} =$$
$$\sum_z \frac{t!(c+e-b+f+t-v)!}{z!(t-z)!(c+e-b+f+t-v-z)!(-t+b+d-f-u-z)!}$$

(C.37)

Similarly, with,

$$a' \rightarrow u+v-t-w$$
$$b' \rightarrow v-t-w$$
$$c' \rightarrow a+c-f-t$$

(C.38)

we get

$$\frac{(u+v-t-w)!}{(v-t-w)!(a+c-f-t)!}$$
$$=\sum_x \frac{u!(u+v-w-a-c+f)!}{x!(u-x)!(u+v-w-a-c+f-x)!(-u+a+c-f-t-x)!} \tag{C.39}$$

Putting the results of (C.37) and (C.39) into (C.35), our fragment becomes

$$F' =\sum_{z,x}\frac{(c+e-b+f+t-v)!}{z!(t-z)!(c+e-b+f+t-v-z)!(-t+b+d-f-u-z)!}$$
$$\times\frac{(u+v-w-a-c+f)!}{x!(u-x)!(u+v-w-a-c+f-x)!(-u+a+c-f-t-x)!} \tag{C.40}$$

Let $t-z\to z''-x$ and $u-x\to x''$, and re-sum:

$$F' =\sum_{z'',x''}\frac{(c+e-b+f+t-v)!(u+v-w-a-c+f)!}{\begin{bmatrix}x''!(u-x'')!(z''-x'')!(t-z''+x'')!(c+e-b+f-v+z''-x'')!\\(b+d-f-u-z''+x'')!(a+c-f-t-x'')!(f-a-c+v-w+x'')!\end{bmatrix}} \tag{C.41}$$

Substituting (C.35) and (C.41) into (C.34) gives (after dropping the primes)

$$\sum_{\alpha,\beta,\gamma,\delta,\phi}(-1)^{f+\phi}v(abe;\alpha,\beta,-e)v(acf;-\alpha,\gamma,\phi)v(bdf;-\beta,\delta,-\phi)v(cde;\gamma,\delta,-e)$$

$$=\sum_{\substack{t,u,v\\w,x,z}}\frac{(-1)^{a-c-f+t+u}\begin{bmatrix}2e!(a+b+e+1)!(c+d+e+1)!(v-w)!(d+e-v+c)!\\(c+e-b+f+t-v)!(u+v-w-a-c+f)!\end{bmatrix}}{\begin{bmatrix}(e+a-b)!(e+b-a)!(e+c-d)!(e+d-c)!w!(a-c-d+b+w)!(e+c+d-w+1)!^2\\(-c+f-b-e+u+v+w)!(f-a+c+t-v)!x!(u-x)!(z-x)!(t-z+x)!\\(c+e-b+f-v+z-x)!(b+d-f-u-z+x)!(a+c-f-t-x)!(f-a-c+v-w+x)!\end{bmatrix}} \tag{C.42}$$

We let $t-z+x\to t''$, which means

$$\sum_t\frac{(-1)^{a-c-f-t}(c+e+f-b+t-v)!}{(f-a+c+t-v)!(t-z+x)!(a+c-f-t-x)!}$$
$$\to\sum_{t''}\frac{(-1)^{a-c-f+t''+z-x}(c+e+f-b+t''+z-x-v)!}{(f-a+c+t''+z-x-v)!t''!(a+c-f-t''-z)!} \tag{C.43}$$

Apply (C.10) to (C.43) with the following identifications,

$$s \to t''$$
$$r' \to a + c - f - z$$
$$b' \to 2c - x - v \qquad \text{(C.44)}$$
$$c' \to 2c + e - b + a - x - v$$

resulting in

$$\sum_{t''} \frac{(-1)^{a-c-f+t''+z-x}(c+e+f-b+t''+z-x-v)!}{(f-a+c+t''+z-x-v)!t''!(a+c-f-t''-z)!}$$
$$= \frac{(-1)^{a-c-f+z-x}(-1)^{a-c-f-z}(e-b+a)!(c+e-b-x-v+f+z)!}{(a+c-f-z)!(2c-x-v)!(e-b-c+f+z)!} \qquad \text{(C.45)}$$

Let $u - x \to u''$:

$$\sum_{u} \frac{(-1)^{u}(f-a-c+u+v-w)!}{(f-b-c-e+u+v-w)!(u-x)!(b+d-f-u-z+x)!}$$
$$\to (-1)^{x}\sum_{u''} \frac{(-1)^{u''}(f-a-c+u''+x+v-w)!}{(f-b-c-e+u''+x+v-w)!u''!(b+d-f-u''-z)!} \qquad \text{(C.46)}$$

We now apply (C.9) to (C.46) with the following identifications:

$$s \to u''$$
$$r' \to b + d - f - z$$
$$b' \to -c - e + x + v - w + d - z \qquad \text{(C.47)}$$
$$c' \to -a - c + x + v - w + b + d - z$$

which gives

$$(-1)^{x}\sum_{u''} \frac{(-1)^{u''}(f-a-c+u''+x+v-w)!}{(f-b-c-e+u''+x+v-w)!u''!(b+d-f-u''-z)!}$$
$$= \frac{(-1)^{x+b+d-f-z}(-a+b+e)!(f-a-c+x+v-w)!}{(b+d-f-z)!(-c-e+x+v-w+d-z)!(-a+e-d+f+z)!} \qquad \text{(C.48)}$$

Substituting the results of (C.45) and (C.48) into (C.42) gives (after dropping the primes)

$$\sum_{\alpha,\beta,\gamma,\delta,\phi} (-1)^{f+\phi} v(abe;\alpha,\beta,-e)v(acf;-\alpha,\gamma,\phi)v(bdf;-\beta,\delta,-\phi)v(cde;\gamma,\delta,-e)$$

$$= \sum_{v,w,x,z} \frac{(-1)^{2e+f+d-b+z}2e!(a+b+e+1)!(c+d+e+1)!(v-w)!(d+e-v+c)!}{\begin{bmatrix}(e+c-d)!(e+d-c)!w!(a-c-d+b+w)!(e+c+d-w+1)!^2x!(z-x)! \\ (a+c-f-z)!(2c-x-v)!(e-b-c+f+z)! \\ (b+d-f-z)!(-c-e+x+v-w+d-z)!(-a+e-d+f+z)!\end{bmatrix}}$$

$$(C.49)$$

where we also saw that $(-1)^{2a+4f-2b+2z} = 1$. Now we modify (C.10) as follows:

$$\begin{aligned}
\sigma &\to v \\
a' &\to -w \\
b' &\to c+d+e \\
c' &\to d-c-e-w+x-z \\
d' &\to 2c-x
\end{aligned} \qquad (C.50)$$

to get

$$\sum_v \frac{(c+d+e-v)!(v-w)!}{(2c-v-x)!(d-c-e+v-w+x-z)!}$$
$$= \frac{(-w+c+d+e+1)!(-d+c+e-x+z)!(-c+d+e+x)!}{(d+c-e-w-z)!(2e+z+1)!} \qquad (C.51)$$

Substituting this result into (C.49) gives

$$\sum_{\alpha,\beta,\gamma,\delta,\phi} (-1)^{f+\phi} v(abe;\alpha,\beta,-e)v(acf;-\alpha,\gamma,\phi)v(bdf;-\beta,\delta,-\phi)v(cde;\gamma,\delta,-e)$$

$$= \sum_{w,x,z} \frac{(-1)^{2e+f+d-b+z}2e!(a+b+e+1)!(c+d+e+1)!(-d+c+e-x+z)!(-c+d+e+x)!}{\begin{bmatrix}(e+c-d)!(e+d-c)!w!(a-c-d+b+w)!(e+c+d-w+1)!x!(z-x)! \\ (a+c-f-z)!(e-b-c+f+z)!(b+d-f-z)!(-a+e-d+f+z)! \\ (d+c-e-w-z)!(2e+z+1)!\end{bmatrix}}$$

$$(C.52)$$

Let us now extract the following fragment from (C.52) and manipulate it as follows:

$$\sum_w \frac{(c+d+e+1)!}{[w!(a-c-d+b+w)!(d+c-e-w-z)!(e+c+d-w+1)!]}$$

$$=\sum_w \frac{(c+d+e+1)!}{[w!(a-c-d+b+w)!(d+c-e-w-z)!(e+c+d-w+1)!]}\frac{(2e+1)(c+d-e-z)!}{(2e+1)(c+d-e-z)!}$$

$$(C.53)$$

We operate on the expression in (C.53) with (C.8), using the following identifications:

$$s \rightarrow w$$
$$a' \rightarrow a+b+c+d+1-z$$
$$b' \rightarrow a+b-e-z \qquad\qquad (C.54)$$
$$c' \rightarrow a+b+e+1$$

to get

$$\sum_w \frac{(c+d+e+1)!}{[w!(a-c-d+b+w)!(d+c-e-w-z)!(e+c+d-w+1)!]}$$

$$=\sum_w \frac{(c+d+e+1)!(c+d-e-z)!}{[w!(a-c-d+b+w)!(d+c-e-w-z)!(e+c+d-w+1)!]}\frac{(2e+1)}{(2e+1)(c+d-e-z)!}$$

$$=\frac{(2e+1)}{(2e+1)(c+d-e-z)!}\frac{(a+b+c+d+1-z)!}{(a+b-e-z)!(a+b+e+1)!}$$

$$(C.55)$$

Substituting this result into (C.52) gives

$$\sum_{\alpha,\beta,\gamma,\delta,\phi} (-1)^{f+\phi}v(abe;\alpha,\beta,-e)v(acf;-\alpha,\gamma,\phi)v(bdf;-\beta,\delta,-\phi)v(cde;\gamma,\delta,-e)$$

$$=\frac{(2e+1)!}{(2e+1)}\sum_{x,z} \frac{(-1)^{2e+f+d-b+z}(-d+c+e-x+z)!(-c+d+e+x)!(a+b+c+d+1-z)!}{\left[\begin{array}{l}(e+c-d)!(e+d-c)!x!(z-x)!\\ (a+c-f-z)!(e-b-c+f+z)!(b+d-f-z)!(-a+e-d+f+z)!\\ (2e+z+1)!(c+d-e-z)!(a+b-e-z)!\end{array}\right]}$$

$$(C.56)$$

Operating on the relevant portion of (C.56) using (C.10) and the following identifications,

$$\sigma \to x$$
$$a' \to d + e - c$$
$$b' \to c + e - d + z \tag{C.57}$$
$$c' \to 0$$
$$d' \to z$$

gives us

$$\sum_x \frac{(d + e - c + x)!(c + e - d + z - x!)}{x!(z - x)!}$$
$$= \frac{(2e + 1 + z)!(d + e - c)!(c + e - d)!}{z!(2e + 1)!} \tag{C.58}$$

which, when substituted into (C.56), leaves

$$\sum_{\alpha,\beta,\gamma,\delta,\phi} (-1)^{f+\phi} v(abe; \alpha, \beta, -e) v(acf; -\alpha, \gamma, \phi) v(bdf; -\beta, \delta, -\phi) v(cde; \gamma, \delta, -e)$$

$$= \frac{(-1)^{2e+f+d-b}}{(2e+1)} \sum_z \frac{(-1)^z (a+b+c+d+1-z)!}{\left[\begin{array}{c} (a+c-f-z)!(e+f-b-c+z)!(b+d-f-z)!(e+f-a-d+z)! \\ z!(c+d-e-z)!(a+b-e-z)! \end{array} \right]} \tag{C.59}$$

At this point, we define the following quantity, again, after Racah [2]:

$$V(abc; \alpha\beta\gamma) \equiv \Delta(abc) v(abc; \alpha\beta\gamma) \tag{C.60}$$

where

$$\Delta(abc) \equiv \left[\frac{(a+b-c)!(a-b+c)!(-a+b+c)!}{(a+b+c+1)!} \right] \tag{C.61}$$

In the context of angular momentum theory, the $\Delta(abc)$ vanish unless the triangle condition of equations (2.122) are met for a, b, c. Note also that (B.133), (2.142), (C.11), (C.60), and (C.61) imply the following relationships:

$$C^{j,m}_{j_1,m_1;j_2,m_2} = (-1)^{j+m}(2j+1)^{\frac{1}{2}} V(j_1 j_2 j; m_1, m_2, -m) \tag{C.62a}$$

$$V(j_1 j_2 j; m_1, m_2, m) = (-1)^{j_1 - j_2 - j} \begin{pmatrix} j_1 & j_2 & j \\ m_1 & m_2 & m \end{pmatrix} \qquad \text{(C.62b)}$$

Using (C.59) and (C.60), we can build the following sum:

$$\sum_{\alpha, \beta, \gamma, \delta, \phi} (-1)^{f + \phi} V(abe; \alpha, \beta, -e) V(acf; -\alpha, \gamma, \phi) V(bdf; -\beta, \delta, -\phi) V(cde; \gamma, \delta, -e)$$

$$= \Delta(abe) \Delta(acf) \Delta(bdf) \Delta(cde)$$

$$\times \frac{(-1)^{2e + f + d - b}}{(2e + 1)} \sum_z \frac{(-1)^z (a + b + c + d + 1 - z)!}{\left[\begin{array}{l} (a + c - f - z)!(e + f - b - c + z)!(b + d - f - z)!(e + f - a - d + z)! \\ z!(c + d - e - z)!(a + b - e - z)! \end{array} \right]}$$

$$= \frac{(-1)^{2e + f + d - b}}{(2e + 1)} W(abcd; ef)$$

$$\qquad \text{(C.63)}$$

where

$$W(abcd; ef) \equiv \Delta(abe) \Delta(acf) \Delta(bdf) \Delta(cde)$$

$$\times \sum_z \frac{(-1)^z (a + b + c + d + 1 - z)!}{\left[\begin{array}{l} (a + c - f - z)!(e + f - b - c + z)!(b + d - f - z)!(e + f - a - d + z)! \\ z!(c + d - e - z)!(a + b - e - z)! \end{array} \right]}$$

$$\qquad \text{(C.64)}$$

are called the Racah coefficients.

Some make the transformation $z \rightarrow a + b + c + d - z'$ to rewrite (C.64) in the following equivalent form[6]:

$$W(abcd; ef) \equiv \Delta(abe) \Delta(acf) \Delta(bdf) \Delta(cde)$$

$$\times \sum_{z'} \frac{(-1)^{z' + a + b + c + d} (z' + 1)!}{\left[\begin{array}{l} (z' - a - b - e)!(z' - c - d - e)!(z' - a - c - f)!(z' - b - d - f)! \\ (a + b + c + d - z')!(a + d + e + f - z')!(b + c + e + f - z')! \end{array} \right]}$$

$$\qquad \text{(C.65)}$$

The summations in (C.64) and (C.65) have $\sigma + 1$ terms, where σ is the smallest of the numbers:

---

[6] A slightly shorter version of the derivation of equation (C.65) can be found in [7], Chap. 3, App. A.

$$a+b-e; \; a+c-f; \; b+d-f; \; c+d-e$$
$$e+a-b; \; f+a-c; \; f+b-d; \; e+c-d$$
$$b+e-a; \; c+f-a; \; d+f-b; \; d+e-c$$

Equations (C.62) and (C.63) imply

$$\frac{(-1)^{2e+f+d-b}}{(2e+1)}W(abcd;ef)$$

$$= \sum_{\alpha,\beta,\delta,\gamma,\phi} (-1)^{f+\phi} \left\{ \begin{array}{l} \left[C^{e,*}_{a,\alpha;b,\beta}(2e+1)^{-\frac{1}{2}}\right]\left[C^{f,-\phi}_{a,-\alpha;c,\gamma}(-1)^{-f+\phi}(2f+1)^{-\frac{1}{2}}\right] \\ \left[C^{f,\phi}_{b,-\beta;d,\delta}(-1)^{-f-\phi}(2f+1)^{-\frac{1}{2}}\right]\left[C^{e,*}_{c,\gamma;d,\delta}(2e+1)^{-\frac{1}{2}}\right] \end{array} \right\}$$

$$= \sum_{\alpha,\beta,\delta,\gamma,\phi} (-1)^{f+\phi}\left[C^{e,*}_{a,\alpha;b,\beta}(2e+1)^{-\frac{1}{2}}\right]\left[\frac{(2f+1)^{\frac{1}{2}}}{(2c+1)^{\frac{1}{2}}}C^{c,\gamma}_{a,\alpha;f,\phi}(-1)^{-f+\phi+a-\alpha}(2f+1)^{-\frac{1}{2}}\right]$$

$$\times \left[C^{f,\phi}_{b,\beta;d,\delta}(-1)^{-f-\phi}(2f+1)^{-\frac{1}{2}}\right]\left[\frac{(2c+1)^{\frac{1}{2}}}{(2e+1)^{\frac{1}{2}}}(-1)^{d-\delta}C^{c,\gamma}_{e,*;d,\delta}(2e+1)^{-\frac{1}{2}}\right]$$

$$\Rightarrow \frac{(-1)^{2e+f+d-b}}{(2e+1)}W(abcd;ef) = \frac{(-1)^{a-\alpha-f+d-\delta+\phi}}{(2e+1)^{\frac{3}{2}}(2f+1)^{\frac{1}{2}}}\sum_{\alpha,\beta,\delta,\gamma,\phi}C^{e,*}_{a,\alpha;b,\beta}C^{c,\gamma}_{a,\alpha;f,\phi}C^{f,\phi}_{b,\beta;d,\delta}C^{c,\gamma}_{e,*;d,\delta}$$

$$\Rightarrow W(abcd;ef) = [(2e+1)(2f+1)]^{-\frac{1}{2}}\sum_{\alpha,\beta,\delta,\gamma,\phi}C^{e,*}_{a,\alpha;b,\beta}C^{c,\gamma}_{a,\alpha;f,\phi}C^{f,\phi}_{b,\beta;d,\delta}C^{c,\gamma}_{e,*;d,\delta}$$

$$(C.66)$$

where we used the appropriate permutations of (2.142) and noted that $(-1)^{2e+2f-(a+b-\delta+\phi-\alpha)} = 1$. We also see that the signs of α, β, δ, γ, ϕ in any of the C-G coefficients in (C.66) are irrelevant, due to the summations over those quantities. We note finally that the total magnetic quantum numbers in some of the C-G coefficients (indicated by a $*$) are irrelevant, since they are fixed by the triangle rule.

Comparing (C.66) and (2.170) leads to the following result:

$$W(j_1 j_2 j j_3; j_{12} j_{23}) = [(2j_{12}+1)(2j_{23}+1)]^{-\frac{1}{2}}\langle(j_1,j_{23})j|(j_{12},j_3)j\rangle \qquad (C.67)$$

As a final remark, note that the vanishing of the Racah coefficients is given by the triangle conditions seen in equation (C.65), something which can also be deduced by a close examination of (C.66) or (C.67).

We now make the following definition of the 6-j symbols, which have a higher symmetry than that of the Racah coefficients:

$$\left\{ \begin{matrix} j_1\, j_2\, j_{12} \\ j_3\, j\, j_{23} \end{matrix} \right\} \equiv (-1)^{j_1+j_2+j_3+j}\, W(j_1 j_2 j j_3; j_{12} j_{23}) \tag{C.68}$$

It is seen immediately from (C.65) and (C.68) that the general expression for the 6-j symbols is given by

$$\left\{ \begin{matrix} a\, b\, e \\ d\, c\, f \end{matrix} \right\} \equiv \Delta(abe)\Delta(acf)\Delta(bdf)\Delta(cde)$$

$$\times \sum_{z'} \frac{(-1)^{z'}(z'+1)!}{\left[\begin{array}{c} (z'-a-b-e)!(z'-c-d-e)!(z'-a-c-f)!(z'-b-d-f)! \\ (a+b+c+d-z')!(a+d+e+f-z')!(b+c+e+f-z')! \end{array} \right]} \tag{C.69}$$

where we have noted that $(-1)^{2(a+b+c+d)} = 1$. Obviously the 6-j symbols are subject to the same triangle conditions as the Racah coefficients, and the summation in (C.69) has the same number of terms in its expansion as does the summation in (C.65) [see remarks below (C.65)]. In addition, it can be seen directly from (C.69) that the 6-j symbols are invariant under an interchange of any two columns, or under the interchange of any two numbers in the bottom row with the numbers immediately above them [3, 7, 10, 89].

The 6-j symbols can be written as a contraction of four 3-j symbols, which is derived and shown here in standard notation as follows:

$$\left\{ \begin{matrix} j_1\, j_2\, j_3 \\ j_4\, j_5\, j_6 \end{matrix} \right\} = (-1)^{j_1+j_2+j_4+j_5}\, W(j_1 j_2 j_5 j_4; j_3 j_6)$$

$$= (-1)^{j_1+j_2+j_4+j_5} \frac{(2j_3+1)}{(-1)^{2j_3+j_6+j_1-j_2}}$$

$$\times \sum_{\substack{m_1,m_2,m_4 \\ m_5,m_6}} (-1)^{j_6+m_6} \left[\begin{array}{c} V(j_1,j_2,j_3;m_1,m_2,-m_3)V(j_1,j_5,j_6;-m_1,m_5,m_6) \\ V(j_2,j_4,j_6;-m_2,m_4,-m_6)V(j_5,j_4,j_3;m_5,m_4,-m_3) \end{array} \right]$$

$$= (-1)^{j_1+j_2+j_4+j_5} \frac{(2j_3+1)}{(-1)^{2j_3+j_6+j_1-j_2}}$$

$$\times \sum_{\substack{m_1,m_2,m_4 \\ m_5,m_6}} (-1)^{j_6+m_6} \left\{ \begin{array}{c} \left[(-1)^{j_1+j_2-j_5}\begin{pmatrix} j_1 & j_2 & j_3 \\ m_1 & m_2 & -m_3 \end{pmatrix} \right]\left[(-1)^{j_1+j_5-j_6}\begin{pmatrix} j_1 & j_5 & j_6 \\ -m_1 & m_5 & m_6 \end{pmatrix} \right] \\ \left[(-1)^{j_2+j_4-j_6}\begin{pmatrix} j_2 & j_4 & j_6 \\ -m_2 & m_4 & -m_6 \end{pmatrix} \right]\left[(-1)^{j_5+j_4-j_3}\begin{pmatrix} j_5 & j_4 & j_3 \\ m_5 & m_4 & m_3 \end{pmatrix} \right] \end{array} \right\}$$

$$
\begin{aligned}
&= \left(2j_3+1\right) \sum_{\substack{m_1,m_2,m_4 \\ m_5,m_6}} (-1)^{[2j_1+4j_2-4j_3+3j_4} {}^{+3j_5-2j_6+m_6]} \left\{ \underbrace{\left[\begin{pmatrix} j_1 & j_2 & j_3 \\ m_1 & m_2 & -m_3 \end{pmatrix}\right] \left[\begin{pmatrix} j_1 & j_5 & j_6 \\ -m_1 & m_5 & m_6 \end{pmatrix}\right]}_{(-1)^{j_1+j_5+j_6}\begin{pmatrix} j_5 & j_1 & j_6 \\ m_5 & -m_1 & m_6 \end{pmatrix}} \right. \\
&\qquad\qquad \times \left. \left[\begin{pmatrix} j_2 & j_4 & j_6 \\ -m_2 & m_4 & -m_6 \end{pmatrix}\right] \underbrace{\left[\begin{pmatrix} j_5 & j_4 & j_3 \\ m_5 & m_4 & m_3 \end{pmatrix}\right]}_{(-1)^{j_5+j_4+j_3}\begin{pmatrix} j_4 & j_5 & j_3 \\ m_4 & m_5 & m_3 \end{pmatrix}} \right\}
\end{aligned}
$$

$$
= (2j_3+1)\underbrace{(-1)^{[2j_1+3j_2-2j_3+3j_4+4j_5-2j_6]}}_{=1}
$$

$$
\times \sum_{\substack{m_1,m_2,m_4 \\ m_5,m_6}} (-1)^{[j_1+j_2-j_3+j_4 \,+j_5+j_6+m_6]} \begin{pmatrix} j_1 & j_2 & j_3 \\ m_1 & m_2 & -m_3 \end{pmatrix} \begin{pmatrix} j_5 & j_1 & j_6 \\ m_5 & -m_1 & m_6 \end{pmatrix} \begin{pmatrix} j_2 & j_4 & j_6 \\ -m_2 & m_4 & -m_6 \end{pmatrix} \begin{pmatrix} j_4 & j_5 & j_3 \\ m_4 & m_5 & m_3 \end{pmatrix}
$$

$$
= \left(2j_3+1\right) \sum_{\substack{m_1,m_2,m_4 \\ m_5,m_6}} (-1)^{[j_1+j_2-j_3+j_4 \,+j_5+j_6+m_6]} \begin{pmatrix} j_1 & j_2 & j_3 \\ m_1 & m_2 & -m_3 \end{pmatrix} \begin{pmatrix} j_5 & j_1 & j_6 \\ m_5 & m_1 & m_6 \end{pmatrix} \begin{pmatrix} j_2 & j_4 & j_6 \\ m_2 & m_4 & m_6 \end{pmatrix} \begin{pmatrix} j_4 & j_5 & j_3 \\ m_4 & m_5 & m_3 \end{pmatrix}
$$

$$(C.70)$$

where we have used (C.62b), (C.63), and (C.68), as well as the comment preceding (2.143). Also, for the last line in (C.70), we changed the sign on some of the magnetic quantum numbers because, again, as long as those numbers are being summed over, their sign is irrelevant. Finally, in (C.70), if the sum over any of the magnetic quantum numbers ζ is added/dropped, the expression must be divided/ (multiplied) by the term $(2j_\zeta + 1)$. So, for example, if we added the sum over m_3 to the RHS of (C.70), we would also have to divide the RHS by $(2j_3 + 1)$ to get

$$
\begin{aligned}
\begin{Bmatrix} j_1 & j_2 & j_3 \\ j_4 & j_5 & j_6 \end{Bmatrix} &= \sum_{all\,m} (-1)^{[j_1+j_2-j_3+j_4 \,+j_5+j_6+m_6]} \begin{pmatrix} j_1 & j_2 & j_3 \\ m_1 & m_2 & -m_3 \end{pmatrix} \\
&\qquad \times \begin{pmatrix} j_5 & j_1 & j_6 \\ m_5 & m_1 & m_6 \end{pmatrix} \begin{pmatrix} j_2 & j_4 & j_6 \\ m_2 & m_4 & m_6 \end{pmatrix} \\
&\qquad \times \begin{pmatrix} j_4 & j_5 & j_3 \\ m_4 & m_5 & m_3 \end{pmatrix}
\end{aligned}
$$

$$(C.71)$$

Many orthogonality properties and symmetry relations for the 6-j symbols can be deduced from (C.71) and the properties of the 3-j symbols. As stated earlier, the 6-j symbols are invariant with the interchange of any two columns or to the exchange of the upper and lower entries in any two columns. Consequently, we must also have [3]

$$(-1)^{j_1+j_2+j_3} = (-1)^{-j_1-j_2-j_3} \text{ and } (-1)^{2(j_1+j_2+j_3)} = 1 \qquad (C.72)$$

Many sum rules and tabulations for specific 6-j symbols can also be deduced from (C.71), or directly from (C.69). See, for example, [3, 7, 10, 89].

Some useful algebraic formulas involving 6-j symbols are offered below [3, 7, 16, 89]:

$$
\begin{Bmatrix} j_1 & j_2 & j_3 \\ j_4 & j_5 & j_6 \end{Bmatrix}
\begin{pmatrix} j_1 & j_6 & j_5 \\ m_1 & m_6 & m_5 \end{pmatrix}
= \sum_{m_2, m_3, m_4} (-1)^{j_1+j_2-j_3+j_4+j_5+j_6-m_1-m_4}
$$
$$
\times
\begin{pmatrix} j_1 & j_2 & j_3 \\ m_1 & m_2 & -m_3 \end{pmatrix}
\begin{pmatrix} j_4 & j_5 & j_3 \\ m_4 & m_5 & m_3 \end{pmatrix}
\begin{pmatrix} j_2 & j_4 & j_6 \\ m_2 & m_4 & -m_6 \end{pmatrix}
\qquad (C.73)
$$

Note that a version of this identity is derived in problem 7.1. Continuing,

$$
\sum_{j_6} (2j_6 + 1)(-1)^{j_1+j_2-j_3+j_4+j_5+j_6-m_1-m_4}
\begin{Bmatrix} j_1 & j_2 & j_3 \\ j_4 & j_5 & j_6 \end{Bmatrix}
\begin{pmatrix} j_1 & j_6 & j_5 \\ m_1 & m_6 & m_5 \end{pmatrix}
\begin{pmatrix} j_2 & j_4 & j_6 \\ m_2 & m_4 & -m_6 \end{pmatrix}
$$
$$
=
\begin{pmatrix} j_1 & j_2 & j_3 \\ m_1 & m_2 & -m_3 \end{pmatrix}
\begin{pmatrix} j_4 & j_5 & j_3 \\ m_4 & m_5 & m_3 \end{pmatrix}
\qquad (C.74)
$$

$$
\begin{Bmatrix} j_1 & j_2 & 0 \\ j_4 & j_5 & j_6 \end{Bmatrix}
= (-1)^{j_1+j_4+j_6}[(2j_1 + 1)(2j_4 + 1)]^{-1/2}\delta_{j_1 j_2}\delta_{j_4 j_5}
\qquad (C.75)
$$

$$
\begin{Bmatrix} j_1 & j_2 & j_3 \\ 1 & j_3 & j_2 \end{Bmatrix}
= (-1)^{j_1+j_2+j_3}2\frac{j_1(j_1 + 1) - j_2(j_2 + 1) - j_3(j_3 + 1)}{\sqrt{2j_2(2j_2 + 1)(2j_2 + 2)2j_3(2j_3 + 1)(2j_3 + 2)}}
$$
$$
\qquad (C.76)
$$

$$
\begin{Bmatrix} j_1 & j_2 & j_3 \\ j_4 & j_5 & j_6 \end{Bmatrix} \begin{Bmatrix} j_1 & j_2 & j_3 \\ j_7 & j_8 & j_9 \end{Bmatrix} = \sum_j (-1)^{j_1+j_2+j_3+j_4+j_5+j_6+j_7+j_8+j_9+j}(2j+1)
$$

$$
\times \begin{Bmatrix} j_2 & j_4 & j_6 \\ j & j_9 & j_7 \end{Bmatrix} \begin{Bmatrix} j_1 & j_5 & j_6 \\ j & j_9 & j_8 \end{Bmatrix} \begin{Bmatrix} j_3 & j_4 & j_5 \\ j & j_8 & j_7 \end{Bmatrix}
$$

$$\tag{C.77}$$

Equation (C.77) is known as the Biedenharn-Elliot sum rule.

And finally, the orthonormality relation[7]

$$
\sum_{j_6} (2j_3+1)(2j_6+1) \begin{Bmatrix} j_1 & j_2 & j_3 \\ j_4 & j_5 & j_6 \end{Bmatrix} \begin{Bmatrix} j_1 & j_2 & j_3 \\ j_4 & j_5 & j_6' \end{Bmatrix} = \delta_{j_6 j_6'} \tag{C.78}
$$

Appendix D: The 9-*j* Symbols

The 9-*j* symbols are used when we are coupling together four angular momenta. In this appendix, we derive some key results involving these important objects.

Let us start by looking at the reduced matrix element of (2.181) and write some different (but equivalent) expressions:

$$
\langle \alpha j \| T^{(k)} \| \alpha' j' \rangle \equiv (-1)^{k-j'+j}(2j+1)^{-\frac{1}{2}} \sum_{\mu,q',\mu'} \langle \alpha j \mu | T_{q'}^{(k)} | \alpha' j' \mu' \rangle C_{k,q';j',\mu'}^{j,\mu}
$$

$$
= (-1)^{k-j'+j}(2j+1)^{-\frac{1}{2}} \sum_{\mu,q',\mu'} \langle \alpha j \mu | T_{q'}^{(k)} | \alpha' j' \mu' \rangle [(-1)^{j'+\mu'} \frac{(2j+1)^{\frac{1}{2}}}{(2k+1)^{\frac{1}{2}}} C_{j,-\mu;j',\mu'}^{k,-q'}] \tag{D.1}
$$

$$
= (2k+1)^{-\frac{1}{2}} \sum_{\mu,q',\mu'} (-1)^{k+j+\mu'} C_{j,-\mu;j',\mu'}^{k,-q'} \langle \alpha j \mu | T_{q'}^{(k)} | \alpha' j' \mu' \rangle
$$

To continue, note that (2.125) implies

$$
\delta_{k,k'} \delta_{q,q'} \langle \alpha j \| T^{(k)} \| \alpha' j' \rangle = (2k+1)^{-\frac{1}{2}} \sum_{\mu,q',\mu'} (-1)^{k+j+\mu'} C_{j,-\mu;j',\mu'}^{k,-q'} \langle \alpha j \mu | T_{q'}^{(k)} | \alpha' j' \mu' \rangle
$$

$$\tag{D.2}$$

We now perform some further work on (D.2):

---

[7]For an exhaustive collection of formulas involving the 6-*j* symbols, see [10].

$$\delta_{k,k'}\delta_{q,q'}\langle\alpha j\|T^{(k)}\|\alpha'j'\rangle$$

$$= (2k+1)^{-\frac{1}{2}} \sum_{\mu,q',\mu'} (-1)^{k+j+\mu'}(-1)^{j-j'-q'}(2k+1)^{\frac{1}{2}}\begin{pmatrix} j & j' & k \\ -\mu & \mu' & q' \end{pmatrix}\langle\alpha j\mu|T_{q'}^{(k)}|\alpha'j'\mu'\rangle$$

$$= \sum_{\mu,q',\mu'} (-1)^{k+2j-j'+\mu'-q'}(-1)^{j+j'+k}\begin{pmatrix} j & k & j' \\ -\mu & q' & \mu' \end{pmatrix}\langle\alpha j\mu|T_{q'}^{(k)}|\alpha'j'\mu'\rangle$$

$$= \sum_{\mu,q',\mu'} (-1)^{j-\mu}\begin{pmatrix} j & k & j' \\ -\mu & q' & \mu' \end{pmatrix}\langle\alpha j\mu|T_{q'}^{(k)}|\alpha'j'\mu'\rangle$$

$$\text{(D.3)}$$

where we have used (2.122), (2.144), and (2.145).

Because of the orthogonality of the C-G coefficients (or similarly, the 3-j symbols), the summation(s) over μ and μ' yield a value independent of q'. Thus, the summation over q' has the effect of yielding the term $(2k + 1)$ [3, 16] [see also comment after (C.70)]. That is,

$$\delta_{k,k'}\delta_{q,q'}\langle\alpha j\|T^{(k)}\|\alpha'j'\rangle = (2k+1)\sum_{\mu,\mu'}(-1)^{j-\mu}\begin{pmatrix} j & k & j' \\ -\mu & q & \mu' \end{pmatrix}\langle\alpha j\mu|T_q^{(k)}|\alpha'j'\mu'\rangle$$

$$\Rightarrow (2k+1)^{-1}\sum_{q}\langle\alpha j\|T^{(k)}\|\alpha'j'\rangle = \langle\alpha j\|T^{(k)}\|\alpha'j'\rangle$$

$$\text{(D.4)}$$

Or equivalently

$$(2j'+1)^{-1}\sum_{m'}\langle\alpha j\|T^{(k)}\|\alpha'j'\rangle = \langle\alpha j\|T^{(k)}\|\alpha'j'\rangle \qquad \text{(D.5)}$$

If we recall (2.119), it is not hard to imagine that we could construct a relation between two spherical tensor operators, $T_{q_1}^{(k_1)}$ and $T_{q_2}^{(k_2)}$, which together contract to form the spherical tensor $T_q^{(k)}$ as follows:

$$T_q^{(k)} = \sum_{q_1} C_{k_1,q_1;k_2,q_2}^{k,q} T_{q_1}^{(k_1)} T_{q_2}^{(k_2)} \qquad \text{(D.6)}$$

As a reminder, to show that $T_q^{(k)}$ is, in fact, a spherical tensor, it suffices to note that $T_{q_1}^{(k_1)}$ and $T_{q_2}^{(k_2)}$ operate on different spaces (and hence commute), and to show that the RHS of (D.6) transforms according to (2.174) [3, 6]. This was done in Chap. 2.

We form the matrix elements of $T_q^{(k)}$ as follows. First, rewrite (2.119)

$$|j,m\rangle = \sum_{m_1,m_2}(-1)^{j_1-j_2+m}\sqrt{2j+1}\begin{pmatrix} j_1 & j_2 & j \\ m_1 & m_2 & -m \end{pmatrix}|j_1,m_1\rangle|j_2,m_2\rangle \quad (D.7)$$

and also rewrite (D.6)

$$T_q^{(k)} = \sum_{q_1,q_2}(-1)^{k_1-k_2+q}\sqrt{2k+1}\begin{pmatrix} k_1 & k_2 & k \\ q_1 & q_2 & -q \end{pmatrix}T_{q_1}^{(k_1)}T_{q_2}^{(k_2)} \quad (D.8)$$

Apply the W-E theorem to the general matrix element:

$$\langle \alpha jm|T_q^{(k)}|\alpha' j'm'\rangle = (-1)^{j-m}\begin{pmatrix} j & k & j' \\ -m & q & m' \end{pmatrix}\langle \alpha,j\|T^{(k)}\|\alpha',j'\rangle \quad (D.9)$$

We could also apply (D.7) and (D.8) to the same matrix element:

$$\langle \alpha jm|T_q^{(k)}|\alpha' j'm'\rangle = \sum_{\alpha''}\sum_{m_1,m_2 m_1',m_2'}\sum_{q_1,q_2}(-1)^{k_1-k_2+q+j_1-j_2+m+j_1'-j_2'+m'}$$

$$\times \sqrt{(2j+1)(2j'+1)(2k+1)}\begin{pmatrix} j_1 & j_2 & j \\ m_1 & m_2 & -m \end{pmatrix}\begin{pmatrix} j_1' & j_2' & j' \\ m_1' & m_2' & -m' \end{pmatrix}\begin{pmatrix} k_1 & k_2 & k \\ q_1 & q_2 & -q \end{pmatrix}$$

$$\times \langle \alpha j_1 m_1|T_{q_1}^{(k_1)}|\alpha'' j_1'' m_1''\rangle\langle \alpha'' j_2 m_2|T_{q_2}^{(k_2)}|\alpha' j_2' m_2'\rangle$$

$$(D.10)$$

Applying the W-E theorem to the last two terms on the RHS of (D.10),

$$\langle \alpha jm|T_q^{(k)}|\alpha' j'm'\rangle = \sum_{\alpha''}\sum_{m_1,m_2 m_1',m_2'}\sum_{q_1,q_2}(-1)^{k_1-k_2+q+j_1-j_2+m+j_1'-j_2'+m'}$$

$$\times \sqrt{(2j+1)(2j'+1)(2k+1)}\begin{pmatrix} j_1 & j_2 & j \\ m_1 & m_2 & -m \end{pmatrix}\begin{pmatrix} j_1' & j_2' & j' \\ m_1' & m_2' & -m' \end{pmatrix}\begin{pmatrix} k_1 & k_2 & k \\ q_1 & q_2 & -q \end{pmatrix}$$

$$\times (-1)^{j_1-m_1+j_2-m_2}\begin{pmatrix} j_1 & k_1 & j_1' \\ -m_1 & q_1 & m_1' \end{pmatrix}\begin{pmatrix} j_2 & k_2 & j_2' \\ -m_2 & q_2 & m_2' \end{pmatrix}$$

$$\times \langle \alpha j_1\|T^{(k_1)}\|\alpha'' j_1'\rangle\langle \alpha'' j_2\|T^{(k_2)}\|\alpha' j_2'\rangle$$

$$(D.11)$$

Equations (D.9) and (D.11) are two expressions for the same matrix element. We now equate them, multiply both sides by $(-1)^{j-m}\begin{pmatrix} j & k & j' \\ -m & q & m' \end{pmatrix}$, and sum over m, m', q

$$\sum_{m'}\langle\alpha j\|T^{(k)}\|\alpha'j'\rangle(2j'+1)^{-1} = \sum_{\alpha''}\sum_{m_1,m_2}\sum_{m_1',m_2'}\sum_{q_1,q_2m,m',q}(-1)^{k_1-k_2+q+2j_1+j+j_1'-j_2'+m'-m_1-m_2}$$

$$\times\sqrt{(2j+1)(2j'+1)(2k+1)}\begin{pmatrix}j_1 & j_2 & j\\ m_1 & m_2 & -m\end{pmatrix}\begin{pmatrix}j_1' & j_2' & j'\\ m_1' & m_2' & -m'\end{pmatrix}\begin{pmatrix}k_1 & k_2 & k\\ q_1 & q_2- & q\end{pmatrix}$$

$$\times\begin{pmatrix}j_1 & k_1 & j_1'\\ -m_1 & q_1 & m_1'\end{pmatrix}\begin{pmatrix}j_2 & k_2 & j_2'\\ -m_2 & q_2 & m_2'\end{pmatrix}\begin{pmatrix}j & k & j'\\ -m & q & m'\end{pmatrix}$$

$$\times\langle\alpha j_1\|T^{(k_1)}\|\alpha''j_1'\rangle\langle\alpha''j_2\|T^{(k_2)}\|\alpha'j_2'\rangle$$

$$(D.12)$$

where we have used the 3-j version of (2.125)

$$\sum_{m,q}\begin{pmatrix}j & k & j'\\ -m & q & m'\end{pmatrix}\begin{pmatrix}j & k & j''\\ -m & q & m''\end{pmatrix} = (2j'+1)^{-1}\delta_{j'j''}\delta_{m'm''} \qquad (D.13)$$

Now use (D.5) on the LHS of (D.12) to get

$$\langle\alpha j\|T^{(k)}\|\alpha'j'\rangle = \sum_{\alpha''}\sum_{m_1,m_2}\sum_{m_1',m_2'}\sum_{q_1,q_2m,m',q}(-1)^{k_1-k_2+q+2j_1+j+j_1'-j_2'+m'-m_1-m_2}$$

$$\times\sqrt{(2j+1)(2j'+1)(2k+1)}\begin{pmatrix}j_1 & j_2 & j\\ m_1 & m_2 & -m\end{pmatrix}\begin{pmatrix}j_1' & j_2' & j'\\ m_1' & m_2' & -m'\end{pmatrix}\begin{pmatrix}k_1 & k_2 & k\\ q_1 & q_2 & -q\end{pmatrix}$$

$$\times\begin{pmatrix}j_1 & k_1 & j_1'\\ -m_1 & q_1 & m_1'\end{pmatrix}\begin{pmatrix}j_2 & k_2 & j_2'\\ -m_2 & q_2 & m_2'\end{pmatrix}\begin{pmatrix}j & k & j'\\ -m & & m'\end{pmatrix}$$

$$\times\langle\alpha j_1\|T^{(k_1)}\|\alpha''j_1'\rangle\langle\alpha''j_2\|T^{(k_2)}\|\alpha'j_2'\rangle$$

$$(D.14)$$

It would not be surprising given that the 6-j symbol (which was involved in the coupling of three angular momenta) was expressed as a sum over the product of four 3-j symbols, that we could, by extension, define the 9-j symbol as a sum over the product of six 3-j symbols, and that this 9-j symbol would be involved in the coupling of four angular momenta. With this in mind, we write (D.14) as

$$\langle\alpha, [j_1j_2]j\|[T^{(k_1)}\otimes T^{(k_2)}]^{(k)}\|\alpha', [j_1'j_2']j'\rangle =$$

$$\sqrt{(2j+1)(2j'+1)(2k+1)}\begin{Bmatrix}j_1 & j_2 & j\\ j_1' & j_2' & j'\\ k_1 & k_2 & k\end{Bmatrix}\sum_{\alpha''}\langle\alpha j_1\|T^{(k_1)}\|\alpha''j_1'\rangle\langle\alpha''j_2\|T^{(k_2)}\|\alpha'j_2'\rangle$$

$$(D.15)$$

where we have defined the 9-j symbol as

$$
\begin{Bmatrix} j_1 & j_2 & j \\ j_1' & j_2' & j' \\ k_1 & k_2 & k \end{Bmatrix} = \sum_{all\,m} \begin{pmatrix} j_1 & j_2 & j \\ m_1 & m_2 & m \end{pmatrix} \begin{pmatrix} j_1' & j_2' & j' \\ m_1' & m_2' & m' \end{pmatrix} \begin{pmatrix} k_1 & k_2 & k \\ q_1 & q_2 & q \end{pmatrix}
$$

$$
\times \begin{pmatrix} j_1 & j_1' & k_1 \\ m_1 & m_1' & q_1 \end{pmatrix} \begin{pmatrix} j_2 & j_2' & k_2 \\ m_2 & m_2' & q_2 \end{pmatrix} \begin{pmatrix} j & j' & k \\ m & m' & q \end{pmatrix} \quad \text{(D.16)}
$$

and where, in coming up with (D.15), we have used the symmetry properties of the 3-j symbols [(2.145) and its variants] to eliminate the phase factor and the negative signs on the magnetic quantum numbers in the 3-j symbols of (D.14). Those same symmetry properties also combine to produce the symmetry properties of the 9-j symbols. In fact, we see that the 9-j symbols are invariant for a reflection about any diagonal, and for an even permutation of rows or columns [3, 7, 89]. One important property of the 9-j symbols bears mentioning. If one of the arguments vanishes, we can reformulate in terms of a 6-j symbol [3, 7, 10, 16, 89]:

$$
\begin{Bmatrix} j_1 & j_2 & j_3 \\ j_4 & j_5 & j_6 \\ j_7 & j_8 & 0 \end{Bmatrix} = (-1)^{j_2+j_3+j_4+j_7} [(2j_3+1)(2j_7+1)]^{-1/2} \begin{Bmatrix} j_1 & j_2 & j_3 \\ j_5 & j_4 & j_7 \end{Bmatrix} \delta_{j_3 j_6} \delta_{j_7 j_8}
$$

$$
\text{(D.17)}
$$

Some additional algebraic formulas involving 9-j symbols are offered below [3, 7, 10, 16, 89]:

$$
\begin{Bmatrix} j_1 & j_2 & j_3 \\ j_4 & j_5 & j_6 \\ j_7 & j_8 & j_9 \end{Bmatrix} = \sum_k (-1)^{2k}(2k+1) \begin{Bmatrix} j_1 & j_4 & j_7 \\ j_8 & j_9 & k \end{Bmatrix} \begin{Bmatrix} j_2 & j_5 & j_8 \\ j_4 & k & j_6 \end{Bmatrix}
$$

$$
\times \begin{Bmatrix} j_3 & j_6 & j_9 \\ k & j_1 & j_2 \end{Bmatrix} \quad \text{(D.18)}
$$

$$
\sum_{j_{12}} (2j_{12}+1) \begin{Bmatrix} j_1 & j_2 & j_{12} \\ j_3 & j_4 & j_{34} \\ j_{13} & j_{24} & j \end{Bmatrix} \begin{Bmatrix} j_1 & j_2 & j_{12} \\ j_{34} & j & j' \end{Bmatrix}
$$

$$
= (-1)^{2j'} \begin{Bmatrix} j_3 & j_4 & j_{34} \\ j_2 & j' & j_{24} \end{Bmatrix} \begin{Bmatrix} j_{13} & j_{24} & j \\ j' & j_1 & j_3 \end{Bmatrix} \quad \text{(D.19)}
$$

And finally, the orthonormality relation,[8]

---

[8]For an exhaustive collection of formulas involving the 9-j symbols, see [10].

$$\sum_{j_{12},j_{34}} (2j_{12}+1)(2j_{34}+1)(2j_{14}+1)(2j_{23}+1)$$

$$\times \begin{Bmatrix} j_1 & j_2 & j_{12} \\ j_4 & j_3 & j_{34} \\ j_{14} & j_{23} & j \end{Bmatrix} \begin{Bmatrix} j_1 & j_2 & j_{12} \\ j_4 & j_3 & j_{34} \\ j'_{14} & j'_{23} & j \end{Bmatrix} = \delta_{j_{14}j'_{14}} \delta_{j_{23}j'_{23}} \tag{D.20}$$

In general, the coupling of n angular momenta requires the introduction of $(n-2)$ intermediate angular momenta vectors. The coefficients of such a coupling can be put in the form of $3(n-1)$-j symbols [89], which in turn are sums over the projection quantum numbers of $2(n-1)$ C-G coefficients. So, for example, the coupling of four angular momenta requires the introduction of $4-2=2$ intermediate angular momenta vectors. The coupling is represented by the $3(4-1)$-$j = 9$-j symbols. The 9-j symbols are decomposed into a sum over the projection quantum numbers of $2(4-1) = 6$ C-G coefficients [or equivalently, six 3-j symbols, as seen in (D.16)].

Appendix E: Hamiltonian for the Interaction of an Electron with an Electromagnetic Field

From (3.2), we know that the electric field $\vec{E}$ and magnetic field $\vec{B}$ can be derived from the scalar potential $\Phi\left(\vec{r},t\right)$ and the vector potential $\vec{A}\left(\vec{r},t\right)$, as follows [11, 12]:

$$\vec{E} = -\vec{\nabla}\Phi - \frac{1}{c}\frac{\partial \vec{A}}{\partial t} \tag{E.1}$$

$$\vec{B} = \vec{\nabla} \times \vec{A} \tag{E.2}$$

We want to show that the electromagnetic (EM) force on a particle of charge e can be described by the velocity-dependent potential:

$$V = e\Phi - \frac{e}{c}\vec{A} \cdot \vec{v} \tag{E.3}$$

where $\vec{v}$ is the velocity of a particle of charge e and mass m moving through the EM field. To demonstrate this, all we need to do is construct the Lagrangian from the given potential, and by applying the Euler-Lagrange dynamical equations, arrive at the correct form of the Lorentz force law. From the given potential, the Lagrangian is [11]

$$L = T - V = \frac{1}{2}mv^2 - e\Phi + \frac{e}{c}\vec{A} \cdot \vec{v} \tag{E.4}$$

We form the Euler-Lagrange (E-L) equation for the x-coordinate only, realizing that the E-L equations for the other two coordinates will be the same:

$$\frac{d}{dt}\frac{\partial L}{\partial v_x} - \frac{\partial L}{\partial x} = 0 \tag{E.5}$$

where

$$\frac{\partial L}{\partial x} = -e\frac{\partial \Phi}{\partial x} + \frac{e}{c}\frac{\partial \vec{A}}{\partial x} \cdot \vec{v} \tag{E.6}$$

and the x-component of the canonical momentum is

$$p_x \equiv \frac{\partial L}{\partial v_x} = mv_x + \frac{e}{c}A_x \tag{E.7}$$

Combining (E.5, E.6, and E.7),

$$\frac{d}{dt}mv_x = -e\frac{\partial \Phi}{\partial x} + \frac{e}{c}\frac{\partial \vec{A}}{\partial x} \cdot \vec{v} - \frac{e}{c}\frac{dA_x}{dt} \tag{E.8}$$

where

$$\frac{dA_x(\vec{r},t)}{dt} = \frac{\partial A_x}{\partial t} + \frac{\partial A_x}{\partial x}\frac{dx}{dt} + \frac{\partial A_x}{\partial y}\frac{dy}{dt} + \frac{\partial A_x}{\partial z}\frac{dz}{dt}$$

$$= \frac{\partial A_x}{\partial t} + \frac{\partial A_x}{\partial x}v_x + \frac{\partial A_x}{\partial y}v_y + \frac{\partial A_x}{\partial z}v_z \tag{E.9}$$

and

$$\frac{\partial \vec{A}}{\partial x} \cdot \vec{v} = \frac{\partial A_x}{\partial x}v_x + \frac{\partial A_y}{\partial x}v_y + \frac{\partial A_z}{\partial x}v_z \tag{E.10}$$

Inserting the results from (E.9) and (E.10) into equation (E.8)

$$\frac{d}{dt}mv_x = -e\left(\frac{\partial \Phi}{\partial x} + \frac{1}{c}\frac{\partial A_x}{\partial t}\right) + \frac{e}{c}\left(\frac{\partial A_y}{\partial x} - \frac{\partial A_x}{\partial y}\right)v_y - \frac{e}{c}\left(\frac{\partial A_x}{\partial z} - \frac{\partial A_z}{\partial x}\right)v_z$$

$$= eE_x + \frac{e}{c}(B_xv_y - B_yv_z) = e\left(\vec{E} + \frac{1}{c}\vec{v} \times \vec{B}\right)_x \tag{E.11}$$

Adding the corresponding expressions for the y- and z-components, we get

$$\frac{d}{dt}m\vec{v} = e\left(\vec{E} + \frac{1}{c}\vec{v} \times \vec{B}\right) \tag{E.12}$$

which is indeed the correct force law for the interaction of a charged particle with the EM field. We now proceed to find the Hamiltonian using the definition

$$H \equiv \vec{p} \cdot \vec{v} - L = mv^2 + \frac{e}{c}\vec{A} \cdot \vec{v} - \frac{1}{2}mv^2 + e\Phi - \frac{e}{c}\vec{A} \cdot \vec{v} = \frac{1}{2}mv^2 + e\Phi$$

$$= \frac{\left(\vec{p} - \frac{e}{c}\vec{A}\right)^2}{2m} + e\Phi$$

(E.13)

Notice that the Hamiltonian is written in terms of the canonical momentum:

$$\vec{p} = m\vec{v} + \frac{e}{c}\vec{A}$$

(E.14)

On a final note, we remind the reader that in this convention, the electron has a charge of -e.

Appendix F: Integral Representation of the Spherical Bessel Functions and the Expansion of Plane Waves in Terms of Spherical Functions[9]

Integral Representation of the Spherical Bessel Functions

The purpose of this section is to derive an integral representation (in the complex plane) of the spherical Bessel functions j_l and then solve that integral to find a series representation for those functions. In doing so, we will also be able to find the form of the spherical Bessel functions in the limit as the argument gets very large, thus verifying the portion of (4.14) taken in that limit. Because the spherical Bessel functions play such a prominent role in describing free particles in quantum mechanics, it is worth taking the time to get to know these functions.

We want to show that

$$j_l(x) = \frac{1}{2\pi} \frac{(-2)^l l!}{x^{l+1}} \oint_{C_j} \frac{e^{-ixz}}{(z+1)^{l+1}(z-1)^{l+1}} dz$$

(F.1)

where C_j is any (positively oriented) closed contour in the complex plane enclosing the poles at $z = \pm 1$ (see Fig. F.1).

One way to do this is to show that the integral expression given above solves the differential equation for the spherical Bessel functions:

---

[9]Primary references for App. F: [9, 172, 173, 175, 176].

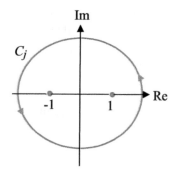

Fig. F.1 Contour in the complex plane for evaluating the integral representation of the spherical Bessel functions

$$j_l''(x) + \frac{2}{x}j_l'(x) + \left[1 - \frac{l(l+1)}{x^2}\right]j_l(x) = 0 \tag{F.2}$$

To demonstrate, first form the expression

$$f_l(x) = \frac{C_l}{x^{l+1}} \oint \frac{e^{-ixz}}{(z+1)^{l+1}(z-1)^{l+1}} dz \tag{F.3}$$

The contour in this case is any closed curve that contains at least one of the points $z = \pm 1$ (otherwise $f_l(x)$ would be zero by the Cauchy Theorem).

Next, form the derivatives that we will need (noting that the constant C_l is immaterial to this exercise):

$$f_l'(x) = -\frac{(l+1)}{x^{l+2}} \oint \frac{e^{-ixz}}{(z^2-1)^{l+1}} dz - \frac{i}{x^{l+1}} \oint \frac{ze^{-ixz}}{(z^2-1)^{l+1}} dz \tag{F.4}$$

$$f_l''(x) = \frac{(l+1)(l+2)}{x^{l+3}} \oint \frac{e^{-ixz}dz}{(z^2-1)^{l+1}} + \frac{2i(l+1)}{x^{l+2}} \oint \frac{ze^{-ixz}dz}{(z^2-1)^{l+1}} - \frac{1}{x^{l+1}}$$
$$\times \oint \frac{z^2 e^{-ixz}dz}{(z^2-1)^{l+1}} \tag{F.5}$$

Substitute (F.3), (F.4), and (F.5) into (F.2) and see if (F.2) is satisfied:

$$0 \stackrel{?}{=} \frac{(l+1)(l+2)}{x^{l+3}} \oint \frac{e^{-ixz}dz}{(z^2-1)^{l+1}} + \frac{2i(l+1)}{x^{l+2}} \oint \frac{ze^{-ixz}dz}{(z^2-1)^{l+1}}$$
$$-\frac{1}{x^{l+1}} \oint \frac{z^2 e^{-ixz}dz}{(z^2-1)^{l+1}} - \frac{2(l+1)}{x^{l+3}} \oint \frac{e^{-ixz}dz}{(z^2-1)^{l+1}} - \frac{2i}{x^{l+2}} \oint \frac{ze^{-ixz}dz}{(z^2-1)^{l+1}} \tag{F.6}$$
$$+\frac{1}{x^{l+1}} \oint \frac{e^{-ixz}dz}{(z^2-1)^{l+1}} - \frac{l(l+1)}{x^{l+3}} \oint \frac{e^{-ixz}dz}{(z^2-1)^{l+1}}$$

Simplifying,

$$0 \overset{?}{=} \frac{2il}{x^{l+2}} \oint \frac{ze^{-ixz}dz}{(z^2-1)^{l+1}} - \frac{1}{x^{l+1}} \oint \frac{e^{-ixz}dz}{(z^2-1)^{l}} \tag{F.7}$$

Now note

$$\frac{2lze^{-ixz}}{(z^2-1)^{l+1}} = -\frac{d}{dz}\left[\frac{e^{-ixz}}{(z^2-1)^{l}}\right] - ix\frac{e^{-ixz}}{(z^2-1)^{l}} \tag{F.8}$$

so that

$$0 \overset{?}{=} -\frac{i}{x^{l+2}} \oint \left\{\frac{d}{dz}\left[\frac{e^{-ixz}}{(z^2-1)^{l}}\right] + ix\frac{e^{-ixz}}{(z^2-1)^{l}}\right\}dz - \frac{1}{x^{l+1}} \oint \frac{e^{-ixz}dz}{(z^2-1)^{l}}$$

$$\Rightarrow 0 = -\frac{i}{x^{l+2}} \oint \frac{d}{dz}\left[\frac{e^{-ixz}}{(z^2-1)^{l}}\right]dz \tag{F.9}$$

The integrand is now an exact differential. Because the integral of an exact differential around a closed contour vanishes, we find that the integral expression given above does solve the differential equation for the spherical Bessel functions.

Now that we know our integral expression solves the differential equation for the spherical Bessel functions, we can use residue calculus to solve the integral expression directly. Before starting, remember that Cauchy's theorem states the remarkable fact that the shape of the closed contour does not matter. What matters is only whether the contour encloses any singularities of the integrand. Therefore, as long as we do not change the number of singularities that are enclosed, we are free to deform the contour any way we want (as long as the enclosed region(s) remain simply connected). With this in mind, we may imagine that we could gradually deform the contour C_j in such a way as to "pinch off" the contour until it makes two separate closed contours in a manner suggested by Fig. F.2.

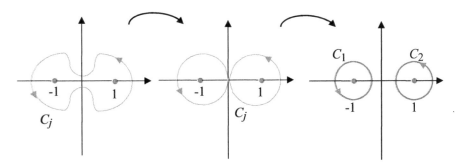

Fig. F.2 Deformed contour for evaluating the integral representation of the spherical Bessel functions

Because both poles are still enclosed, the value of the integral has not changed, and we can now say

$$
\begin{aligned}
j_l(x) = {} & \frac{1}{2\pi}\frac{(-2)^l l!}{x^{l+1}} \oint_{C_1} \frac{e^{-ixz}}{(z+1)^{l+1}(z-1)^{l+1}}\,dz \\
& + \frac{1}{2\pi}\frac{(-2)^l l!}{x^{l+1}} \oint_{C_2} \frac{e^{-ixz}}{(z+1)^{l+1}(z-1)^{l+1}}\,dz; \quad C_j = C_1 + C_2
\end{aligned}
\tag{F.10}
$$

Each of these two integrands has a pole of order $l+1$ contained inside the contour, so the integrals can easily be evaluated using the residue theorem. First, we must find the residues at each of the poles. We use the formula [172]

$$
\mathrm{Res}f(z_0) = \lim_{z\to z_0} \frac{1}{(m-1)!}\frac{d^{m-1}}{dz^{m-1}}[(z-z_0)f(z)]
\tag{F.11}
$$

For the integral around the contour C_2, we find

$$
\begin{aligned}
\mathrm{Res}\,\{f(z); z_0 = 1\} = {} & \lim_{z\to 1}\frac{1}{l!}\frac{d^l}{dz^l}\left[(z-1)^{l+1}\frac{e^{-ixz}}{(z+1)^{l+1}(z-1)^{l+1}}\right] \\
= {} & \lim_{z\to 1}\frac{1}{l!}\frac{d^l}{dz^l}\left[\frac{e^{-ixz}}{(z+1)^{l+1}}\right]
\end{aligned}
\tag{F.12}
$$

We will have to evaluate this derivative using Leibnitz's rule [172]:

$$
(f\cdot g)^{(n)} = \sum_{k=0}^{n}\frac{n!}{k!(n-k)!}f^{(k)}g^{(n-k)}
\tag{F.13}
$$

$$
\Rightarrow \mathrm{Res}\,\{f(z); z_0 = 1\} = \frac{1}{l!}\sum_{m=0}^{l}\left[\frac{l!}{m!(l-m)!}\frac{d^{l-m}}{dz^{l-m}}e^{-ixz}\frac{d^m}{dz^m}\frac{1}{(z+1)^{l+1}}\right]_{z=1}
\tag{F.14}
$$

Similarly, for the integral around the contour C_1, we have for the residue of the enclosed pole:

$$
\mathrm{Res}\,\{f(z); z_0 = -1\} = \frac{1}{l!}\sum_{m=0}^{l}\left[\frac{l!}{m!(l-m)!}\frac{d^{l-m}}{dz^{l-m}}e^{-ixz}\frac{d^m}{dz^m}\frac{1}{(z-1)^{l+1}}\right]_{z=-1}
\tag{F.15}
$$

Now evaluate the integrals using the residue theorem:

$$j_l(x) = 2\pi i \left(\frac{1}{2\pi} \frac{(-2)^l l!}{x^{l+1} l!} \right) \sum_{m=0}^{l} \frac{l!}{m!(l-m)!}$$

$$\times \left\{ \left[\frac{d^{l-m}}{dz^{l-m}} e^{-ixz} \frac{d^m}{dz^m} \frac{1}{(z+1)^{l+1}} \right]_{z=1} + \left[\frac{d^{l-m}}{dz^{l-m}} e^{-ixz} \frac{d^m}{dz^m} \frac{1}{(z-1)^{l+1}} \right]_{z=-1} \right\}$$

$$(F.16)$$

Evaluating the derivatives and simplifying

$$j_l(x) = \left(\frac{i(-2)^l}{x^{l+1}} \right) \sum_{m=0}^{l} \frac{l!}{m!(l-m)!} (-ix)^{l-m} (-1)^m \frac{(l+m)!}{l!} \left[\frac{e^{-ix}}{(2)^{l+m+1}} + \frac{e^{ix}}{(-2)^{l+m+1}} \right]$$

$$= \sum_{m=0}^{l} \frac{(l+m)!}{m!(l-m)!2^m} \frac{1}{x^{m+1}} (-i)^{l-m+1} \left[\frac{e^{-ix} + (-1)^{-(l-m+1)} e^{ix}}{2} \right]$$

$$(F.17)$$

where we have used the fact that, for $l=$ integer,

$$(-1)^{l+l-m+m} = (-1)^{2l} = 1 \tag{F.18}$$

And, for l, $m=$integer,

$$\frac{1}{(-2)^{l+m+1}} = \frac{1}{2^{l+m+1}(-1)^{l+m+1}} = \frac{(-1)^{-(l+m+1)}}{2^{l+m+1}} = \frac{(-1)^{-(l-m+1)}}{2^{l+m+1}} \tag{F.19}$$

Noting that $(-1) = e^{i\pi}$ and $i = e^{i\pi/2}$, we can further simplify

$$j_l(x) = \sum_{m=0}^{l} \frac{(l+m)!}{m!(l-m)!2^m} \frac{1}{x^{m+1}} \left\{ \frac{\exp\left[x - \frac{\pi(l-m+1)}{2} \right] + \exp - \left[x - \frac{\pi(l-m+1)}{2} \right]}{2} \right\}$$

$$= \sum_{m=0}^{l} \frac{(l+m)!}{m!(l-m)!2^m} \frac{\cos\left(x - \frac{\pi(l-m+1)}{2} \right)}{x^{m+1}}$$

$$(F.20)$$

This particular series expression has advantages in that it makes clear not only the relation of the spherical Bessel functions to the trigonometric functions but also the asymptotic behavior of the spherical Bessel functions. Specifically, as x gets very large, (F.20) shows the dominant term in the sum will be the $m = 0$ term, so that

$$\lim_{x\to\infty} j_l(x) \to \frac{\cos\left[x - \frac{\pi(l+1)}{2}\right]}{x} = \frac{\sin\left(x - \frac{l\pi}{2}\right)}{x} \tag{F.21}$$

On the other hand, this particular form obscures the fact that the leading term in the series for small x is of order x^l. Let us see if we can find another series representation that makes this fact clear. We revisit (F.1) and make the change of variable $\xi \equiv xz$,

$$\Rightarrow j_l(x) = \frac{1}{2\pi} \frac{(-2)^l l!}{x^{l+2}} \oint_{C_j} \frac{e^{-i\xi}}{\left(\frac{\xi^2}{x^2} - 1\right)^{l+1}} d\xi \tag{F.22}$$

We pause here and note that the contour C_j is a positively oriented, closed contour that contains the singularities at $\xi = \pm x$.

Rewriting,

$$j_l(x) = \frac{1}{2\pi}(-2)^l l! x^l \oint_{C_j} \frac{e^{-i\xi}}{\xi^{2l+2}} \left[1 - \frac{x^2}{\xi^2}\right]^{-l-1} d\xi \tag{F.23}$$

For any region in the complex $\xi-$ plane outside the circle $|\xi| = x$, the quantity in the brackets is uniformly convergent and can be expanded in a power series:

$$\left[1 - \frac{x^2}{\xi^2}\right]^{-l-1} = 1 + (l+1)\frac{x^2}{\xi^2} + \frac{1}{2}(l+1)(l+2)\frac{x^4}{\xi^4} + \cdots$$

$$= \sum_{m=0}^{\infty} \frac{(l+m)!}{m!l!} \left(\frac{x^2}{\xi^2}\right)^m \tag{F.24}$$

If we deform the contour C_j so that all parts of it lie outside the disk(s) $|\xi| = x$, then we can evaluate the integral using the residue theorem. First, we interchange the order of summation and integration (which we can do because of the uniform convergence) and then evaluate

$$\begin{aligned}
j_l(x) &= \frac{1}{2\pi}(-2)^l l! x^l \left[\oint_{C_j} \frac{e^{-i\xi}}{\xi^{2l+2}} d\xi + (l+1)x^2 \oint_{C_j} \frac{e^{-i\xi}}{\xi^{2l+4}} d\xi + \cdots\right] \\
&= i(-2)^l l! x^l \left[\frac{(-i)^{2l+1}}{(2l+1)!} + (l+1)x^2 \frac{(-i)^{2l+3}}{(2l+3)!} + \cdots\right] \\
&= \frac{2^l l!}{(2l+1)!} \left[x^l - \frac{x^{l+2}}{2(2l+3)} + \cdots\right]
\end{aligned} \tag{F.25}$$

where (F.11) was used to evaluate the residues of all the poles at $\xi = 0$. Equation (F.25) makes it abundantly clear that x^l is the leading term in this expansion. We finish by writing equation (F.25) in summation form,

$$
j_l(x) = 2^l x^l \sum_{m=0}^{\infty} (-1)^m \frac{(l+m)! x^{2m}}{m!(2l+2m+1)!}
$$

$$
= \frac{\sqrt{\pi}}{2} \sum_{m=0}^{\infty} (-1)^m \frac{(x/2)^{2m+l}}{m! \Gamma(l+m+3/2)} \tag{F.26}
$$

In a similar fashion, we can construct

$$
n_l(x) = \frac{1}{2\pi i} \frac{(-2)^l l!}{x^{l+1}} \oint_{C_n} \frac{e^{-ixz}}{(z+1)^{l+1}(z-1)^{l+1}} dz; \, C_n = C_1 - C_2 \tag{F.27}
$$

This construction allows us to realize the series representation of the spherical Neumann functions $n_l(x)$ right away. All we have to do is change the sign of the term e^{-ix} in (F.20) to $-e^{-ix}$ and then divide by i to get

$$
n_l(x) = \sum_{m=0}^{\infty} \frac{(l+m)!}{2^m m!(l-m)!} \frac{\sin\left(x - \frac{l-m+1}{2}\pi\right)}{x^{m+1}} \tag{F.28}
$$

Expansion of Plane Waves in Terms of Spherical Functions

In this section, we derive the formula for the expansion of a plane wave in terms of spherical waves; a formulation that first appears in (4.14). We start by assuming

$$
e^{ikrx} = \sum_{l=0}^{\infty} a_l j_l(kr) P_l(x) \tag{F.29}
$$

Our task will be to find the coefficients a_l. Multiplying both sides of (F.29) by $P_n(x)$ and integrating

$$
\int_{-1}^{1} e^{ikrx} P_n(x) dx = \int_{-1}^{1} \sum_{l=0}^{\infty} a_l j_l(kr) P_l(x) P_n(x) dx \tag{F.30}
$$

Using the orthogonality of the Legendre polynomials (2.71),

$$\int_{-1}^{1} e^{ikrx} P_n(x)dx = \sum_{l=0}^{\infty} a_l j_l(kr) \frac{2}{2l+1} \delta_{nl} = \frac{2}{2n+1} a_n j_n(kr) \tag{F.31}$$

and differentiating both sides n times with respect to kr, we get

$$\frac{d^n}{d(kr)^n} \int_{-1}^{1} e^{ikrx} P_n(x)dx = \frac{2}{2n+1} a_n \frac{d^n}{d(kr)^n} j_n(kr)$$

$$\Rightarrow (i)^n \int_{-1}^{1} x^n e^{ikrx} P_n(x)dx = \frac{2}{2n+1} a_n \frac{d^n}{d(kr)^n} j_n(kr) \tag{F.32}$$

To evaluate the derivative on the RHS of (F.32), we appeal to (F.26):

$$\begin{aligned}
\frac{d^n}{d(kr)^n} j_n(kr) &= \frac{\sqrt{\pi}}{2} \frac{d^n}{d(kr)^n} \sum_{m=0}^{\infty} \frac{(-1)^m}{m!\Gamma(n+m+3/2)} \left(\frac{kr}{2}\right)^{2m+n} \\
&= \frac{\sqrt{\pi}}{2} \sum_{m=0}^{\infty} \frac{(-1)^m (1/2)^n (2m+n)(2m+m-1)\cdots(2m+1)}{m!\Gamma(n+m+3/2)} \left(\frac{kr}{2}\right)^{2m} \\
&= \frac{\sqrt{\pi}}{2} \sum_{m=0}^{\infty} \frac{(-1)^m (1/2)^n (2m+n)!}{m!\Gamma(n+m+3/2)(2m)!} \left(\frac{kr}{2}\right)^{2m}
\end{aligned} \tag{F.33}$$

So now we have

$$(i)^n \int_{-1}^{1} x^n e^{ikrx} P_n(x)dx = \frac{2}{2n+1} a_n \frac{\sqrt{\pi}}{2} \sum_{m=0}^{\infty} \frac{(-1)^m (1/2)^n (2m+n)!}{m!\Gamma(n+m+3/2)(2m)!} \left(\frac{kr}{2}\right)^{2m} \tag{F.34}$$

Taking the limit of both sides as $kr \to 0$,

$$\lim_{kr\to 0} (i)^n \int_{-1}^{1} x^n e^{ikrx} P_n(x)dx = \lim_{kr\to 0} \frac{2}{2n+1} a_n \frac{\sqrt{\pi}}{2} \sum_{m=0}^{\infty} \frac{(-1)^m (1/2)^n (2m+n)!}{m!\Gamma(n+m+3/2)(2m)!} \left(\frac{kr}{2}\right)^{2m}$$

$$\Rightarrow (i)^n \int_{-1}^{1} x^n P_n(x)dx = \lim_{kr\to 0} \frac{2}{2n+1} a_n \frac{\sqrt{\pi}}{2} \sum_{m=0}^{\infty} \frac{(-1)^m (1/2)^n (2m+n)!}{m!\Gamma(n+m+3/2)(2m)!} \left(\frac{kr}{2}\right)^{2m} \tag{F.35}$$

To evaluate the LHS, we use (A.32)

$$\int\limits_{-1}^{1} x^n P_n(x)dx = \frac{2^{n+1}(n!)^2}{(2n+1)!} \tag{A.32}$$

To evaluate the RHS, we note that, in the limit, only the first term of the summation is non-negligible:

$$\Rightarrow \lim_{kr \to 0} \frac{2}{2n+1} a_n \frac{\sqrt{\pi}}{2} \sum_{m=0}^{\infty} \frac{(-1)^m (1/2)^n (2m+n)!}{m!\Gamma(n+m+3/2)(2m)!} \left(\frac{kr}{2}\right)^{2m} \to \frac{a_n \sqrt{\pi}}{2n+1} \frac{(1/2)^n n!}{\Gamma(n+3/2)} \tag{F.36}$$

Making use of the following identity [172],

$$\Gamma\left(n+\frac{3}{2}\right) = \frac{\sqrt{\pi}(2n+1)!}{2^{2n+1}n!} \tag{F.37}$$

which, after mating the RHS and the LHS back together gives

$$(i)^n \frac{2^{n+1}(n!)^2}{(2n+1)!} = \frac{a_n(n!)^2 2^{n+1}}{(2n+1)(2n+1)!} \Rightarrow a_n = (i)^n(2n+1) \tag{F.38}$$

The derivation of (F.38) was based, in part, on (A.31), which we never proved. For that reason, the previous derivation may be unsatisfactory to some. In that case, we offer the following alternative derivation. Here we assume

$$e^{i\vec{k}\cdot\vec{r}} = \sum_{l=0}^{\infty} \sum_{m=-l}^{l} a_{lm} j_l(kr) Y_{lm}(\theta, \phi) \tag{F.39}$$

Let the z-axis lie along the direction of $\vec{k}$:

$$\Rightarrow e^{i\vec{k}\cdot\vec{r}} = e^{ikr\cos\theta} = \sum_{l=0}^{\infty} \sum_{m=-l}^{l} a_{lm} j_l(kr) Y_{lm}(\theta, \phi) \tag{F.40}$$

The quantity on the LHS has no ϕ-dependance; therefore, we must have $m = 0$ on the RHS:

$$\Rightarrow e^{i\vec{k}\cdot\vec{r}} = e^{ikr\cos\theta} = \sum_{l=0}^{\infty} a_l j_l(kr) Y_{l0}(\theta, \phi) \tag{F.41}$$

Using (2.60),

$$\Rightarrow e^{i\vec{k}\cdot\vec{r}} = \sum_{l=0}^{\infty} a_l j_l(kr) \left(\frac{2l+1}{4\pi}\right)^{1/2} P_l(\cos\theta) \tag{F.42}$$

Again, we use the orthogonality of the Legendre polynomials (2.71) to establish

$$a_l j_l(kr) = \frac{1}{2}[4\pi(2l+1)]^{1/2} \int_{-1}^{1} dz P_l(z) e^{ikrz}$$

As before, we take the limit of both sides as $kr \to 0$. To evaluate the LHS, we again note that, in the limit, only the first term of the summation for $j_l(kr)$ in (F.26) is non-negligible. Using (F.37), we get

$$a_l \frac{\sqrt{\pi}}{2} \frac{2^l l!}{(2l+1)!} \frac{2^{2l+1}}{\sqrt{\pi}} \left(\frac{kr}{2}\right)^l = \frac{1}{2}[4\pi(2l+1)]^{1/2} \int_{-1}^{1} dz P_l(z) e^{ikrz}$$

$$\Rightarrow a_l \frac{2^l l!}{(2l+1)!} (kr)^l = \frac{1}{2}[4\pi(2l+1)]^{1/2} \int_{-1}^{1} dz P_l(z) e^{ikrz} \tag{F.43}$$

The oscillatory nature of the exponential term on the RHS of (F.43) makes it hard to evaluate (even in the limit as $kr \to 0$), but we can proceed by equating the correct power of kr on the RHS with the corresponding power of kr on the LHS:

$$\Rightarrow a_l \frac{2^l l!}{(2l+1)!} (kr)^l = \frac{1}{2}[4\pi(2l+1)]^{1/2} (ikr)^l \int_{-1}^{1} dz P_l(z) \frac{z^l}{l!}$$

$$\Rightarrow a_l \frac{2^l (l!)^2}{(2l+1)!} = \frac{1}{2}[4\pi(2l+1)]^{1/2} (i)^l \int_{-1}^{1} dz P_l(z) z^l \tag{F.44}$$

We can evaluate the integral by remembering that $P_l(z)$ is an lth-degree polynomial in z. The coefficient of the leading power, z^l, can be obtained from (A.3)

$$P_l(z) = \frac{1}{2^l l!} \frac{d^l}{dz^l} (z^2 - 1)^l = \frac{2l(2l-1)(2l-2)\cdots(l+1)}{2^l l!} z^l + O(z^{l-1})$$

$$\Rightarrow z^l = \frac{2^l l!}{2l(2l-1)(2l-2)\cdots(l+1)} P_l(z) + \text{terms involving } P_{l+1}(z) \text{ and higher} \tag{F.45}$$

Substituting this into the integral of (F.44) and again using (2.71) gives us (to leading order)

$$a_l \frac{2^l(l!)^2}{(2l+1)!} = \frac{1}{2}[4\pi(2l+1)]^{1/2}(i)^l \frac{2^l l!}{2l(2l-1)(2l-2)\cdots(l+1)} \frac{2}{2l+1}$$

$$\Rightarrow a_l \frac{l!}{(2l)!} = [4\pi(2l+1)]^{1/2}(i)^l \frac{1}{2l(2l-1)(2l-2)\cdots(l+1)}$$

$$\Rightarrow a_l \frac{l!}{(2l)(2l-1)(2l-2)\cdots(l+1)l!} = [4\pi(2l+1)]^{1/2}(i)^l \frac{1}{2l(2l-1)(2l-2)\cdots(l+1)}$$

$$\Rightarrow a_l = [4\pi(2l+1)]^{1/2}(i)^l$$

$$\text{(F.46)}$$

Finally resulting in,

$$e^{i\vec{k}\cdot\vec{r}} = \sum_{l=0}^{\infty} [4\pi(2l+1)]^{1/2}(i)^l j_l(kr)\left(\frac{2l+1}{4\pi}\right)^{1/2} P_l(\cos\theta)$$

$$\Rightarrow e^{i\vec{k}\cdot\vec{r}} = \sum_{l=0}^{\infty} (2l+1)(i)^l j_l(kr) P_l(\cos\theta) \qquad\text{(F.47)}$$

as before.

If we want to account for the spin of the photoelectron (e.g., in relativistic applications), then our continuum photoelectron plane wave would have the form

$$\psi_{p\kappa m_j}(pr) = e^{i\vec{p}\cdot\vec{r}/\hbar}\chi^{m_s} \qquad\text{(F.48)}$$

We now expand in the usual way [73]:

$$\psi_{p\kappa m_j}(pr) = e^{i\vec{p}\cdot\vec{r}/\hbar}\chi^{m_s} = \sum_{\kappa,m_j} a_{\kappa m_j} j_l(kr)\chi_\kappa^{m_j}(\hat{r}) \qquad\text{(F.49)}$$

where $k = p/\hbar$ is the wave number of the lth partial wave of the continuum photoelectron. To find the coefficients $a_{\kappa m_j}$, we start by multiplying both sides of (F.49) by $\chi_{\kappa'}^{m_j'\dagger}(\hat{r})$ and integrating over the solid angle. Using (6.3) we get

$$a_{\kappa m_j} j_l(kr) = \int \chi_\kappa^{m_j\dagger}(\hat{r})e^{i\vec{k}\cdot\vec{r}}\chi^{m_s}\,d\hat{r} \qquad\text{(F.50)}$$

Expand the exponential in terms of spherical harmonics using (F.47), (6.1), and (2.79):

$$a_{\kappa m_j} j_l(kr) = 4\pi \int \chi_\kappa^{m_j\dagger}(\hat{r}) \sum_{l',m'} (i)^{l'} j_{l'}(kr) Y_{l'm'}^*(\hat{k}) Y_{l'm'}(\hat{r}) \chi^{m_s} d\hat{r}$$

$$= 4\pi \int \left[\begin{array}{c} \sum_{m'_s} (-1)^{l-\frac{1}{2}-m_j} \sqrt{2j+1} \begin{pmatrix} l & \frac{1}{2} & j \\ m_j - m'_s & m'_s & -m_j \end{pmatrix} Y_{l,m_j-m'_s}(\hat{r}) \underbrace{\chi^{m'_s}_s \chi^{m_s}}_{\delta_{m'_s m_s}} \\ \times \sum_{l',m'}^{\infty} (i)^{l'} j_{l'}(kr) Y_{l'm'}^*(\hat{k}) Y_{l'm'}(\hat{r}) d\hat{r} \end{array} \right]$$

$$= 4\pi \sum_{m'_s} (-1)^{l-\frac{1}{2}-m_j} \sqrt{2j+1} \begin{pmatrix} l & \frac{1}{2} & j \\ m_j - m'_s & m'_s & -m_j \end{pmatrix}$$

$$\times \sum_{l',m'}^{\infty} (i)^{l'} j_{l'}(kr) Y_{l'm'}^*(\hat{k}) \delta_{ll'} \delta_{m_j-m'_s,m'} \delta_{m'_s m_s}$$

$$= 4\pi (-1)^{l-\frac{1}{2}-m_j} \sqrt{2j+1} \begin{pmatrix} l & \frac{1}{2} & j \\ m_j - m_s & m_s & -m_j \end{pmatrix} (i)^l Y_{l,m_j-m_s}^*(\hat{k}) j_l(kr)$$

$$(F.51)$$

where we also used (6.2) and (2.64). Comparing the LHS of (F.51) with the RHS of (F.51),

$$\Rightarrow a_{\kappa m_j} = 4\pi(i)^l(-1)^{l-\frac{1}{2}-m_j} \sqrt{2j+1} \begin{pmatrix} l & \frac{1}{2} & j \\ m_j - m_s & m_s & -m_j \end{pmatrix} Y_{l,m_j-m_s}^*(\hat{k}) \quad (F.52)$$

Giving us, finally [73],

$$\psi_{p\kappa m_j}(pr) = e^{i\vec{p}\cdot\vec{r}/\hbar} \chi^{m_s} = 4\pi \sum_{\kappa,m_j} (i)^l (-1)^{l-\frac{1}{2}-m_j} \sqrt{2j+1}$$

$$\times \begin{pmatrix} l & \frac{1}{2} & j \\ m_j - m_s & m_s & -m_j \end{pmatrix} Y_{l,m_j-m_s}^*(\hat{k}) j_l(kr) \chi_\kappa^{m_j}(\hat{r}) \quad (F.53)$$

This form is equivalent to the one given in (6.16).

Appendix G: Basic Theory of the Design of the COLTRIMS Reaction Microscope[10]

Recoil Ion Detection

This appendix details the basic kinematic ideas behind the design of the COLTRIMS device. A schematic of the COLTRIMS (or C-REMI) spectrometer is shown in the figure below. Photons are introduced into the spectrometer along the x-axis in the positive x-direction. A well-collimated jet of cold molecules is introduced along the y-axis in the positive y-direction. The intersection of the photon beam and the gas jet is called the interaction region (IR). For now, we assume that the IR is small enough that it can be approximated by a point in space which we place at the origin of a set of lab-fixed axes (see Fig. G.1).

Magnetic and electrostatic fields are used to accelerate positive ion (recoil) fragments and electrons onto their respective position-sensitive detectors via the Lorentz force:

$$\vec{F} = q\left[\vec{E} + \vec{v} \times \vec{B}\right] \tag{G.1}$$

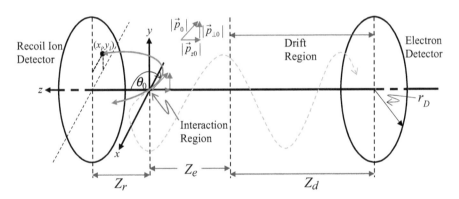

Fig. G.1 Schematic of COLTRIMS device. The solid (blue) curve shows two typical ion trajectories (in the left portion of the diagram). The impact coordinates (x, y) of one of the trajectories onto the recoil ion detector is also shown. The initial momentum vector triangle (for one of the trajectories) is shown near the origin. The initial momentum vector triangle is reproduced again at the top of the diagram for clarity. A typical electron trajectory is shown as a (green) dashed curve. The recoil ion detector is a fixed distance Z_r along the z-axis from the IR. Both detectors are flat and lie in planes parallel to the x-y plane. The distances Z_e and Z_d indicated in the figure are also constants (and are, along with the $\vec{E}$ and $\vec{B}$ fields, controlled by the experimenter). Adapted from [152] and republished with permission of Elsevier; permission conveyed through Copyright Clearance Center, Inc

---

[10]Primary references for App. G: [150–152, 177].

where electric and magnetic fields fill the entire volume of the spectrometer according to

$$\vec{E} = \left\{ \begin{array}{l} E_0\widehat{z}; Z_r \geq z \geq -Z_e \\ 0 \ ; z \leq -Z_e \end{array} \right\} \quad \vec{B} = B_0\widehat{z}; Z_r \geq z \geq -(Z_e + Z_d) \quad E_0, B_0$$

$$= \text{const} \tag{G.2}$$

By measuring the time of flight (TOF) of the ion, and the x- and y-coordinates of the ion's impact on the recoil ion detector, it is possible to reconstruct the components of the ion's initial momentum vector (and by extension, the orientation of the molecule in space at the time of its break-up) with respect to the lab-fixed axes. That is, it is possible to find and separately express each of the three individual Cartesian components of the recoil ion's initial momentum vector as a function of the time of flight of the ion, the x- and y-coordinates of its impact on the recoil ion detector (as appropriate), and other constants of the problem.

The initial momentum vector of the recoil ion can be written in component form as

$$\vec{P}_0 = \vec{P}_{x0} + \vec{P}_{y0} + \vec{P}_{z0} \tag{G.3}$$

where $p_{\perp 0} = \sqrt{p_{x0}^2 + p_{y0}^2}$ is the magnitude of the transverse component of the initial ion momentum.

If the time of flight t_i of the ion is an integer multiple of the cyclotron period, $T = 2\pi/\omega$, where $\omega = qB_0/m$ is the cyclotron frequency for a particle of charge q and mass m, then the impact coordinates of the ion onto the detector will be $(0, 0, Z_r)$.

To start, we assume that, right before the collision with the photon (and the resulting Coulomb explosion), the molecule is at rest with respect to the lab frame. That is, we ignore the mechanical momentum imparted to the molecule by the photon (because it is negligible) and the constant initial velocity in the positive y-direction that all the particles will have by their being part of a collimated, fast-moving jet (we do this because it is easy to account for this initial velocity in post-experiment analysis, and so it is not essential to the main concept at this point). We may also assume that the trajectory of the ion does not cross into the drift region of the spectrometer. All motions are nonrelativistic.

A couple of quick notes:

- The high-energy photons needed for these types of experiments are generally produced by a synchrotron. The synchrotron facility provides an electronic timing marker when photons are produced to assist the experimenter in measuring times of flight.
- The drift region and the fields in the drift region are designed in such a way as to confine and capture fast-moving electrons (to be discussed below).

We start with the Lorentz force equation for the force on a particle of charge q and mass m, moving with velocity $\vec{v}$,

$$\vec{F} = q\left[\vec{E} + \vec{v} \times \vec{B}\right]$$

$$\Rightarrow m\vec{a} = q\left[E_0\hat{z} + \begin{vmatrix} \hat{x} & \hat{y} & \hat{z} \\ v_x & v_y & v_z \\ 0 & 0 & B_0 \end{vmatrix}\right] = q\left[\hat{x}B_0v_y - \hat{y}B_0v_x + E_0\hat{z}\right] \qquad (G.4)$$

$$\Rightarrow \dot{v}_x = \frac{qB_0}{m}v_y; \quad \dot{v}_y = -\frac{qB_0}{m}v_x; \quad \dot{v}_z = \frac{qE_0}{m}$$

Examining the z-component:

$$\frac{dv_z}{dt} = \frac{qE_0}{m} \Rightarrow m\int_{v_{z0}}^{v_z} dv_z' = \frac{qE_0}{m}\int_{t_0}^{t} dt' \Rightarrow (v_z - v_{z0}) = \frac{qE_0}{m}\left(t - \underbrace{t_0}_{=0}\right)$$

$$\Rightarrow \frac{dz}{dt} = \frac{qE_0}{m}t + v_{z0} \Rightarrow \int_{z_0}^{z} dz' = \frac{qE_0}{m}\int_0^t t' dt' + v_{z0}\int_0^t dt'$$

$$\Rightarrow z - \underbrace{z_0}_{=0} = \frac{qE_0}{2m}t^2 + v_{z0}t \Rightarrow mz = \frac{qE_0}{2}t^2 + p_{z0}t$$

$$\Rightarrow p_{z0} = \frac{1}{t_i}\left(m_iZ_r - \frac{qE_0}{2}t_i^2\right)$$

$$(G.5)$$

where t_i is the time of flight of the ion, m_i is the mass of the ion, and we have noted that the z-coordinate of all ions striking the detector will be Z_r. Note also that we have started the clock at the instant of the Coulomb explosion of the molecule.

Although the z-equation is decoupled, the x- and y-equations are not:

$$\dot{v}_x = \frac{qB_0}{m}v_y; \quad \dot{v}_y = -\frac{qB_0}{m}v_x \qquad (G.6)$$

These two coupled equations can be easily solved:

$$\dot{v}_x = \frac{qB_0}{m}v_y \Rightarrow \ddot{v}_x = \frac{qB_0}{m}\dot{v}_y = -\left(\frac{qB_0}{m}\right)^2 v_x = -\omega^2 v_x \text{ etc.} \qquad (G.7)$$

$$\Rightarrow v_x = A\cos(\omega t + \varphi) \text{ and } v_y = -A\sin(\omega t + \varphi)$$

where $\omega = qB_0/m$ is the cyclotron frequency for a particle of charge q and mass m, and A and φ are constant to be determined from the initial conditions.

Continuing,

$$
\begin{aligned}
v_x &= A[\cos \omega t \cos \varphi - \sin \omega t \sin \varphi] \\
&\Rightarrow \int_0^x dx' = A\left[\cos \varphi \int_0^t \cos \omega t' dt' - \sin \varphi \int_0^t \sin \omega t' dt'\right] \\
&\Rightarrow x = \frac{A}{\omega}[\cos \varphi \sin \omega t + \sin \varphi(\cos \omega t - 1)]
\end{aligned}
\tag{G.8}
$$

Similarly,

$$
\begin{aligned}
v_y &= -A[\sin \omega t \cos \varphi + \cos \omega t \sin \varphi] \\
&\Rightarrow \int_0^y dy' = -A\left[\cos \varphi \int_0^t \sin \omega t' dt' + \sin \varphi \int_0^t \cos \omega t' dt'\right] \\
&\Rightarrow y = -\frac{A}{\omega}[\cos \varphi(1 - \cos \omega t) + \sin \varphi \sin \omega t]
\end{aligned}
\tag{G.9}
$$

To find A and φ, we use (G.7):

$$
v_x^2 + v_y^2 = A^2 \Rightarrow A = \sqrt{v_x^2 + v_y^2} \equiv |\vec{v}_\perp|
\tag{G.10}
$$

Since A is a constant, we must have

$$
A = |\vec{v}_\perp| = |\vec{v}_{\perp 0}|
\tag{G.11}
$$

which makes sense, because the magnetic field does no work and hence cannot change the transverse momentum of the particle. Now evaluate (G.7) at $t_0 = 0$:

$$
v_{x0} = v_{\perp 0} \cos \varphi = v_{\perp 0} \cos(-\varphi) \quad \text{and} \quad v_{y0} = -v_{\perp 0} \sin \varphi = v_{\perp 0} \sin(-\varphi)
\tag{G.12}
$$

But,

$$
v_{x0} = v_{\perp 0} \cos \phi_0 \quad \text{and} \quad v_{y0} = v_{\perp 0} \sin \phi_0
\tag{G.13}
$$

Thus,

$$
\phi_0 = -\varphi
\tag{G.14}
$$

So, the phase angle is the negative of the initial azimuthal angle ϕ_0 at which $v_{\perp 0}$ was directed in the x-y plane.

Equations (G.4) can now be written as

$$v_x = v_{\perp 0} \cos(\omega t - \phi_0) \tag{G.15a}$$

$$v_y = -v_{\perp 0} \sin(\omega t - \phi_0) \tag{G.15b}$$

Equations (G.8) and (G.9) can be written as

$$
\begin{aligned}
x &= \frac{A}{\omega} \sin\varphi(\cos\omega t - 1) + \frac{A}{\omega}\cos\varphi\sin\omega t = \frac{v_{\perp 0}\sin(-\phi)}{\omega}(\cos\omega t - 1) \\
&\quad + \frac{v_{\perp 0}\cos(-\phi)}{\omega}\sin\omega t \\
&= \frac{-p_{y0}}{m\omega}(\cos\omega t - 1) + \frac{p_{x0}}{m\omega}\sin\omega t
\end{aligned} \tag{G.16}
$$

and

$$
\begin{aligned}
y &= -\frac{A}{\omega}\cos\varphi(1 - \cos\omega t) - \frac{A}{\omega}\sin\varphi\sin\omega t = -\frac{v_{\perp 0}\cos(-\phi)}{\omega}(1 - \cos\omega t) \\
&\quad - \frac{v_{\perp 0}\sin(-\phi)}{\omega}\sin\omega t \\
&= -\frac{p_{x0}}{m\omega}(1 - \cos\omega t) + \frac{p_{y0}}{m\omega}\sin\omega t
\end{aligned} \tag{G.17}
$$

Now combine (G.16) and (G.17):

$$
\begin{aligned}
p_{x0} &= \frac{1}{2}\left[\frac{\sin(\omega t/2)}{\sin(\omega t/2)} + \frac{\sin(\omega t/2)}{\sin(\omega t/2)}\right]p_{x0} \\
&= \frac{1}{2}\left\{-p_{y0}\left[\frac{\cos(\omega t/2)}{\sin(\omega t/2)} - \frac{\cos(\omega t/2)}{\sin(\omega t/2)}\right] + p_{x0}\left[\frac{\sin(\omega t/2)}{\sin(\omega t/2)} + \frac{\sin(\omega t/2)}{\sin(\omega t/2)}\right]\right\} \\
&= \frac{1}{2}\left\{
\begin{aligned}
&-p_{y0}\left[\frac{\cos(\omega t/2)\cos(\omega t) + \sin(\omega t/2)\sin(\omega t)}{\sin(\omega t/2)} - \frac{\cos(\omega t/2)}{\sin(\omega t/2)}\right] \\
&+p_{x0}\left[\frac{\cos(\omega t/2)\sin(\omega t) - \sin(\omega t/2)\cos(\omega t)}{\sin(\omega t/2)} + \frac{\sin(\omega t/2)}{\sin(\omega t/2)}\right]
\end{aligned}
\right\} \\
&= \frac{m\omega}{2}\left\{
\begin{aligned}
&\left[-\frac{p_{y0}}{m\omega}\frac{\cos(\omega t/2)(\cos(\omega t) - 1)}{\sin(\omega t/2)} + \frac{p_{x0}}{m\omega}\frac{\cos(\omega t/2)\sin(\omega t)}{\sin(\omega t/2)}\right] \\
&+\left[\frac{p_{x0}}{m\omega}\frac{\sin(\omega t/2)(1 - \cos(\omega t))}{\sin(\omega t/2)} - \frac{p_{y0}}{m\omega}\frac{\sin(\omega t/2)\sin(\omega t)}{\sin(\omega t/2)}\right]
\end{aligned}
\right\} \\
\Rightarrow p_{x0} &= \frac{m_i\omega}{2}\left[x_i\cot\left(\frac{\omega t_i}{2}\right) - y_i\right]
\end{aligned} \tag{G.18}
$$

where x_i and y_i are the x- and y-coordinates of the ion's impact on the recoil ion detector, respectively.

Similarly,

$$
\begin{aligned}
p_{y0} &= \frac{1}{2}\left[\frac{\sin(\omega t/2)}{\sin(\omega t/2)} + \frac{\sin(\omega t/2)}{\sin(\omega t/2)}\right]p_{y0} \\
&= \frac{1}{2}\left\{ p_{y0}\left[\frac{\sin(\omega t/2)}{\sin(\omega t/2)} + \frac{\sin(\omega t/2)}{\sin(\omega t/2)}\right] + p_{x0}\left[\frac{\cos(\omega t/2)}{\sin(\omega t/2)} - \frac{\cos(\omega t/2)}{\sin(\omega t/2)}\right]\right\} \\
&= \frac{1}{2}\left\{ \begin{array}{l} p_{y0}\left[\dfrac{\cos(\omega t/2)\sin(\omega t) - \sin(\omega t/2)\cos(\omega t)}{\sin(\omega t/2)} + \dfrac{\sin(\omega t/2)}{\sin(\omega t/2)}\right] \\[2ex] +p_{x0}\left[\dfrac{\sin(\omega t/2)\sin(\omega t) - +\cos(\omega t/2)\cos(\omega t)}{\sin(\omega t/2)} - \dfrac{\cos(\omega t/2)}{\sin(\omega t/2)}\right] \end{array}\right\} \\
&= \frac{m\omega}{2}\left\{ \begin{array}{l} \left[-\dfrac{p_{x0}}{m\omega}\dfrac{\cos(\omega t/2)(1-\cos(\omega t))}{\sin(\omega t/2)} + \dfrac{p_{y0}}{m\omega}\dfrac{\cos(\omega t/2)\sin(\omega t)}{\sin(\omega t/2)}\right] \\[2ex] -\left[\dfrac{p_{y0}}{m\omega}\dfrac{\sin(\omega t/2)(\cos(\omega t)-1)}{\sin(\omega t/2)} + \dfrac{p_{x0}}{m\omega}\dfrac{\sin(\omega t/2)\sin(\omega t)}{\sin(\omega t/2)}\right] \end{array}\right\} \\
&\Rightarrow p_{y0} = \frac{m_i\omega}{2}\left[y_i \cot\left(\frac{\omega t_i}{2}\right) + x_i\right]
\end{aligned}
$$

$$(G.19)$$

Thus, we have seen that the x- and y- motions are coupled, whereas the motion in the z-direction is characterized by a constant acceleration.

We continue by manipulating (G.16) and (G.17):

$$
\begin{aligned}
r^2 &= x^2 + y^2 \\
&= \left(\frac{A}{\omega}\right)^2\left\{[\cos\varphi\sin\omega t + \sin\varphi(\cos\omega t - 1)]^2 + [\cos\varphi(\cos\omega t - 1)\right. \\
&\quad \left. - \sin\varphi\sin\omega t]^2\right\} \\
&= \left(\frac{A}{\omega}\right)^2\left[(\cos\omega t - 1)^2 + \sin^2\omega t\right] = \left(\frac{A}{\omega}\right)^2[2(1-\cos\omega t)] \\
&= \left(\frac{A}{\omega}\right)^2\left[2\left(1 - \cos\frac{2\pi t}{T}\right)\right] \\
&\Rightarrow r(t) = \frac{v_{\perp 0}}{\omega}\sqrt{2\left(1 - \cos\frac{2\pi t}{T}\right)}
\end{aligned}
$$

$$(G.20)$$

It is clear that whenever $t = nT$ for n=integer, $r(t)$ is zero.[11] Thus, if $t_i = nT$, the x and y impact coordinates are both zero. The experimenter will have to disregard these events, because they are indeterminate (they do not uniquely define an initial momentum vector). Equation (G.20) also implies we must design the spectrometer so that the detector radius $r_D > 2v_{\perp 0}/\omega$ ensures that all the charged particles strike the detector [c.f. (G.23) below].

---

[11] Note that this result could also be obtained by evaluating equations (G.16) and (G.17) at $t = nT$.

Geometric Analysis

Notice that, from (G.20), the projection of the ion trajectory onto the x-y plane is a circle, which can be written as

$$r(t) = \frac{v_{\perp 0}}{\omega}\sqrt{2(1 - \cos \omega t)} = \frac{v_{\perp 0}}{\omega}\sqrt{2 \cdot 2 \sin^2\left(\frac{\omega t}{2}\right)} = 2\frac{v_{\perp 0}}{\omega}\left|\sin\left(\frac{\omega t}{2}\right)\right| \quad \text{(G.21)}$$

A projection of a typical ion trajectory onto the x-y plane is shown in Fig. G.2. The view in Fig. G.2 is from behind the ion detector looking back (in the negative z-direction) toward the IR. From this point of view, the ion is traveling in a clockwise direction around the circle. From the figure (remembering that ϕ is measured clockwise from the x-axis),

$$|\phi_0| = \frac{\omega t}{2} - \phi' \quad \text{(G.22a)}$$

$$\phi + (-\phi') = 2\pi \Rightarrow \phi' = \phi - 2\pi \quad \text{(G.22b)}$$

$$\cos \phi' = \frac{x}{r} \quad \text{(G.22c)}$$

$$\sin \phi' = -\frac{y}{r} \quad \text{(G.22d)}$$

We also have

$$v_\perp = R\omega \Rightarrow R = \frac{v_\perp}{\omega} = \frac{v_{\perp 0}}{\omega} \quad \text{(G.23)}$$

Fig. G.2 Projection of a typical ion trajectory onto the x-y plane. Adapted from [151] and reproduced with permission from IOP publishing; permission conveyed through Copyright Clearance Center, Inc

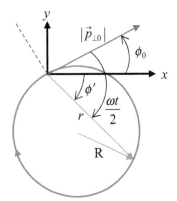

Now we have enough information to proceed

$$
\begin{aligned}
p_{x0} &= p_{\perp 0} \cos \phi_0 = p_{\perp 0}\left[\cos \left(\frac{\omega t}{2} - \phi' \right) \right] \\
&= p_{\perp 0}\left[\cos \left(\cos^{-1} \frac{x}{r} \right) \cos \left(\frac{\omega t}{2} \right) + \sin \left(\sin^{-1} \frac{-y}{r} \right) \sin \left(\frac{\omega t}{2} \right) \right] \\
&= m_i v_{\perp 0}\left[\frac{x}{r} \cos \left(\frac{\omega t}{2} \right) - \frac{y}{r} \sin \left(\frac{\omega t}{2} \right) \right] = \frac{m_i v_{\perp 0}}{r}\left[x\cos \left(\frac{\omega t}{2} \right) - y\sin \left(\frac{\omega t}{2} \right) \right] \\
&= \frac{m_i v_{\perp 0}}{2\frac{v_{\perp 0}}{\omega}\left| \sin \left(\frac{\omega t}{2} \right) \right|}\left[x\cos \left(\frac{\omega t}{2} \right) - y\sin \left(\frac{\omega t}{2} \right) \right] \\
&\Rightarrow p_{x0} = \frac{m_i \omega}{2}\left[x_i \cot \left(\frac{\omega t_i}{2} \right) - y_i \right]
\end{aligned}
$$

$$(G.24)$$

$$
\begin{aligned}
p_{y0} &= p_{\perp 0} \sin \phi_0 = p_{\perp 0}\left[\sin \left(\frac{\omega t}{2} - \phi' \right) \right] \\
&= p_{\perp 0}\left[\cos \left(\cos^{-1} \frac{x}{r} \right) \sin \left(\frac{\omega t}{2} \right) - \sin \left(\sin^{-1} \frac{-y}{r} \right) \cos \left(\frac{\omega t}{2} \right) \right] \\
&= m_i v_{\perp 0}\left[\frac{x}{r} \sin \left(\frac{\omega t}{2} \right) + \frac{y}{r} \cos \left(\frac{\omega t}{2} \right) \right] = \frac{m_i v_{\perp 0}}{r}\left[x\sin \left(\frac{\omega t}{2} \right) + y\cos \left(\frac{\omega t}{2} \right) \right] \\
&= \frac{m_i v_{\perp 0}}{2\frac{v_{\perp 0}}{\omega}\left| \sin \left(\frac{\omega t}{2} \right) \right|}\left[x\sin \left(\frac{\omega t}{2} \right) + y\cos \left(\frac{\omega t}{2} \right) \right] \\
&\Rightarrow p_{y0} = \frac{m_i \omega}{2}\left[y_i \cot \left(\frac{\omega t_i}{2} \right) + x_i \right]
\end{aligned}
$$

$$(G.25)$$

as before.

Because the cyclotron period scales linearly with the mass of the ion, as a practical matter the ion will not be much affected by the magnetic field before it hits the ion detector. Therefore, we may neglect the influence of the magnetic field on the recoil ions and rewrite the last two equations of (G.4) as

$$\dot{v}_y = 0 \quad \text{and} \quad \dot{v}_x = 0 \tag{G.26}$$

If we now include the constant initial velocity in the positive y-direction that all the particles will have by their being part of a collimated, fast-moving jet, v_{jet}, we can solve the first of equations (G.26) as

$$
\dot{v}_y = 0 \Rightarrow \int_{v_{yinital}}^{v_y} dv_y' = 0 \Rightarrow v_y = v_{yinitial} = v_{y0} + v_{jet} \Rightarrow \int_0^y dy' = (v_{y0} + v_{jet})\int_0^t dt'
$$

$$\Rightarrow y(t) = (v_{y0} + v_{jet})t \tag{G.27}$$

which means

$$m_i y = (m_i v_{y0} + m_i v_{jet})t \Rightarrow p_{y0} = \frac{m_i(y_i - v_{jet})}{t_i} \tag{G.28}$$

We also have

$$x(t) = v_{x0}t \tag{G.29}$$

which leads to

$$p_{x0} = \frac{m_i x_i}{t_i} \tag{G.30}$$

Note that (G.27) and (G.29) imply that the ion trajectory lies in the plane given by

$$x(t) = \frac{v_{x0}y}{v_{y0} + v_{jet}} \tag{G.31}$$

Because we now have cylindrical symmetry, we are free to reorient the x-axis so that $v_{x0} = 0$. In that case, we can rewrite (G.27) as

$$y(t) = v_0 \sin \theta_0 t + v_{jet}t \tag{G.32}$$

We combine (G.32) with (G.5), eliminating the common TOF to show that the ion trajectory is a parabola (as suggested in Fig. G.1):

$$z(y) = \frac{v_0 \cos \theta_0}{v_0 \sin \theta_0 + v_{jet}} y + \frac{qE_0}{2m(v_0 \sin \theta_0 + v_{jet})^2} y^2 \tag{G.33}$$

Set $z(y) = Z_r$ in (G.5) to find the TOF of the ion:

$$t_i = \frac{-v_0 \cos \theta_0 + \sqrt{(v_0 \cos \theta_0)^2 + \frac{2qE_0 Z_r}{m_i}}}{qE_0/m_i} \tag{G.34}$$

where we have carefully chosen the positive root of the radical to ensure the TOF is positive. The good news is that all the factors under the radical are inherently positive, so there is no chance of having an imaginary TOF.

The y-coordinate of the ion's impact on the recoil ion detector can now be found by evaluating (G.32) at the time indicated in (G.34):

$$y = v_0 \sin \theta_0 t_i + v_{jet} t_i$$

$$= v_0 \sin \theta_0 \left[\frac{-v_0 \cos \theta_0 + \sqrt{(v_0 \cos \theta_0)^2 + \dfrac{2qE_0 Z_r}{m_i}}}{qE_0/m_i} \right]$$

$$+ v_{jet} \left[\frac{-v_0 \cos \theta_0 + \sqrt{(v_0 \cos \theta_0)^2 + \dfrac{2qE_0 Z_r}{m_i}}}{qE_0/m_i} \right]$$

$$= -\frac{v_0^2 \sin 2\theta_0}{2qE_0/m_i} + \frac{v_0}{qE_0/m_i} \left(\sin \theta_0 - v_{jet} \cos \theta_0 \right) \sqrt{(v_0 \cos \theta_0)^2 + \frac{2qE_0 Z_r}{m_i}}$$

$$\text{(G.35)}$$

Electron Detection

We can apply the same analysis used to find the components of the ion initial momentum vector to find the components of the initial momentum vectors of electrons produced in the Coulomb explosion. For example, from the vantage point of a position behind the electron detector, looking back toward the IR (in the positive z-direction), the electron describes a clockwise circle (see Fig. G.3).

From the figure,

$$\phi_0 = \phi - \frac{\omega t}{2} \qquad\qquad\qquad\qquad \text{(G.36a)}$$

$$\cos \phi = \frac{-x}{r} \qquad\qquad\qquad\qquad \text{(G.36b)}$$

$$\sin \phi = \frac{y}{r} \qquad\qquad\qquad\qquad \text{(G.36c)}$$

The rest of analysis is the same as that which led to (G.24) and (G.25), giving us,

Fig. G.3 Projection of a typical electron trajectory onto the x-y plane. Adapted from [151] and reproduced with permission from IOP publishing; permission conveyed through Copyright Clearance Center, Inc.

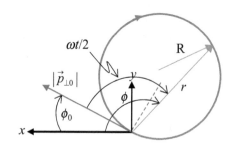

$$P_{x0} = \frac{m_e \omega}{2} \left[x_e \cot \left(\frac{\omega t_e}{2} \right) + y_e \right] \tag{G.37}$$

$$P_{y0} = \frac{m_e \omega}{2} \left[y_e \cot \left(\frac{\omega t_e}{2} \right) - x_e \right] \tag{G.38}$$

where t_e is the time of flight of the electron, m_e is the mass of the electron, and x_e and y_e are the x- and y-coordinates of the electron's impact on the electron detector, respectively.

We cannot neglect the influence of the magnetic field on the electrons (in fact, the magnetic field is there to confine the electrons), but because the magnetic field does no work (i.e., $|\vec{v}_\perp| = |\vec{v}_{\perp 0}|$), we can calculate the TOF for the electrons as we did for the ions above. First, we use (G.5) to find the time it takes the electron to travel from the IR to the coordinate $z = -Z_e$,

$$t_{Z_e} = \frac{-v_0 \cos \theta_0 + \sqrt{(v_0 \cos \theta_0)^2 + \frac{2qE_0 Z_e}{m_e}}}{qE_0/m_e} \tag{G.39}$$

Here q refers to the electron charge, and m_e is the electron mass. From above,

$$(v_z - v_{z0}) = \frac{qE_0}{m} t \Rightarrow v_z = v_{z0} + \frac{qE_0}{m} t \tag{G.40}$$

Therefore, the z-component of the speed of the electron when it reaches the end of the acceleration region will be

$$
\begin{aligned}
v_z|_{t_{Z_e}} &= v_{z0} + \frac{qE_0}{m_e} t_{Z_e} \\
&= v_0 \cos \theta_0 + \frac{qE_0}{m_e} \left[\frac{-v_0 \cos \theta_0 + \sqrt{(v_0 \cos \theta_0)^2 + \frac{2qE_0 Z_e}{m_e}}}{qE_0/m_e} \right] \\
&= \sqrt{(v_0 \cos \theta_0)^2 + \frac{2qE_0 Z_e}{m_e}}
\end{aligned}
\tag{G.41}
$$

The time t_d required for the electron to travel from $z = -Z_e$ to $z = -(Z_e + Z_d)$ is $Z_d/v_z|_{t_{Z_e}}$.

The TOF of the electron is $t_{Z_e} + t_d$:

$$
\begin{aligned}
t_e &= t_{Z_e} + t_d \\
&= \frac{-v_0 \cos \theta_0 + \sqrt{(v_0 \cos \theta_0)^2 + \frac{2qE_0 Z_e}{m_e}}}{qE_0/m_e} + \frac{Z_d}{\sqrt{(v_0 \cos \theta_0)^2 + \frac{2qE_0 Z_e}{m_e}}}
\end{aligned}
\tag{G.42}
$$

Including a drift region in the spectrometer is a way to account for the fact that the IR is not truly a point but as a practical matter has a finite size for which one must account. An electron that originates at a point some small distance h (along the z-axis) from the origin (so that $z = h + Z_e$) will have a TOF given by

$$t_e(z) = \frac{-v_0 \cos \theta_0 + \sqrt{(v_0 \cos \theta_0)^2 + \frac{2qE_0 z}{m_e}}}{qE_0/m_e} + \frac{Z_d}{\sqrt{(v_0 \cos \theta_0)^2 + \frac{2qE_0 z}{m_e}}} \qquad (G.43)$$

Because the distance h is small, we can expand $t_e(z)$ in a Taylor series about the point Z_e:

$$t_e(z) = t_e(Z_e) + h \frac{d[t_e(z)]}{dh}\bigg|_{z=Z_e} + \cdots \qquad (G.44)$$

We wish to ensure that the TOF deviation $\Delta TOF \equiv t_e(z) - t_e(Z_e)$ is zero:

$$\Delta TOF \equiv t_e(z) - t_e(Z_e) = \sum_{n=1}^{\infty} \frac{h^n}{n!} \frac{d^n[t_e(z)]}{dz^n}\bigg|_{z=Z_e} = 0 \qquad (G.45)$$

where we have noted that $dh = dz$. From (G.45), it is clear that, to achieve our aim, all the indicated derivatives must simultaneously be zero. As it turns out, this is not possible (given the spectrometer design shown above), so we settle for an approximate solution by requiring only the first derivative be zero:

$$\Delta TOF \approx \frac{d[t_e(z)]}{dz}\bigg|_{z=Z_e} = 0$$

$$\Rightarrow \left[\frac{1}{qE_0/m_e} \frac{qE_0/m_e}{\sqrt{(v_0 \cos \theta_0)^2 + \frac{2qE_0 z}{m_e}}} - \frac{Z_d qE_0/m_e}{\left[(v_0 \cos \theta_0)^2 + \frac{2qE_0 z}{m_e}\right]^{3/2}} \right]_{z=Z_e} = 0$$

$$\qquad (G.46)$$

We make the further approximation that the z-component of the electron's initial velocity, v_{z0}, is small compared to the velocity the electron gains while in the acceleration region:

$$v_0 \cos \theta_0 << \frac{2qE_0 Z_e}{m_e},$$

$$\Rightarrow \frac{1}{qE_0/m_e} \frac{qE_0/m_e}{\sqrt{\frac{2qE_0 Z_e}{m_e}}} - \frac{Z_d qE_0/m_e}{\left[\frac{2qE_0 Z_e}{m_e}\right]^{3/2}} = 0$$

$$\Rightarrow \frac{1}{\sqrt{\frac{2qE_0 Z_e}{m_e}}} = \frac{Z_d qE_0/m_e}{\left[\frac{2qE_0 Z_e}{m_e}\right]^{3/2}} \Rightarrow \frac{2qE_0 Z_e}{m_e} = \frac{Z_d qE_0}{m_e} \qquad (G.47)$$

$$\Rightarrow 2Z_e = Z_d$$

So, if we construct the drift region of the spectrometer to conform with the requirement in (G.47), electrons that have the same initial velocity but originate at different points within the IR will nevertheless arrive at the electron detector at the same time (to first-order) as long as they start with a small initial velocity as compared to the velocity they gain while in the acceleration region. This is an example of "first-order" space focusing. The beauty of this type of focusing is that it is a function only of the geometry of the spectrometer and is independent of the electric field strength.[12] This type of geometry is called a Wiley-McLaren geometry [178].

Using (G.47), we simplify (G.42) as follows:

$$
\begin{aligned}
t_e &= \frac{\sqrt{(v_0 \cos\theta_0)^2 + 2aZ_a}}{a} + \frac{Z_d}{\sqrt{(v_0 \cos\theta_0)^2 + 2aZ_a}} - \frac{v_0 \cos\theta_0}{a} \\
&= \sqrt{\frac{Z_d}{a}}\left[\sqrt{\frac{(v_0 \cos\theta_0)^2 + 2aZ_a}{aZ_d}} + \sqrt{\frac{aZ_d}{(v_0 \cos\theta_0)^2 + 2aZ_a}}\right] - \frac{v_0 \cos\theta_0}{a} \quad \text{(G.48)} \\
&\approx \sqrt{\frac{Z_d}{a}}\left[\sqrt{\frac{2aZ_a}{aZ_d}} + \sqrt{\frac{aZ_d}{2aZ_a}}\right] - \frac{v_0 \cos\theta_0}{a} \\
\Rightarrow t_e &= 2\sqrt{\frac{Z_d}{a}} - \frac{v_0 \cos\theta_0}{a}
\end{aligned}
$$

where we have defined the constant $a \equiv E_0 q/m_e$ and used the approximation $v_0 \cos\theta_0 \ll 2aZ_a$.

It must be emphasized again that the accuracy of measurement of the initial velocity vector of any charged particle produced in the Coulomb explosion is a function of the size of the IR. Thus, the size of the gas jet and the spot size of the photon beam will play a key role. As an example of how to get around this issue, we briefly examine the case of the Coulomb explosion of a diatomic molecule. For molecular fragments that do not originate at the origin, $(x_0, y_0) \neq (0, 0)$, we must modify (G.28) and (G.30) as follows:

$$
P_{y0} = \frac{m_i(y_i - y_0)}{t_i} - m_i v_{jet} \quad \text{(G.49a)}
$$

$$
P_{x0} = \frac{m_i(x_i - x_0)}{t_i} \quad \text{(G.49b)}
$$

---

[12] Strictly speaking, this type of focusing is only in the direction of the axis of the spectrometer. For focusing in directions perpendicular to the spectrometer, additional electrostatic lensing is required.

In the CM frame (frame of the molecule), the total momentum of the system is zero, and if we ignore the recoil due to any photoelectrons that may be produced in the Coulomb explosion, then the momentum relative to the CM, $\vec{p}_{rel}$, can be written as

$$\vec{p}_{0rel} = \vec{p}_{01} = -\vec{p}_{02} \tag{G.50}$$

where $\vec{p}_{01}$ and $\vec{p}_{02}$ are the initial momenta of the two ion fragments (as measured in the lab frame).

Combining (G.49a) and (G.50),

$$\frac{m_1(y_1 - y_0)}{t_1} - m_1 v_{jet}$$

$$= -\frac{m_2(y_2 - y_0)}{t_2} + m_2 v_{jet}$$

$$\Rightarrow \frac{m_1 y_1}{t_1} - m_1 v_{jet} + \frac{m_2 y_2}{t_2} - m_2 v_{jet} = y_0 \left(\frac{m_1}{t_1} + \frac{m_2}{t_2} \right) \tag{G.51}$$

$$\Rightarrow \frac{m_1 y_1 t_2 + m_2 y_2 t_1}{t_1 t_2} - v_{jet}(m_1 + m_2) = y_0 \frac{m_1 t_2 + m_2 t_1}{t_1 t_2}$$

$$\Rightarrow y_0 = \frac{m_1 y_1 t_2 + m_2 y_2 t_1 - v_{jet}(m_1 + m_2)t_1 t_2}{m_1 t_2 + m_2 t_1}$$

Thus,

$$\left(\vec{p}_{0rel} \right)_y$$

$$= \left(\vec{p}_{01} \right)_y = \frac{m_1 y_1}{t_1} - \frac{m_1}{t_1} \left[\frac{m_1 y_1 t_2 + m_2 y_2 t_1 - v_{jet}(m_1 + m_2)t_1 t_2}{m_1 t_2 + m_2 t_1} \right] - m_1 v_{jet}$$

$$= \frac{m_1}{t_1} \left[\frac{y_1(m_1 t_2 + m_2 t_1) - m_1 y_1 t_2 - m_2 y_2 t_1 + v_{jet}(m_1 + m_2)t_1 t_2 - v_{jet}t_1(m_1 t_2 + m_2 t_1)}{m_1 t_2 + m_2 t_1} \right]$$

$$= \frac{m_1}{t_1} \left[\frac{y_1 m_2 t_1 - m_2 y_2 t_1 + v_{jet}m_2 t_1 t_2 - v_{jet}t_1^2 m_2}{m_1 t_2 + m_2 t_1} \right]$$

$$\Rightarrow \left(\vec{p}_0 \right)_y = m_1 m_2 \left[\frac{(y_1 - y_2) + v_{jet}(t_2 - t_1)t_2}{m_1 t_2 + m_2 t_1} \right] \tag{G.52}$$

From (G.52), we can immediately write the expression for $\left(\vec{p}_{0rel} \right)_x$ by setting $v_{jet} = 0$, since $\left(\vec{v}_{jet} \right)_x = 0$,

$$\left(\vec{p}_{0rel} \right)_x = \left(\vec{p}_{01} \right)_x = m_1 m_2 \left[\frac{(x_1 - x_2)}{m_1 t_2 + m_2 t_1} \right] \tag{G.53}$$

From (G.50) and (G.5),

$$\frac{m_1 Z_r}{t_1} - \frac{q_1 E_0 t_1}{2} = -\frac{m_2 Z_r}{t_2} + \frac{q_2 E_0 t_2}{2} \Rightarrow Z_r \left(\frac{m_1}{t_1} + \frac{m_2}{t_2} \right) = \frac{E_0}{2} (q_1 t_1 + q_2 t_2)$$

$$\Rightarrow Z_r = \frac{E_0}{2} (q_1 t_1 + q_2 t_2) \left(\frac{t_1 t_2}{m_1 t_2 + m_2 t_1} \right)$$

$$(G.54)$$

Thus,

$$\left(\vec{P}_{0rel} \right)_z = \left(\vec{P}_{01} \right)_z = \frac{m_1}{t_1} \left[\frac{E_0}{2} (q_1 t_1 + q_2 t_2) \left(\frac{t_1 t_2}{m_1 t_2 + m_2 t_1} \right) \right] - \frac{q_1 E_0 t_1}{2} \left(\frac{m_1 t_2 + m_2 t_1}{m_1 t_2 + m_2 t_1} \right)$$

$$= \frac{E_0}{2} \left[\frac{m_1 q_1 t_1 t_2 + m_1 q_2 t_2^2 - q_1 t_1 m_1 t_2 - q_1 m_2 t_1^2}{m_1 t_2 + m_2 t_1} \right]$$

$$\Rightarrow \left(\vec{P}_0 \right)_z = \frac{E_0}{2} \left[\frac{m_1 q_2 t_2^2 - m_2 q_1 t_1^2}{m_1 t_2 + m_2 t_1} \right]$$

$$(G.55)$$

So, in the case of a diatomic molecule, the components of the initial ion fragment momentum vectors can be found by measuring the TOF of both fragments and the coordinates (x_1, y_1) and (x_2, y_2) of the impact of the two fragments on the ion detector, regardless of any uncertainty as to where (within the IR) the fragments originated.

For molecules that break up into two charged ionic fragments via several breakup channels, (G.54) can be used to plot the dependence of t_2 on t_1:

$$Z_r = \frac{E_0}{2} (q_1 t_1 + q_2 t_2) \left(\frac{t_1 t_2}{m_1 t_2 + m_2 t_1} \right) \Rightarrow \frac{2 Z_r}{E_0} (m_1 t_2 + m_2 t_1) = q_1 t_1^2 t_2 + q_2 t_2^2 t_1$$

$$\Rightarrow (q_2 t_1) t_2^2 + \left(q_1 t_1^2 - \frac{2 Z_r m_1}{E_0} \right) t_2 + \left(-\frac{2 Z_r m_2 t_1}{E_0} \right) = 0$$

$$\Rightarrow t_2 = \frac{-q_1 t_1^2 + \frac{2 Z_r m_1}{E_0} + \sqrt{\left(q_1 t_1^2 - \frac{2 Z_r m_1}{E_0} \right)^2 + 4 q_2 t_1 \left(\frac{2 Z_r m_2 t_1}{E_0} \right)}}{2 q_2 t_1}$$

$$\Rightarrow t_2 = \frac{Z_r m_1}{E_0 q_2 t_1} - \frac{q_1 t_1}{2 q_2} + \sqrt{\frac{q_1^2 t_1^4 - 4 q_1 Z_r m_1 t_1^2 / E_0 + 4 Z_r^2 m_1^2 / E_0^2 + 8 q_2 Z_r m_2 t_1^2 / E_0}{4 q_2^2 t_1^2}}$$

$$= \frac{Z_r m_1}{E_0 q_2 t_1} - \frac{q_1 t_1}{2 q_2} + \sqrt{\frac{q_1^2 t_1^2}{4 q_2^2} - \frac{q_1 Z_r m_1}{q_2^2 E_0} + \frac{Z_r^2 m_1^2}{q_2^2 t_1^2 E_0^2} + \frac{2 Z_r m_2}{q_2 E_0}}$$

$$\Rightarrow t_2 = \frac{Z_r m_1}{E_0 q_2 t_1} - \frac{q_1 t_1}{2 q_2} + \sqrt{\left(\frac{Z_r m_1}{q_2 t_1 E_0} - \frac{q_1 t_1}{2 q_2} \right)^2 + \frac{2 Z_r m_2}{q_2 E_0}}$$

$$(G.56)$$

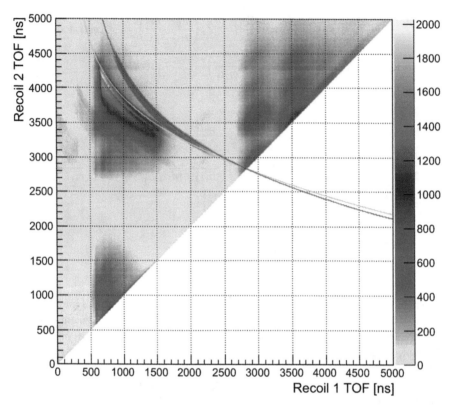

Fig. G.4 PIPICO plot for the photodissociation of methane. The (simulated) solid PIPICO lines are calculated via (G.56) for the reactions (1) $CH_4 + \gamma \rightarrow CH_4^* + e_\gamma^- \rightarrow CH_2^+ + H_2^+ + e_\gamma^- + e_{Auger}^-$ and (2) $CH_4 + \gamma \rightarrow CH_4^* + e_\gamma^- \rightarrow CH_3^+ + H^+ + e_\gamma^- + e_{Auger}^-$. The simulated (red and green) lines are overlaid on top of actual experimental data. The number of experimentally-recorded events is indicated by the different colors appearing in the diagram (see the vertical scale to the right of the plot) [179]. Adapted and reproduced from [179] under the Creative Commons Attribution-ShareAlike 3.0 International License. To view a copy of this license, see https://creativecommons.org/licences/by-sa/3.0/legalcode

Equation (G.56) yields unique curves for plots of t_2 vs. t_1 for different breakup channels of the molecule (as long as there are exactly two ionic fragments). These plots are called PIPICO (photoion/photoion coincidence) plots (see Figs. G.4 and G.6). Recorded events that do not lie along these curves can be eliminated as noise. Recorded events that lie outside the IR can also be discarded [c.f. (G.51)].

With the COLTRIMS, one can detect all fragments emitted in a molecular breakup event in coincidence, which allows for the creation of a fixed-in-space MFPAD to be reconstructed after the fact. A typical COLTRIMS-generated MFPAD with error bars is shown in Fig. G.5, and the accompanying PIPICO is shown in Fig. G.6. The current technology that supports the COLTRIMS technique allows the impact location of individual particles to be known to within $50\mu m$ and

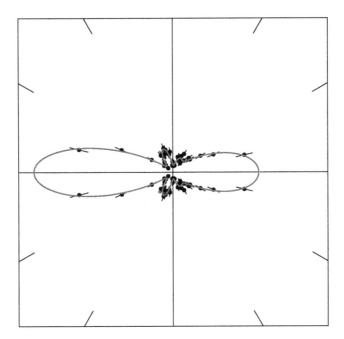

Fig. G.5 MFPAD of the photoionization of a valence electron in CO. A 48.4 eV, linearly polarized photon was used to ionize the molecule, the axis of which was parallel to the polarization vector. The "dots" represent experimental data (along with the associated error bars) and the solid (red) line is a theoretical fit to the data. The MFPAD shows the rich features corresponding to strong d- and f-wave contributions that are typical when the transition is close to a shape resonance of the molecule. (If the MFPAD was due to circularly polarized photons, the forward- and rear-facing lobes would not lie in the direction of the molecular axis as they do here but would instead be tilted at an angle that depends on the photoelectron kinetic energy—see Chap. 10) [154, 180]. Reproduced from [180] with permission from the author

Fig. G.6 PIPICO for the MFPAD of Fig. G.5. The faint images that are present in the upper part of the figure are "echoes" of the actual PIPICO trace and are caused by random temporal fluctuations on the part of the collection electronics [180]. Reproduced from [180] with permission from the author

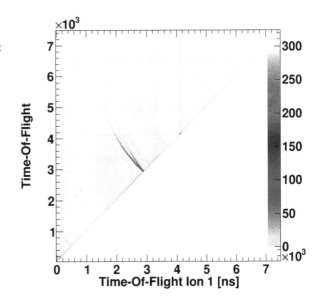

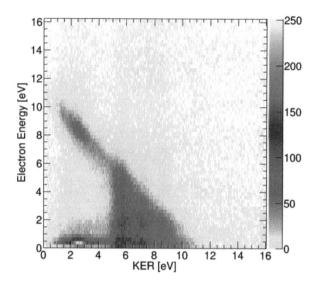

Fig. G.7 Histogram depicting the kinetic energy release (KER) of the recoil ions verses photo-electron kinetic energy (when a molecule fragments, the internal energy of the molecule is often released as kinetic energy of the ionic recoil fragments in a process called KER). These types of diagrams are useful in determining which photodissociation reaction channel is being measured. In this diagram, the diagonal structure located at 6 eV <EE < 11 eV and 1 eV < KER < 5 eV corresponds to the photoelectron MFPAD of Fig. G.5. Also present is a low-energy autoionizing electron (near 0 eV), which is clearly distinguishable from the photoelectron. The triangular feature at higher KER is most likely the result of a mixture of photoionization + autoionization and direct double ionization, the latter of which yields two electrons with broad energy sharing. It is this sharing that results in the filled triangular shape at the bottom center of the figure [180]. Reproduced from [180] with permission from the author

the arrival time to within less than 100 ps, allowing for a single-particle momentum resolution of below 0.01 a.u. As mentioned in Chap. 10, one can then infer relative timing of the entangled dynamical events of photon-molecule interactions to within one attosecond or less. Advances in the design and manufacture of position-sensitive detectors, the development of PC-based, multiparameter data storage systems which can store vast amounts of data (from which many different types of analyses can be conducted—see, e.g., Figs. 10.10, 10.11, and G.7), the introduction of specialized techniques in electrostatic focusing of ions and magnetic confinement of high-energy electrons (allowing for a near-unity collection efficiency over the entire 4π solid angle), along with the introduction of supersonic gas jets were all necessary before these types of resolutions could be achieved. It is conceivable, with the available resolutions, that molecular dynamics on the order of zeptoseconds (10^{-21} s) could be measured. For more information and for an in-depth discussion of the history of the development of the COLTRIMS, see [170] and the references therein.

Bibliography

1. C.N. Yang, Phys. Rev. **74**, 764 (1948)
2. G. Racah, Phys. Rev. **61**, 186 (1942), Phys. Rev. **62**, 438 (1942)
3. R.N. Zare, *Angular Momentum: Understanding Spatial Aspects in Chemistry and Physics* (Wiley, New York, 1988)
4. M.E. Rose, *Elementary Theory of Angular Momentum* (Wiley, New York, 1957)
5. A.R. Edmonds, *Angular Momentum in Quantum Mechanics* (Princeton University Press, New Jersey, 1960)
6. J.J. Sakurai, J. Napolitano, *Modern Quantum Mechanics*, 3rd edn. (Cambridge University Press, Cambridge, 2021)
7. L.C. Biedenharn, J.D. Louck, *Angular Momentum in Quantum Physics; Theory and Application [Encyclopedia of Mathematics, Vol 8, Gian-Carlo Rota (ed.)]* (Addison-Wesley Publishing Company, London, 1981)
8. C. Cohen-Tannoudji, B. Diu, F. Laloe, *Quantum Mechanics*, vol I–II, 2nd edn. (Wiley, New York, 1977)
9. G.B. Arfken, H.J. Weber, F.E. Harris, *Mathematical Methods for Physicists*, 7th edn. (Elsevier, San Diego, 2013)
10. D.A. Varshalovich, A.N. Moskalev, V.K. Khersonskii, *Quantum Theory of Angular Momentum* (World Scientific Publishing, New Jersey, 1988)
11. J.D. Jackson, *Classical Electrodynamics*, 3rd edn. (Wiley, New York, 1999)
12. D.J. Griffiths, *Introduction to Electrodynamics*, 4th edn. (Cambridge University Press, Cambridge, 2017)
13. J. Cooper, R.N. Zare, Photoelectron Angular Distributions, in *Lectures in Theoretical Physics*, ed. by S. Geltman, K. T. Mahanthappa, W. E. Britten, vol. XI-c, (Gordon and Breach, New York, 1969)
14. R.N. Zare, Mol. Photochem. **4**, 1–37 (1972)
15. S. Gasiorowicz, *Quantum Physics*, 2nd edn. (Wiley, New York, 1996)
16. I. Lindgren, J. Morrison, *Atomic Many-Body Theory*, 2nd edn. (Springer, Berlin, 1986)
17. H. Hotop, T.A. Paterson, W.C. Lineberger, Phys. Rev. A **8**, 762 (1973)
18. E.P. Wigner, Phys. Rev. **73**, 1002 (1948)
19. W. Demtroder, *Atoms, Molecules and Photons; An Introduction to Atomic-, Molecular- and Quantum-Physics* (Springer, Berlin, 2006)
20. D. Hanstorp, C. Bengtsson, D.J. Larson, Phys. Rev. A **40**, 670 (1989)
21. E.R. Grumbling, A. Sanov, J. Chem. Phys. **135**, 164302 (2011)
22. F. Breyer, P. Frey, H. Hotop, Z. Phys. A **286**, 133 (1978)

© The Editor(s) (if applicable) and The Author(s), under exclusive license to
Springer Nature Switzerland AG 2022
V. T. Davis, *Introduction to Photoelectron Angular Distributions*, Springer Tracts in
Modern Physics 286, https://doi.org/10.1007/978-3-031-08027-2

23. K. Chartkunchand, Dissertation (University of Nevada, Reno, 2015)
24. H. Hotop, W.C. Lineberger, J. Phys. Chem. Ref. Data **14**, 731 (1985)
25. T. Andersen, H.K. Haugen, H. Hotop, J. Phys. Chem. Ref. Data **28**, 1511 (1999)
26. U. Fano, Phys. Rev. **124**, 1866 (1961), Phys. Rev. **140**, A67 (1965)
27. J. Cooper, R. N. Zare, J. Chem. Phys. **48**, 942 (1968), J. Chem. Phys. **49**, 4252 (1968)
28. C.M. Oana, A.I. Krylov, J. Chem. Phys. **131**, 124114 (2009)
29. S.T. Manson, J. Electron Spectrosc. Relat. Phenom. **37**, 37–56 (1985)
30. D.J. Kennedy, S.T. Manson, Phys. Rev. A **5**, 227 (1972)
31. M.Y. Amusa, N.A. Cherepkov, L.V. Chernysheva, Phys. Lett. **40A**, 15 (1972)
32. G.R. Branton, C.E. Brion, J. Electron Spectrosc. Relat. Phenom. **3**, 123–128 (1974)
33. M. J. Van der Wiel and C. E. Brion, J. Electron Spectrosc. Relat. Phenom. 1, 439 (1972/73)
34. R.G. Houlgate, J.B. West, K. Codling, G.V. Marr, J. Electron Spectrosc. Relat. Phenom. **9**, 205–209 (1976)
35. G. Aravind, A.K. Gupta, M. Krishnamurthy, E. Krishnakumar, Phys. Rev. A **75**, 042714 (2007)
36. G. Aravind, N. Bhargava Ram, A.K. Gupta, E. Krishnakumar, Phys. Rev. A **79**, 043411 (2009)
37. W.W. Williams, D.L. Carpenter, A.M. Covington, J.S. Thompson, Phys. Rev. A **59**, 4368 (1999)
38. S.J. Cavanagh, S.T. Gibson, M.N. Gale, C.J. Dedman, E.H. Roberts, B.R. Lewis, Phys. Rev. A **76**, 052708 (2007)
39. D. Calabrese, A.M. Covington, W.W. Williams, D.L. Carpenter, J.S. Thompson, J. Kvale, Phys. Rev. A **71**, 042708 (2005)
40. D. Calabrese, A.M. Covington, D.L. Carpenter, J.S. Thompson, T.J. Kvale, R. Collier, J. Phys. B **30**, 4791 (1997)
41. V.T. Davis, J. Ashokkumar, J.S. Thompson, Phys. Rev. A **65**, 024702 (2002)
42. See Ref. 37
43. S.J. Cavanagh, S.T. Gibson, M.N. Gale, C.J. Dedman, E.H. Roberts, B.L. Lewis, Phys. Rev. A **76**, 052708 (2007)
44. See Ref. 35
45. J.S. Thompson, D.J. Pegg, R.N. Compton, G.D. Alton, J. Phys. B **23**, L15–L19 (1990)
46. A.M. Covington, D. Calabrese, W.W. Williams, J.S. Thompson, T.J. Kvale, Phys. Rev. A **56**, 4746 (1997)
47. A.M. Covington, S.S. Duvvuri, E.D. Emmons, R.G. Krauss, W.W. Williams, J.S. Thompson, D. Calabrese, D.L. Carpenter, R.D. Collier, T.J. Kvale, V.T. Davis, Phys. Rev. A **75**, 022711 (2007)
48. J. Cooper, Phys. Rev. A **42**, 6942 (1990), Phys. Rev. A **45**, 3362 (1992)
49. J. Cooper, Phys. Rev. A **47**, 1841 (1993)
50. A. Derivianko, W.R. Johnson, Data Nucl. Data Tables **73**, 153–211 (1999)
51. H.K. Tseng, R.H. Pratt, S. Yu, A. Ron, Phys. Rev. A **17**, 1061 (1978)
52. P.S. Shaw, U. Arp, S.H. Southworth, Phys. Rev. A **54**, 1463 (1996)
53. M. Ya Amus'ya, A.S. Baltenkov, A.A. Grinberg, S.G. Shapiro, Sov. Phys. JETP **41**, 14 (1975)
54. M. Jung, B. Krassig, D.S. Gremmel, E.P. Kanter, T. LeBrun, S.H. Southworth, L. Young, Phys. Rev. A **54**, 2127 (1996)
55. O. Hemmers, G. Fisher, P. Glans, D.L. Hanesn, H. Wang, S.B. Whitfield, R. Wehlitz, J.C. Levin, I.A. Sellin, R.C.C. Perera, E.W.B. Dias, H.S. Chakraborty, P.C. Desmukh, S.T. Manson, D.W. Lindle, J. Phys. B **30**, L727 (1997)
56. M. Ya Amus'ya, A.S. Baltenkov, Z. Felfli, A.Z. Msezane, Phys. Rev. A **63**, 052506 (2001)
57. M. Ya Amus'ya, A.S. Baltenkov, L.V. Chernysheva, Z. Felfli, A.Z. Msezane, Phys. Rev. A **59**, R2544 (1999), Phys. Rev. A **72**, 032727 (2005)
58. V.K. Domatov, A.S. Baltenkov, S.T. Manson, Phys. Rev. A **67**, 062714 (2003)
59. V.K. Domatov, A.S. Baltenkov, S.T. Manson, Phys. Rev. A **64**, 042718 (2001)

60. A Derivianko, O. Hemmers, S. Oblad, P. Glans, H. Wang, S.B. Whitfield, R. Wehlitz, I.A. Sellin, W.R. Johnson and D.W. Lindle, Phys. Rev. Lett. 84, 2116 (2000)
61. M.B. Trzhaskovskaya, V.K. Nikulin, V.I. Nefedov, V.G. Yarzhemsky, Data Nucl. Data Tables **92**, 245–304 (2006)
62. L. Argenti, R. Moccia, J. Phys. B **43**, 235006 (2010)
63. P.C. Deshmukh, T. Banargee, H.R. Varma, O. Hemmers, R. Guillemin, D. Rolles, A. Wolska, S.W. Yu, D.W. Lindle, W.R. Johnson, S.T. Manson, J. Phys. B **41**, 021002 (2008)
64. A. Bechler, R.H. Pratt, Phys Rev A **39**, 1774 (1989), Phys Rev A **42**, 6400 (1990)
65. M. Ya Amusia, L.V. Chernysheva, J. Phys. B **39**, 4627–4636 (2006)
66. T.E.H. Walker, J.T. Waber, J. Phys. B **6**, 1165 (1973), Phys. Rev. Lett. **30**, 307 (1973)
67. T.E.H. Walker, J.T. Waber, J. Phys. B **7**, 674 (1974)
68. T.E.H. Walker, J. Berkowitz, J.L. Dehmer, J.T. Weber, Phys. Rev. Lett. **31**, 678 (1973)
69. W.R. Johnson, C.D. Lin, Phys. Rev. A **20**, 964 (1979)
70. O. Hemmers, S.T. Manson, M.M. Sant'Anna, P. Focke, H. Wang, I.A. Sellin, D.W. Lindle, Phys. Rev. A **64**, 022507 (2001)
71. H. Harrison, J. Chem. Phys. **19**, 901–905 (1970)
72. W.R. Alling, W.R. Johnson, Phys. Rev. **139**, A1050 (1965)
73. P. Strange, *Relativistic Quantum Mechanics with Applications in Condensed Matter and Atomic Physics* (Cambridge University Press, Cambridge, 1998)
74. I.P. Grant, Adv. Phys. **19**, 747–811 (1970), Proc. R. Soc. Lond. Ser. A Math. Phys. Sci. **262**, 555–576 (1961)
75. W.R. Johnson, *Lectures on Atomic Physics* (Notre Dame, Indiana, 2006)
76. D.M. Brink, G.R. Satchler, *Angular Momentum*, 2nd edn. (Clarendon Press, Oxford, 1968)
77. I.P. Grant, *Relativistic Quantum Theory of Atoms and Molecules; Theory and Computation* (Springer, Oxford, 2007)
78. W.R. Johnson, K.T. Chang, Phys. Rev. A **20**, 978 (1979)
79. L.A. Lajohn, S.T. Manson, R.H. Pratt, Nucl. Instrum. Methods Phys. Res. A **619**, 7–9 (2010)
80. M.B. Trzhaskovskaya, V.K. Nikulin, V.I. Nefedov, V.G. Yarzhemsky, J. Phys. B **34**, 3221–3237 (2001)
81. A. Ron, R.H. Pratt, H.K. Tseng, Chem. Phys. Lett. **47**, 377 (1977)
82. H. Harrison, J. Chem. Phys. **52**, 901 (1970)
83. J.L. Dehmer, Phys. Rev. A **12**, 1966 (1975)
84. Y.-Y. Yin, D.S. Eliot, Phys. Rev. A **45**, 281 (1992)
85. S. Shahabi, Phys. Lett. **72A**, 212 (1979)
86. P.C. Deshmukh, V. Radojevic, S.T. Manson, Phys. Rev. A **45**, 6339 (1992)
87. U. Fano, D. Dill, Phys. Rev. A **6**, 185 (1972)
88. D. Dill, U. Fano, Phys. Rev. Lett. **29**, 1203 (1972)
89. M. Rotenberg, R. Bivins, N. Metropolis, J.K. Wooten Jr., *The 3-j and 6-j symbols* (The Technology Press, Cambridge, MA, 1959)
90. S.T. Manson, A.F. Starace, Rev. Mod. Phys. **54**, 389 (1982)
91. D. Dill, Phys. Rev. A **7**, 1976 (1973)
92. D. Dill, A.F. Starace, S.T. Manson, Phys. Rev. A **11**, 1596 (1975)
93. D. Dill, S.T. Manson, A.F. Starace, Phys. Rev. Lett. **32**, 971 (1974)
94. A.F. Starace, R.H. Rast, S.T. Manson, Phys. Rev. Lett. **38**, 1522 (1977)
95. S.T. Manson, in *Many-body Atomic Physics*, ed. by J. J. Boyle, M. S. Pindzola, (Cambridge University Press, Cambridge, 1998), p. 167
96. W. Ong, S.T. Manson, Phys. Rev. A **20**, 2364 (1979)
97. C.J. Dai, Phys. Rev. A **53**, 3237 (1996)
98. S.T. Manson, J. Electron Spectrosc. Relat. Phenom. **66**, 117–123 (1993)
99. E.S. Chang, K.T. Taylor, J. Phys. B **11**, L507 (1978)
100. C. Sinanis, Y. Komminos, C.A. Nicolaides, Phys. Rev. A **51**, R2672 (1995)
101. S. Shahabi, A.F. Starace, Phys. Rev. A **33**, 2111 (1986)
102. A.F. Starace, S.T. Manson, D.J. Kennedy, Phys. Rev. A **9**, 2453 (1974)

103. M. Eypper, F. Innocenti, A. Morris, S. Stranges, J.B. West, G.C. King, J.M. Dyke, J. Chem. Phys. **132**, 244304 (2010)

104. M. Eypper, F. Innocenti, A. Morris, J.M. Dyke, S. Stranges, J.B. West, G.C. King, J. Chem. Phys. **133**, 084302 (2010)

105. Y. Liu, D.J. Pegg, J.S. Thompson, J. Dellwo, G.D. Alton, J. Phys. B **24**, L1–L5 (1991)

106. B. Langer, J. Viefhaus, O. Hemmers, A. Menzel, R. Wehlitz, U. Becker, Phys. Rev. A **51**, R882 (1995)

107. V.P. Krainov, H.R. Reiss, B.M. Smirnov, *Radiative Processes in Atomic Physics* (Wiley, New York, 1997)

108. P.C. Engelking, W.C. Lineberger, Phys. Rev. A **19**, 149 (1979)

109. D. Dill, J. Chem. Phys. **65**, 1130 (1976)

110. S. Wallace, D. Dill, Phys. Rev. B **17**, 1692 (1978)

111. D. Dill, J.L. Dehmer, J. Chem. Phys. **61**, 692 (1974)

112. R.R. Lucchese, J. Electron Spectrosc. Relat. Phenom. **141**, 201–210 (2004)

113. M. Tadjeddine, G. Bouchoux, L. Malegat, J. Durup, C. Pernot, J. Weiner, Chem. Phys. **69**, 229–246 (1982)

114. S. Yang, R. Bersohn, J. Chem. Phys. **61**, 4400 (1974)

115. C. Miron, Q. Miao, C. Nicolas, J.D. Bozek, W. Andralojc, M. Patanen, G. Simoes, O. Travnikova, H. Agren, F. Gelmukhanov, Nat. Commun. **5**, 3816 (2014). https://doi.org/10.1038/ncomms4816

116. M. Schidbauer, A.L.D. Kilcoyne, H.-M. Koppe, J. Feldhaus, A.M. Bradshaw, Chem. Phys. Lett. **199**, 119 (1992)

117. E. Shigemasa, Nucl. Instrum. Methods B **99**, 132–135 (1995)

118. N. Chandra, Phys. Rev. A **36**, 3163 (1987), J. Phys. B **20**, 3405–3415 (1987)

119. J.L. Dehmer, D. Dill, J. Chem. Phys. **65**, 5327 (1976)

120. E.S. Chang, J. Phys. B **11**, L293 (1978)

121. J.C. Tully, R.S. Berry, B.J. Dalton, Phys. Rev. **176**, 95 (1968)

122. A.N. Grum-Grzhimalo, J. Phys. B **36**, 2385–2407 (2003)

123. A.N. Grum-Grzhimalo, R.R. Lucchese, X.J. Liu, G. Prumper, Y. Morishita, N. Saito, K. Ueda, J. Electron. Spectrosc. Relat. Phenom. **155**, 100–103 (2007)

124. G.M. Seabra, I.G. Kaplan, J.V. Ortiz, J. Chem. Phys. **123**, 114105 (2005)

125. R. Guillemin, O. Hemmers, D.W. Lindle, E. Shigemasa, K. Le Guen, D. Ceolin, C. Miron, N. Leclercq, P. Morin, M. Simon, P.W. Langhoff, Phys. Rev. Lett. **89**, 033002–033001 (2002)

126. P.W. Langhoff, J. Electron. Spectrosc. Relat. Phenom. **114–116**, 23–32 (2001)

127. N.A. Cherepkov, J. Phys. B **14**, 2165–2177 (1981)

128. A.D. Buckingham, Philos. Trans. R. Soc. Lond. A **268**, 147–157 (1970)

129. C. Jonah, J. Chem. Phys. **55**, 1915 (1971)

130. F.A. Grimm, Chem. Phys. **53**, 71–75 (1980)

131. E. Kukk, D. Ayuso, T.D. Thomas, P. Declava, M. Patanen, L. Argenti, E. Plesiat, A. Palacios, K. Kooser, O. Travnioka, S. Mondal, M. Kimura, K. Sakai, C. Miron, F. Martin, K. Ueda, Phys. Rev. A **88** (2013)

132. S. Southworth, W.D. Brewer, C.M. Truesdale, P.H. Korbin, D.W. Lindle, D.A. Shirley, J. Electron. Spectrosc. Relat. Phenom. **26**, 43–51 (1982)

133. L. Chow Chu, S.R. Samanta, J. Quant. Spectrosc. Radiat. Transf. **25**, 253–264 (1981)

134. C.M. Truesdale, S.H. Southworth, P.H. Korbin, U. Becker, D.W. Lindle, H.G. Kerkhoff, D.A. Shirley, Phys. Rev. A **50**, 1265 (1983)

135. T. Weber, O. Jagutzki, A. Staudte, A. Nauert, L. Schmidt, M.H. Prior, A.L. Landers, A. Brauning-Demian, H. Brauning, C.L. Cocke, T. Osipov, I. Ali, R.D. Muino, D. Rolles, F.J.G. de Abajo, C.S. Fadley, M.A. Van Hove, A. Cassimi, H. Schmidt-Bocking, R. Dorner, J. Phys. B **34**, 3669–3678 (2001)

136. R.R. Lucchese, R. Monturo, A.N. Grum-Grzhimailo, X.-J. Liu, G. Prumper, Y. Morishita, N. Saito, K. Ueda, J. Electron. Spectrosc. Relat. Phenom. **155**, 95–97 (2007)

137. S. Katsumata, Y. Achiba, K. Kimura, J. Electron. Spectrosc. Relat. Phenom. **17**, 229–236 (1979)
138. D. Dill, S. Wallace, J. Siegel, J.L. Dehmer, Phys. Rev. Lett. **41**, 1230 (1978)
139. X.-J. Liu, R.R. Lucchese, A.N. Grum-Grzhimailo, Y. Morishita, N. Saito, G. Prumper, K. Ueda, J. Phys. B **40**, 485–496 (2007)
140. See Ref. 128
141. E. Surber, R. Mabbs, A. Sanov, J. Phys. Chem. **107**, 8215–8224 (2003)
142. H. Fukuzawa, X.-J. Liu, R. Montuoro, R.R. Lucchese, Y. Morishita, N. Saito, M. Kato, I.H. Suzuki, Y. Tamenori, T. Teranishi, T. Lischke, G. Prumper, K. Ueda, J. Phys. B **41**, 045102 (2008)
143. M. Ilchen, S. Deinert, L. Glaser, F. Scholz, J. Seltmann, P. Walter, J. Viefhaus, J. Phys. B **45**, 225102 (2012)
144. B. Ritchie, Phys Rev A **13**, 1411 (1976), Phys Rev A **14**, 359, Phys Rev A **14**, 1396 (1976)
145. A. Einstein, Ann. Phys. Leipzig **16**, 132 (1899), Ann. Phys. Leipzig **20**, 199 (1906)
146. D. Hanstorp, Nucl. Instrum. Methods Phys. Res. B **100**, 165–175 (1995)
147. O. Windelius, J. Welander, A. Aleman, D.J. Pegg, K.V. Jayaprasad, S. Ali, D. Hanstorp, Photoelectron angular distributions in photodetachment from P. Phys. Rev. A **103**, 033108 (2021)
148. A.T.J.B. Eppink, D.H. Parker, Rev. Sci. Instrum. **68**, 3477 (1997)
149. K. Chartkunchand, K.R. Carpenter, V.T. Davis, P.A. Neill, J.S. Thompson, A.M. Covington, T.J. Kvale, Unpublished Research Document
150. R. Dorner, V. Mergel, O. Jagutzki, L. Spielberger, J. Ullrich, R. Moshammer, H. Schmidt-Bocking, Phys. Rep. **330**, 95–192 (2000)
151. J. Ullrich, R. Moshammer, A. Dorn, R. Dorner, L.P.H. Schmidt, H. Schmidt-Bocking, Recoil-ion and electron momentum spectroscopy: reaction-microscopes. Rep. Prog. Phys. **66**, 1463–1545 (13 Aug 2003)
152. T. Jahnke, T. Weber, T. Osipov, A.L. Landers, O. Jagutzki, L.P.H. Schmidt, C.L. Cocke, M.H. Prior, H. Schmidt-Bocking, R. Dorner, J. Electron. Spectrosc. Relat. Phenom. **141**, 229–238 (2004)
153. N. Melzer, F. Trinter, J. Rist, L. Kaiser, K. Klyssek, C. Schwarz, N. Anders, J. Siebert, D. Tsitsonis, I. Vela Perez, D. Trabert, M. Kircher, S. Grundmann, L. Schmidt, V. Davis, J.B. Williams, T. Jahnke, To be Published
154. I. Vela-Perez, F. Ota, A. Mhamdi, Y. Tamura, J. Rist, N. Melzer, S. Uerken, G. Nalin, N. Anders, D. You, M. Kircher, C. Janke, M. Waitz, F. Trinter, R. Guillemin, M. N. Piancastelli, M. Simon, V.T. Davis, J. B. Williams, R. Dorner, K. Hatada, K. Yamazaki, K. Fehre, Ph.V. Demekhin, K. Ueda, M.S. Schoffler, T. Jahnke, To be Published
155. D. Call, M. Weller, G. Kastirke, G. Panelli, R.A. Strom, S. Burrows, K. Larsen, N. Melzer, W. Iskander, R. Dorner, T. Jahnke, A.L. Landers, J.B. Williams, To be Published
156. C.S. Trevisan, C.W. McCurdy, T.N. Rescigno, J. Phys. B **45**, 194002 (2012)
157. J.B. Williams, C.S. Trevisan, M.S. Schoffler, T. Jahnke, I. Bocharova, H. Kim, B. Ulrich, R. Wallauer, F. Sturm, T.N. Rescigno, A. Belkacem, R. Dorner, Th. Weber, C.W. McCurdy, A.L. Landers, Phys. Rev. Lett. **108**, 233002 (2012), J. Phys. B **45**, 194003 (2012)
158. A. Landers, T. Weber, I. Ali, A. Cassimi, M. Hattass, O. Jagutzki, A. Nauert, T. Osipov, A. Staudte, M.H. Prior, H. Schmidt-Bocking, C.L. Cocke, R. Dorner, Phys. Rev. Lett. **87**, 013002 (2001)
159. T. Jahnke, T. Weber, A.L. Landers, A. Knapp, S. Schossler, J. Nickles, S. Kammer, O. Jagutzki, L. Schmidt, A. Czasch, T. Osipov, E. Arenholz, A.T. Young, R. Diez Muino, D. Rolles, F.J. Garcia de Abajo, C.S. Fadley, M.A. Van Hove, S.K. Semenov, N.A. Cherepkov, J. Rosch, M.H. Prior, H. Schmidt-Bocking, C.L. Cocke, R. Dorner, Phys. Rev. Lett. **88**, 073002 (2002)
160. J.A.R. Sampson, A.F. Starace, J. Phys. B **8**, 1806 (1975)
161. J.H. Morre, C.C. Davis, M.A. Coplan, *Building Scientific Apparatus*, 4th edn. (Cambridge University Press, Cambridge, 2021)

162. W.C. Lineberger, B.W. Woodward, Phys. Rev. Lett. **25**, 424 (1970)
163. P. Oliver, A. Hibbert, J. Phys. B **41**, 165003 (2008)
164. X. Fu, Z. Lou, X. Chen, J. Li, C. Ning, J. Chem. Phys. **145**, 164307 (2016)
165. Z. Luo, X. Chen, J. Li, C. Ning, Phys. Rev. A **93**, 020501 (2016)
166. R.C. Bilodeau, M. Scheer, H. Haugen, Phys. Rev. A **61**, 012505 (1999)
167. X. Chen, C. Ning, Phys. Rev. A **93**, 052508 (2016)
168. O. Windelius, A. Aguilar, R.C. Bilodeau, A.M. Juarez, I. Rebolledo-Salgado, D.J. Pegg, J. Rohlen, T. Castel, J. Welander, D. Hanstorp, Nucl. Instrum. Methods B **410**, 144–152 (2017)
169. J.B. Williams, Unpublished Research Document
170. T. Jahnke, V. Mergel, O. Jagutzki, A. Czasch, U. Ullmann, R. Ali, V. Frohne, T. Weber, L.P. Schmidt, S. Eckart, M. Schoffler, S. Schosler, S. Voss, A. Landers, D. Fischer, M. Schulz, A. Dorn, L. Spielberger, R. Moshammer, R. Olsen, M. Prior, R. Dorner, J. Ullrich, C.L. Cocke, H. Schmidt-Bocking, High-Resolution Momentum Imaging-From Stern's Molecular Beam Method to the COLRIMS Reaction Microscope, in *Molecular Beams in Physics and Chemistry-from Otto Stern's Pioneering Exploits to Present-Day Feats*, ed. by B. Friedrich, H. Schmidt-Bocking, (Springer, Cham, 2021). https://doi.org/10.1007/978-3-030-63963-1
171. J. Rist, K. Klyssek, N.M. Novikovskiy, M. Kircher, I. Vela-Perez, D. Trabert, S. Grundmann, D. Tsitsonis, J. Siebert, A. Geyer, N. Melzer, C. Schwarz, N. Anders, L. Kaiser, K. Fehre, A. Hartung, S. Eckart, L.P.H. Schmidt, M.S. Schoffler, V.T. Davis, J.B. Williams, F. Trinter, R. Dorner, P.V. Demekhin, T. Jahnke, Nat. Comm. **12**, 6657 (2021). https://doi.org/10.1038/s41467-021-26994-2
172. S.M. Lea, *Mathematics for Physicists* (Thomas Learning, Belmont, 2004)
173. F.B. Hildebrand, *Advanced Calculus for Applications*, 7th edn. (Prentice-Hall Inc., Englewood Cliffs, 1976)
174. E.U. Condon, G.H. Shortley, *The Theory of Atomic Spectra* (Cambridge University Press, Cambridge, 1991)
175. J.T. Cushing, *Applied Analytical Mathematics for Physical Scientists* (Wiley, New York, 1975)
176. F.W. Byron Jr., R.W. Fuller, *Mathematics of Classical and Quantum Physics* (Dover Publications, New York, 1970)
177. Th. Weber, Dissertation (Goethe University, Frankfurt am Main, 2003)
178. W.C. Wiley, I.H. McLaren, Rev. Sci. Instrum. **26**, 1150 (1955)
179. J.B. Williams, *Dissertation* (Auburn University, Auburn, 2011)
180. T. Jahnke, Unpublished Research Document